# MODERN DEVELOPMENTS IN
# FLUID DYNAMICS

# MODERN DEVELOPMENTS IN
# FLUID DYNAMICS

AN ACCOUNT OF THEORY AND EXPERIMENT
RELATING TO
BOUNDARY LAYERS, TURBULENT MOTION
AND WAKES

*Composed by the*

FLUID MOTION PANEL OF THE
AERONAUTICAL RESEARCH COMMITTEE
AND OTHERS

*and edited by*

S. GOLDSTEIN

*in two volumes*
VOLUME II

NEW YORK
DOVER PUBLICATIONS, INC.

Published in Canada by General Publishing Company, Ltd., 30 Lesmill Road, Don Mills, Toronto, Ontario.
Published in the United Kingdom by Constable and Company, Ltd., 10 Orange Street, London W. C. 2.

This Dover edition, first published in 1965, is an unabridged and corrected republication of the work first published by Clarendon Press, Oxford, in 1938, to which has been added a new Preface and Editor's Note.
This edition is published by special arrangement with Oxford University Press.
The publisher is grateful to the Librarian of Duke University for furnishing a copy of this work for reproduction purposes.

Standard Book Number: 486-61358-5

Library of Congress Catalog Card Number: 65-15511

Manufactured in the United States of America
Dover Publications, Inc.
180 Varick Street
New York, N.Y. 10014

*This book has been composed by the Fluid Motion
Panel of the Aeronautical Research Committee
with the collaboration of*

V. M. FALKNER, B.Sc., A.M.I.Mech.E.

L. HOWARTH, B.Sc., M.A., Ph.D.

L. ROSENHEAD, Ph.D., D.Sc.

H. C. H. TOWNEND, D.Sc., F.R.Ae.S.

# CONTENTS

## *VOLUME II*

### VIII. FLOW IN PIPES AND CHANNELS AND ALONG FLAT PLATES (*continued*).

### SECTION III. TURBULENT FLOW.

## XI. FLOW PAST SOLID BODIES OF REVOLUTION.

### SECTION I. SPHERES.

### SECTION II. AIRSHIP SHAPES.

## XV. HEAT TRANSFER (TURBULENT FLOW).

# PLATES

## VOL. II

# FLOW IN PIPES AND CHANNELS AND ALONG FLAT PLATES (*continued*)

## SECTION III
### TURBULENT FLOW

### 152. Calculations of velocity distributions of mean flows.

FOR the calculation of velocity distributions of mean flows in turbulent motion the various methods described in Chap. V, §§ 81, 82, 83, and 85' may be used. We consider investigations carried out first on Prandtl's momentum transfer theory, then on Taylor's vorticity transfer theory, and finally on Kármán's 'similarity theory'.

### 153. The velocity distribution near a wall: momentum transfer theory. Experimental results from flow through smooth circular pipes.

In a two-dimensional mean motion in which the mean velocity $U$ is in the $x$-direction and is a function of $y$ only (where the axes of $x$ and $y$ are at right angles), the shearing stress $\tau$ $(= \overline{p_{xy}})$ at any point in the fluid is given by

$$\tau = \rho l' v \frac{dU}{dy} \tag{1}$$

on the momentum transfer theory† if purely viscous stresses are neglected. $l'$ is a certain length, $v$ is the root-mean-square velocity fluctuation parallel to the axis of $y$, and $\rho$ is the density. Following Prandtl, we put† $l'v = l^2|dU/dy|$ and

$$\tau = \rho l^2 \left|\frac{dU}{dy}\right| \frac{dU}{dy}. \tag{2}$$

But

$$\frac{\partial \tau}{\partial y} = \frac{\partial p}{\partial x}, \tag{3}$$

so that

$$\frac{1}{\rho}\frac{\partial p}{\partial x} = \frac{\partial}{\partial y}\left(l^2\left|\frac{dU}{dy}\right|\frac{dU}{dy}\right). \tag{4}$$

Before (2) or (4) can be solved for any particular problem it is necessary to make some assumption regarding the form of $l$.

In a turbulent flow along a wall there is, near the wall, a thin so-called 'laminar sub-layer' in which the rate of shear is very

---

† Chap. V, equations (22), (26) (pp. 206, 207).

great, while the turbulent velocity component perpendicular to the wall is very small,† so that the apparent Reynolds stress is insignificant compared with the viscous stress. Farther from the wall there is another region in which the Reynolds stresses and the viscous stresses are of the same order of magnitude. This region and the laminar sub-layer together form the region in which viscosity is of importance, and the two together may be called the 'viscous layer'. Finally (for sufficiently high Reynolds numbers) there is a region in which the viscous stresses may be neglected in comparison with the Reynolds stresses.

For sufficiently high Reynolds numbers the viscous layer is very thin, and the value of $dU/dy$ at any point within it (where $U$ is the velocity parallel to the wall, and $y$ is distance from the wall) is very large. For a first approximation for very large Reynolds numbers we may therefore neglect its finite thickness, and simply solve the equations for turbulent flow with viscosity neglected, if we suitably modify the boundary conditions at the wall. We can no longer retain the condition of zero slip: we replace it by the condition that $dU/dy$ must be very large (since there is no abrupt change in $dU/dy$ as we pass out of the viscous layer). The exact value of $dU/dy$ depends on the thickness of the viscous layer; for a first approximation we take it to be infinite.

Consider now the flow along an infinitely long plane wall in the absence of a pressure gradient. We can apply dimensional arguments to determine a form for $l$. If we suppose that $l$ is determined by the configuration of the boundaries and by the physical constants (and not by the local distribution of mean velocity as in the similarity theory (see p. 347 *et seq.*)), then $l$ and $y$ are the only lengths which enter directly into the problem. The other quantities which enter are $\mu$, $\rho$, and the constant value $\tau_0$ of the shearing stress.‡ Only two independent dimensionless combinations can then be formed, which we may take to be $l/y$ and $yU_\tau/\nu$, where $U_\tau = \sqrt{(\tau_0/\rho)}$ and has the dimensions of a velocity. When the influence of viscosity is neglected (as in the equations for the 'fully turbulent' region) then $l/y$ must be a constant. We therefore put

$$l = Ky, \tag{5}$$

so that, from (2),    $\tau_0 = \rho K^2 y^2 (dU/dy)^2.$    (6)

(For positive values of $\tau_0$, $dU/dy$ is positive and equal to its modulus.) We find by integrating (6) that

$$U = \frac{U_\tau}{K} \log_e y + \text{const.} \dagger$$    (7)

This expression for $U$ automatically satisfies the condition that $dU/dy$ should be infinite at the wall.

Equation (7) may be written

$$\left. \begin{aligned} \frac{U}{U_\tau} &= A + B \log_{10} \frac{y U_\tau}{\nu} \\ \text{or} \quad \frac{U}{U_\tau} &= A + \frac{1}{K} \log_e \frac{y U_\tau}{\nu}, \end{aligned} \right\}$$    (8)

where $A$ and $B$ are constants, and $B$ is $K^{-1} \log_e 10$.

When there is a pressure gradient along the wall equation (7) or (8) will still give a first approximation to the velocity distribution near the wall. For we may suppose $\tau$ and $l$ expanded in powers of $y$, the distance from the wall, and only the first terms retained. The first term in $\tau$ is $\tau_0$, the intensity of the wall friction; $l$ presumably vanishes at the wall, so that we may expect the first term to be of the form $Ky$;‡ thus for a first approximation equation (6) remains unaltered, so that equation (7) or (8) should give a first approximation to the flow in the neighbourhood of a plane boundary in the presence of a pressure gradient along it.

Further, when the mean motion is in a fixed direction parallel to the solid boundary and the magnitude of the velocity is a function only of the normal distance $y$ from the boundary, then the equation (2) is still valid on the momentum transfer theory (with assumptions similar to those made before) when the boundary is curved in a plane at right angles to the direction of mean motion—when it is cylindrical, for example, as in flow through a circular pipe,—$\tau$ now being the shearing stress over surfaces parallel to the boundary. First approximations to $\tau$ and $l$ near the wall will still be $\tau_0$ (the

† Prandtl, *Zeitschr. des Vereines deutscher Ingenieure*, **77** (1933), 107, 108. The result was first found by Kármán, who used his formula for $l$ (*Göttinger Nachrichten* (1930), pp. 58–76; *Proc. 3rd Internat. Congress for Applied Mechanics, Stockholm*, 1930, **1**, 85–92; *Hydromechanische Probleme des Schiffsantriebs* (Hamburg, 1932), pp. 50–73). See Chap. V, § 82 (pp. 208, 209), and § 158 *infra*.

‡ It is possible that the first term is of the form $K'y^n$ ($n \neq 1$), but it seems unlikely that the presence of the pressure gradient will alter the exponent.

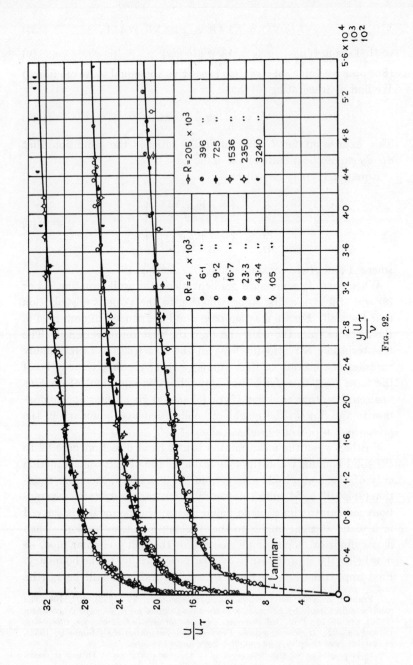

Fig. 92.

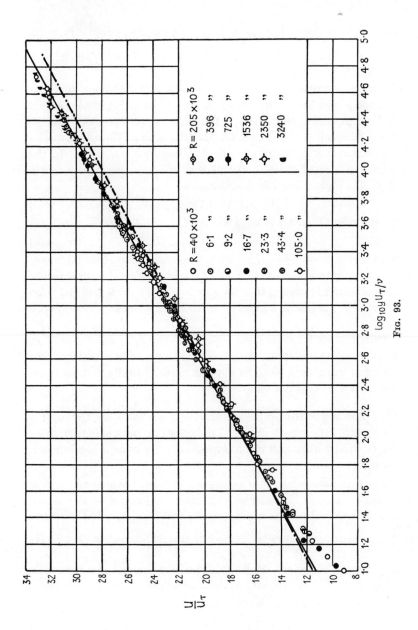

Fig. 93.

intensity of the wall friction) and $Ky$, so that (7) or (8) should still give the velocity distribution near to the wall.

The experimentally observed velocity distributions[†] for flow through a number of circular pipes are shown in Fig. 92 by curves of $U/U_\tau$ plotted against $yU_\tau/\nu$. (The three curves correspond to the three different scales shown for the abscissae.) The same results are shown in Fig. 93 by plotting $U/U_\tau$ against $\log_{10} yU_\tau/\nu$. In the neighbourhood of the wall the plotted points in Fig. 93 lie closely about the broken line whose equation is

$$U/U_\tau = 5 \cdot 8 + 5 \cdot 5 \log_{10} yU_\tau/\nu: \tag{9}$$

the unbroken line, whose equation is[‡]

$$U/U_\tau = 5 \cdot 5 + 5 \cdot 75 \log_{10} yU_\tau/\nu, \tag{10}$$

is in fairly good agreement with the entire experimental curve between the wall and the axis. It is a matter of good fortune rather than of sound reasoning that the equation (10) (which is of the same form as (8)) holds to a good approximation throughout the entire pipe.

Since there is a viscous layer adjacent to the wall we cannot expect an equation of the form (8) to hold right up to the boundary. For values of $y$ less than $30\nu/U_\tau$ it appears experimentally that an equation of this form is not valid.

## 154. The logarithmic resistance formula for flow through a smooth pipe or channel. The relation between the velocity at the axis and the average velocity over a section of a pipe.

For two-dimensional flow through a straight channel or for flow through a straight circular pipe it has been experimentally established[||] (and is also a theoretical result: see §§ 156, 157, and 159) that

$$\frac{U_c - U}{U_\tau} = F\left(\frac{y}{h}\right) \quad \text{or} \quad f\left(\frac{y}{a}\right) \tag{11}$$

respectively, where $h$ is the half-width of the channel or $a$ the radius of the pipe, $U$ is the velocity at the normal distance $y$ from the solid

---

[†] Nikuradse, *Ver. deutsch. Ing., Forschungsheft* 356 (1932).

[‡] The values of $K$ corresponding to equations (9) and (10) are 0·417 and 0·400 respectively, and are taken from Prandtl, *Aerodynamic Theory* (edited by Durand), **3** (Berlin, 1935), 140.

[||] Cf. Darcy, *Mémoires des Savants Étrangers*, **15** (1858), 265–342; Stanton, *Proc. Roy. Soc.* A, **85** (1911), 366–376; Fritsch, *Aachener Abhandlungen*, **8** (1928), 45–62, or *Zeitschr. f. angew Math. u. Mech.* **8** (1928), 199–216.

boundary, $U_c$ is the velocity at the middle of the pipe or channel, $U_\tau = \sqrt{(\tau_0/\rho)}$, and $\tau_0$ is the intensity of the wall friction. For sufficiently high Reynolds numbers $F$ and $f$ are universal functions independent both of the Reynolds number and of any roughness of the walls.

Quite generally $l/y$ is a function of $y/h$ (or $y/a$) and $yU_\tau/\nu$. If the influence of viscosity is neglected, then $l/y$ is a function of $y/h$ (or $y/a$) only. From experimental velocity measurements the relation between $l/a$ and $y/a$ for flow through a pipe can be obtained;† this relation ceases to depend on viscosity for Reynolds numbers greater than $10^5$. Presumably (11) is valid at any rate above the same limit.

With the help of (11) we can obtain a resistance formula for flow through a pipe. Fairly near the wall

$$\frac{U}{U_\tau} = \text{constant} + \frac{1}{K}\log_e\frac{yU_\tau}{\nu} \qquad (12)$$

(compare equation (8)). Since (11) must reduce to (12) for small values of $y$, it must be of the form

$$\frac{U_c-U}{U_\tau} = \text{constant} - \frac{1}{K}\log_e\frac{y}{a} \qquad (13)$$

for small values of $y$. By adding (12) and (13) we find that

$$\frac{U_c}{U_\tau} = \text{constant} + \frac{1}{K}\log_e\frac{aU_\tau}{\nu}. \qquad (14)$$

Since it follows from (11) that

$$\frac{U_c-U_m}{U_\tau} = \text{constant}, \qquad (15)$$

where $U_m$ is the mean value of $U$ over a section, it is equally true that

$$\frac{U_m}{U_\tau} = \text{constant} + \frac{1}{K}\log_e\frac{aU_\tau}{\nu}. \qquad (16)$$

Putting    $R = \dfrac{2aU_m}{\nu}$    and    $\gamma = \dfrac{\tau_0}{\frac{1}{2}\rho U_m^2} = \dfrac{2U_\tau^2}{U_m^2}$,

we find that‡    $\gamma^{-\frac{1}{2}} = \text{constant} + B'\log_{10}R\gamma^{\frac{1}{2}}$, $\qquad (17)$

where    $B' = \dfrac{1}{\sqrt{2}}K^{-1}\log_e 10.$

† Figs. 103 and 104 (p. 357).
‡ Kármán, *Göttinger Nachrichten* (1930), pp. 58–76; Prandtl, *Zeitschr. des Vereines deutscher Ingenieure*, **77** (1933), 105–114.

Similarly, equation (17) with $a$ replaced by $h$ is obtained for two-dimensional flow through a channel.

Equation (11) will not hold in the viscous layer, since it is valid only if viscosity has no effect. The width of this layer decreases when the Reynolds number of the flow increases. The validity of the resistance formula depends on the Reynolds number being sufficiently

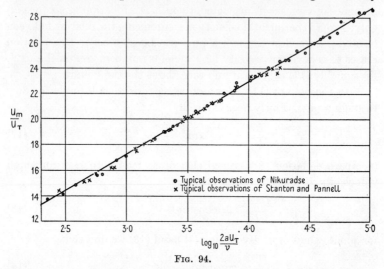

Fig. 94.

large for the following two assumptions to hold:—$(a)$ a region exists in the centre of the pipe or channel where the motion is unaffected by viscosity; $(b)$ this region extends as far as the region where the result (12) for the velocity distribution near a wall is valid.

In Fig. 94 the experimentally obtained values for a circular pipe of $U_m/U_\tau$ $(= \sqrt{(2/\gamma)})$ and $\log_{10} 2aU_\tau/\nu$ $(= \log_{10} R\sqrt{(\tfrac{1}{2}\gamma)})$† are plotted against each other. The linear relation predicted by the theoretical arguments is well substantiated: when we choose the constants to obtain a good fit we find the straight line shown, whose equation is

$$\frac{U_m}{U_\tau} = 0\cdot29 + 5\cdot66 \log_{10} \frac{2aU_\tau}{\nu}. \tag{18}$$

In terms of the resistance coefficient $\gamma$ and the Reynolds number $R$ this equation becomes

$$\gamma^{-\frac{1}{2}} = -0\cdot40 + 4\cdot00 \log_{10} R\gamma^{\frac{1}{2}}. \tag{19}$$

† The observed points are due to Nikuradse, *Ver. deutsch. Ing.*, *Forschungsheft* 356 (1932); Stanton and Pannell, *Phil. Trans.* A, **214** (1914), 199–224.

The value 4·00 for $B'$ corresponds to $K = 0·408$, which should be compared with the value 0·417 derived from equation (9).

It may be mentioned that the value of $U_c - U_m$ for a pipe, as determined from Nikuradse's experiments, is about $4·07U_\tau$.[†] With $U_\tau$ calculated from equation (18), Stanton and Pannell's measurements[‡] lead to the relation $U_c - U_m = 4·7U_\tau$.

## 155. Blasius's resistance formula for smooth pipe flow: power formulae for the velocity distribution.

Blasius‖ gave the formula

$$\gamma = 0·0791(2U_m a/\nu)^{-\frac{1}{4}} = 0·0665(U_m a/\nu)^{-\frac{1}{4}} \qquad (20)$$

as an empirical result for the resistance coefficient for pipe flow. The formula is valid for Reynolds numbers less than $10^5$.

Let us consider first the more general form

$$\gamma = A\left(\frac{U_m a}{\nu}\right)^{-1/n}. \qquad (21)$$

Then since $\tau_0 = \frac{1}{2}\gamma\rho U_m^2 = \rho U_\tau^2$, we have

$$U_\tau^2 = \tfrac{1}{2}A U_m^{2-1/n} a^{-1/n}\nu^{1/n},$$

i.e.

$$\left(\frac{U_m}{U_\tau}\right)^{2-1/n} = \frac{2}{A}\left(\frac{U_\tau a}{\nu}\right)^{1/n}$$

or

$$\frac{U_m}{U_\tau} = B\left(\frac{U_\tau a}{\nu}\right)^{1/(2n-1)} \qquad (22)$$

where $B = (2/A)^{n/(2n-1)}$.

We can, by means of plausible assumptions, deduce a form of the velocity distribution for flow in a pipe corresponding to (22). The mean velocity in a pipe is known to increase rapidly near the wall and not to vary much over the middle of the pipe; and (22) depends largely on the velocity distribution in the neighbourhood of the walls. Near the walls a first approximation is obtained by assuming $\tau$ constant and equal to the intensity of the wall friction; if, instead of simply putting $l$ proportional to the distance $y$ from the wall,

† Nikuradse, *Ver. deutsch. Ing., Forschungsheft* 356 (1932); Prandtl, *Zeitschr. des Vereines deutscher Ingenieure,* **77** (1933), 110.

‡ *Phil. Trans.* A, **214** (1914), 205, Fig. 1.

‖ *Forschungsarbeiten des Ver. deutsch. Ing.,* No. 131 (1913).

we put $l = yf(yU_\tau/\nu)$ in accordance with the dimensional argument in §153 (p. 332), we find from equation (2) that

$$\frac{U}{U_\tau} = \int \frac{dy}{yf(yU_\tau/\nu)} = \phi\left(\frac{yU_\tau}{\nu}\right)$$

say. We try
$$\frac{U}{U_\tau} = C\left(\frac{yU_\tau}{\nu}\right)^m \tag{23}$$

as an approximation to the velocity distribution right across the pipe. We shall find that $m$ is small, so that (23) makes the variation of $U$ small in the middle of the pipe. From (23) it follows that

$$\frac{U_m}{U_\tau} = \left(\frac{aU_\tau}{\nu}\right)^m \frac{2C}{(m+1)(m+2)}. \tag{24}$$

Comparing this result with (22) we see that

$$m = \frac{1}{2n-1}, \qquad C = \frac{n(4n-1)}{(2n-1)^2}\left(\frac{2}{A}\right)^{n/(2n-1)}.$$

When $n = 4$, $m = \frac{1}{7}$, and with the numerical coefficient given in (20) $C = 8\cdot6$. If we slightly alter the numerical value and write

$$\frac{U}{U_\tau} = 8\cdot7\left(\frac{yU_\tau}{\nu}\right)^{\frac{1}{7}}, \tag{25}$$

the result is in rather surprisingly good agreement with measured velocity distributions over the entire pipe for the range of Reynolds numbers for which (20) is valid.

The value of $U_\tau$ given by (25) is

$$\left.\begin{array}{l} U_\tau = 0\cdot150U^{\frac{7}{8}}\nu^{\frac{1}{8}}y^{-\frac{1}{8}}, \\ \tau_0 = \rho U_\tau^2 = 0\cdot0225\rho U^2(\nu/Uy)^{\frac{1}{4}}. \end{array}\right\} \tag{26}$$

so that

For Reynolds numbers greater than $10^5$, $m$ must be taken as $\frac{1}{8}$, $\frac{1}{9}$, etc. successively in order to maintain agreement with experiment.

### 156. Velocity distributions in two-dimensional pressure flow between smooth or rough parallel walls, and in flow through a smooth or rough circular pipe:† momentum transfer theory.

In two-dimensional flow under a pressure gradient between parallel walls, when the flow at any one section is the same as that at any other (i.e. far away from the ends), the velocity $U$ parallel to the central plane is a function only of the distance from that plane, and

† The theory in §§156, 157, and 159 should apply whether the pipes are smooth or (moderately) rough.

equations (2) and (3) hold. $\tau$ vanishes along the central plane, and

$$\frac{\rho U_\tau^2}{h} = \frac{\tau_0}{h} = \frac{\tau}{y'} = -\frac{\partial p}{\partial x}, \tag{27}$$

where $h$ is the half-width of the channel, $y$ is distance measured from a wall and $y'$ is distance measured from the central plane, so that $y = h-y'$, $\tau$ is $\overline{p_{xy}} = -\overline{p_{xy'}}$ (so that it is positive), and $\tau_0$ is the wall friction. Equation (27) is the integral of equation (3), and is the equation of momentum for a symmetrically situated slab of fluid of unit length parallel to the central plane and of width $2y'$.

By dimensional arguments we have already seen that when the influence of viscosity is neglected $l/(h-y')$ is a function of $y'/h$ only. Furthermore, near the wall a first approximation is obtained by taking $l/(h-y')$ constant. The simplest assumption we can make in the present problem is that $l/(h-y')$ is constant throughout, and since we have no further guide to the form of $l$ we shall make this assumption. That is to say we put $l = K_1(h-y')$, so that (2) becomes

$$\frac{y'\tau_0}{\rho h} = K_1^2(h-y')^2\left(\frac{dU}{dy'}\right)^2. \tag{28}$$

The integral of (28) with the condition $U = U_c$ when $y' = 0$ is[†]

$$\frac{U_c-U}{U_\tau} = \frac{1}{K_1}\left[\log_e\frac{1+(y'/h)^{\frac12}}{1-(y'/h)^{\frac12}} - 2(y'/h)^{\frac12}\right]. \tag{29}$$

This integral automatically satisfies the conditions $dU/dy' = 0$ at $y' = 0$, $dU/dy' = \infty$ at $y' = h$. It gives an infinite value for $U$ at $y' = h$. A comparison with the experimental measurements of Dönch[‡] and Nikuradse[||] is shown in Fig. 95, the value 0·23 of the constant $K_1$ being chosen to produce agreement at $y'/h = 0·7$.[††] The crosses reproduce Nikuradse's results, found at $U_m h/\nu$ equal to $6\times 10^4$ approximately ($U_m$ being the average value of $U$ over a cross-section), and the circles Dönch's results at $U_m h/\nu$ equal to $10^5$ approximately. The agreement is good except near the walls.

Consider now flow through a circular pipe. For flow in one direction symmetrical about an axis, with the velocity $U$ a function

---

[†] Taylor, *Proc. Roy. Soc.* A, **159** (1937), 496–506; see especially pp. 504, 505.

[‡] *Forschungsarbeiten des Ver. deutsch. Ing.*, No. 282 (1926).

[||] *Ibid.*, No. 289 (1929).

[††] A more satisfactory method would be to choose $K_1$ by the method of least squares: this method has not been used owing to the labour involved. The same remark applies to all subsequent comparisons of theory and experiment in this section and in §§ 157 and 159.

only of $r$, the distance from the axis, equation (2), when written in the form

$$\tau = -\rho l^2 \left| \frac{dU}{dr} \right| \frac{dU}{dr}, \tag{30}$$

where $\tau = -\overline{p_{rx}}$, is valid with the same assumptions as before on

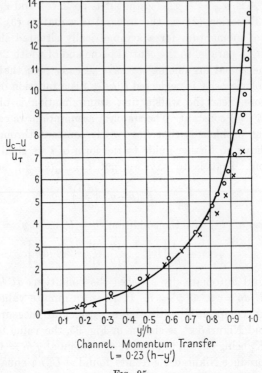

Channel. Momentum Transfer
$l = 0.23\,(h-y')$

Fig. 95.

the momentum transfer theory. (The sign of $\tau$ is chosen in order that it should be positive throughout the fluid for flow in pipes.)

The equation of momentum for a cylinder of fluid of radius $r$ is

$$\frac{2\rho U_\tau^2}{a} = \frac{2\tau_0}{a} = \frac{2\tau}{r} = -\frac{\partial p}{\partial x}. \tag{31}$$

Combining (31) with (30), assuming that $l = K_2(a-r)$, and denoting as before the value of $U$ on the axis by $U_c$, we obtain equation (29) with $K_1$ replaced by $K_2$ and $y'/h$ by $r/a$.

In Fig. 96 this theoretical result is compared with Stanton's experiments† in smooth pipes by choosing $K_2$ to give agreement at $r/a = 0.7$. The value of $K_2$ is 0.20. The crosses, squares, and circles in Fig. 96 refer to Stanton's three series of experiments, the value of $2aU_m/\nu$ being about $4.1 \times 10^4$ for the first two and $9.1 \times 10^4$ for the

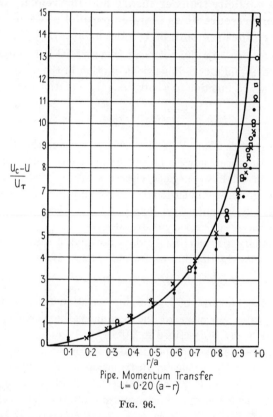

Pipe. Momentum Transfer
$$l = 0.20\,(a-r)$$

FIG. 96.

third. The dots refer to a selection of the results published by Nikuradse in 1932,‡ namely those for $2aU_m/\nu$ equal to $4.34 \times 10^4$ and $1.05 \times 10^5$, the larger value of $(U_c-U)/U_\tau$ corresponding to the larger Reynolds number. Nikuradse's values are definitely lower than Stanton's. The values published by Nikuradse in 1933‖ are still lower, and would give a value 0.24 for $K_2$.††

† *Proc. Roy. Soc.* A, **85** (1911), 366–375.

‡ *Ver. deutsch. Ing., Forschungsheft* 356 (1932).     ‖ *Ibid.* 361 (1933).

†† Stanton's measurements and Nikuradse's early measurements (*Proc. 3rd Internat. Congress for Applied Mechanics, Stockholm*, 1930, **1**, 239–248) are in fairly good

Since two-dimensional flow is a limiting case of flow symmetrical about an axis, the values of $K_1$ and $K_2$ should be identical: they are, in fact, rather different.

### 157. The vorticity transfer theory and the velocity distributions for pressure flows between parallel walls and through circular pipes (smooth or rough surfaces).

If the vorticity of a fluid element is conserved until mixing takes place, we may, for a two-dimensional motion in the $x$-direction with the velocity $U$ a function of $y$ only, take the equation of motion to be

$$\frac{1}{\rho}\frac{\partial p}{\partial x} = l^2 \left|\frac{dU}{dy}\right|\frac{d^2U}{dy^2}† \tag{32}$$

instead of (4) (see Chap. V, §§ 83, 85. If we assume the disturbances as well as the mean flow to be two-dimensional, vorticity is necessarily conserved if we neglect viscosity; alternatively, if the disturbances are three-dimensional, the assumption that vorticity is conserved will be untrue in detail but may lead to satisfactory results for the mean velocity distribution).

The vorticity transfer theory does not give a satisfactory result for the velocity distribution near a plane wall. If $\partial p/\partial x = 0$ in (32), then

$$U = Ay+B \tag{33}$$

whatever be the form of $l$. (An investigation to determine in a particular case how near to the boundary the vorticity transfer theory may apply is described in § 165.)

If, as before, we take $l$ proportional to the distance from the nearer wall, we can find the velocity distribution for flow under pressure between parallel walls at a distance $2h$ apart according to the vorticity transfer theory. When we use the formula

$$-\frac{1}{\rho}\frac{\partial p}{\partial x} = \frac{\tau_0}{\rho h} = \frac{U_\tau^2}{h}, \tag{34}$$

and put

$$l = K_3(h-y'), \tag{35}$$

agreement. Formulae such as (29) should theoretically hold for either smooth or rough pipes. Stanton's measurements and Nikuradse's earlier measurements were made on smooth pipes; Nikuradse's 1933 measurements were made mainly on pipes artificially roughened with a certain kind of sand.

Recent experiments by Fage on smooth pipes at $2aU_m/\nu = 1\cdot9\times10^4$ (*Phil. Mag.* (7), **21** (1936), 80–105) show results in good accord with Stanton's measurements.

† In this expression Prandtl's hypothesis $l'\boldsymbol{\nu} = l^2|dU/dy|$ has been applied. See Chap. V, equation (36) (p. 210).

we find the following integral of (32) with the boundary conditions $U = U_c$ and $dU/dy' = 0$ at $y' = 0$ (where $y' = h-y$ and is distance measured from the central plane):[†]

$$\frac{U_c-U}{U_\tau} = \frac{\sqrt{2}}{K_3}\left[\sin^{-1}\left(\frac{y'}{h}\right)^{\frac{1}{2}} - \left(\frac{y'}{h}\right)^{\frac{1}{2}}\left(1-\frac{y'}{h}\right)^{\frac{1}{2}}\right]. \qquad (36)$$

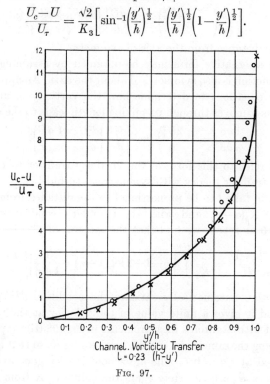

Channel. Vorticity Transfer
$L = 0.23 \ (h-y')$

FIG. 97.

This result is compared with experiment in Fig. 97, $K_3$ being chosen to produce agreement at $y'/h = 0.7$; the value of $K_3$ is $0.23$. The experimental points are the same as in Fig. 95. The agreement is good except in the neighbourhood of the wall. The solution leads to a finite value of $U$ and an infinite value of $dU/dy$ at the wall.

We consider next the velocity distribution in a circular pipe according to the vorticity transfer theory. The mean motion is in the $x$-direction and is symmetrical about the $x$-axis; in such a case, either of the two following forms of the vorticity transfer theory leads to a simple equation of motion.

(i) We may make the assumption of the modified vorticity transfer

[†] Taylor, *Proc. Roy. Soc.* A, **159** (1937), 501.

theory that the vorticity of a fluid element is conserved until mixing takes place; this, together with the assumption that the turbulence is isotropic, leads to the equation[†]

$$\frac{1}{\rho}\frac{\partial p}{\partial x}\left(=-\frac{2U_\tau^2}{a}\right)=l^2\left|\frac{dU}{dr}\right|\left(\frac{d^2U}{dr^2}+\frac{1}{r}\frac{dU}{dr}\right), \qquad (37)$$

where $r$ is distance from the axis of symmetry.

(ii) An alternative form may be obtained by assuming that the eddying motion is also symmetrical about the axis. An application of Cauchy's equations to determine how the vorticity of an element changes as it moves through its mixing length leads to the equation[‡]

$$\frac{1}{\rho}\frac{\partial p}{\partial x}\left(=-\frac{2U_\tau^2}{a}\right)=l^2\left|\frac{dU}{dr}\right|\left(\frac{d^2U}{dr^2}-\frac{1}{r}\frac{dU}{dr}\right). \qquad (38)$$

We shall refer to this form of the vorticity transfer theory as the vorticity transfer theory with symmetrical turbulence.

We again take $l$ to be proportional to the distance from the wall. With $l = K_4(a-r)$, and with the boundary conditions $U = U_c$, $dU/dr = 0$ at $r = 0$, we find that the integral of (37) is[||]

$$\frac{U_c-U}{U_\tau}=\frac{2}{K_4}\int\limits_0^{r/a}\left[\eta-1+2\log(1-\eta)+\frac{1}{1-\eta}\right]^{\frac{1}{2}}\frac{d\eta}{\eta}. \qquad (39)$$

This solution automatically satisfies the condition $dU/dr = \infty$ at $r = a$, and it gives a finite value of the velocity at the wall. It is compared with Stanton's measurements in Fig. 98 (the experimental points being the same as in Fig. 96) by choosing $K_4$ so that agreement is obtained at $r/a = 0.7$. The agreement is very good everywhere; the value of $K_4$ is 0.19. Here again the values of $K$ from pipes and channels are not in agreement. (With Nikuradse's 1933 measurements[††] the value of $K_4$ would be 0.227.)

On the vorticity transfer theory with symmetrical turbulence, with $l$ proportional to $a-r$, $dU/dr$ is imaginary on the axis and no real solution can be obtained. A real solution can be obtained with the flow confined to axial planes if $l$ is allowed to become infinite on the axis. (Cf. § 159 (pp. 354, 355).)

---

† Chap. V, equations (37) and (50) (pp. 210, 214). Prandtl's hypothesis has again been applied.

‡ This result follows easily from equations (38) and (44) of Chap. V. See Goldstein, *Proc. Camb. Phil. Soc.* **31** (1935), 358, equations (43). The result is easily obtained *ab initio*, since for motion symmetrical about an axis $\omega_\theta/r$ is constant for any element, where $\omega_\theta$ is the vorticity.

|| Taylor, *op. cit.*, p. 503.                                          †† *Loc. cit.* (p. 343).

From Figs. 95, 96, 97 and 98 it can be seen that the vorticity transfer theory gives results in better agreement with observation than does the momentum transfer theory for pipe flow, and that there is nothing to choose between the two theories for flow between parallel

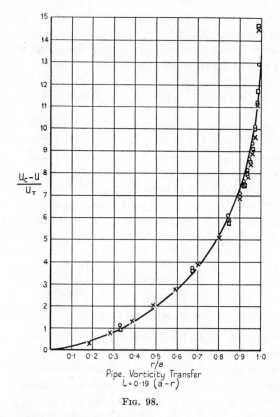

Pipe. Vorticity Transfer
$L = 0.19 \, (a - r)$

Fig. 98.

planes; but in view of the arbitrary nature of the assumption that $l$ is proportional to the distance from the wall at all points in the pipe or the channel, no definite conclusion should be drawn.

## 158. The similarity theory.†

For simplicity we consider the case of a mean motion in the $x$-direction, with the velocity $U$ a function of $y$ only, and we assume

† Kármán, *Göttinger Nachrichten* (1930), *Math.-Phys. Klasse*, pp. 58–76; *Proc. 3rd Internat. Congress for Applied Mechanics, Stockholm*, 1930, **1**, 85–93.

that the eddying motion is also two-dimensional in the $(x, y)$ plane.†
Then the equation for the vorticity is

$$U\frac{\partial \zeta'}{\partial x} + v\frac{d\zeta}{dy} + \frac{\partial \zeta'}{\partial t} + u\frac{\partial \zeta'}{\partial x} + v\frac{\partial \zeta'}{\partial y} = 0, \tag{40}$$

where $\zeta$ and $\zeta'$ are the $z$-components of the mean and the eddy
vorticity respectively (the other components being zero), and $u, v$ are
the turbulent velocity components in the directions of $x$ and $y$.
There is a stream-function $\psi + \psi'$ such that

$$U = \frac{\partial \psi}{\partial y}, \qquad u = \frac{\partial \psi'}{\partial y}, \qquad v = -\frac{\partial \psi'}{\partial x}. \tag{41}$$

We now consider the eddying motion in the neighbourhood of a
certain value of $y$, say $y = y_0$, and assume that the disturbances
due to this eddying motion may be considered confined to a narrow
range of values of $y$ in the neighbourhood of $y_0$. The assumption of
similarity is that, referred to axes moving with the mean velocity at
the value of $y$ under consideration, the eddying motions at all such
points are similar and differ only in the scales of length and time
(or velocity). The scales of length and velocity are introduced by
writing

$$\left.\begin{aligned} y &= y_0 + ly' \\ x &= \text{const.} + lx' \\ t &= \text{const.} + l^2t'/A \\ \psi' &= Af(x', y', t') \end{aligned}\right\} \tag{42}$$

The assumption of dynamical similarity, mathematically expressed,
is that when these substitutions are made the resulting equation
for $f$ is independent of $l$, $A$, $y_0$, $U$ and the derivatives of $U$ at $y_0$.
On the assumption that the eddying motion is limited to a small
range of values of $y$, we take $d\zeta/dy$ equal to its value at $y = y_0$, and
we write

$$U = ly'\left(\frac{dU}{dy}\right)_0 + \tfrac{1}{2}l^2y'^2\left(\frac{d^2U}{dy^2}\right)_0 = -ly'\zeta_0 - \tfrac{1}{2}l^2y'^2\left(\frac{d\zeta}{dy}\right)_0 \tag{43}$$

in this range of values of $y$, where $\zeta_0$ is the value of $\zeta$ at $y = y_0$.
Also we find from (42) that

$$u = \frac{A}{l}\frac{\partial f}{\partial y'}, \qquad v = -\frac{A}{l}\frac{\partial f}{\partial x'}, \qquad \zeta' = -\frac{A}{l^2}\nabla^2 f, \tag{44}$$

† It is not necessary to assume the eddying motion two-dimensional in order to
obtain the results of this paragraph. It is assumed here merely for simplicity.

where $\nabla^2 = \partial^2/\partial x'^2 + \partial^2/\partial y'^2$. The equation for $f$ is therefore, from (40),

$$\frac{A}{l^2}\zeta_0 y'\frac{\partial}{\partial x'}\nabla^2 f - \frac{A}{l}\left(\frac{d\zeta}{dy}\right)_0\left(\frac{\partial f}{\partial x'} - \tfrac{1}{2}y'^2\frac{\partial}{\partial x'}\nabla^2 f\right)$$

$$-\frac{A^2}{l^4}\left[\frac{\partial}{\partial t'} + \frac{\partial f}{\partial y'}\frac{\partial}{\partial x'} - \frac{\partial f}{\partial x'}\frac{\partial}{\partial y'}\right]\nabla^2 f = 0. \quad (45)$$

This equation is independent of $A$, $l$, $\zeta_0$ and $y_0$ provided that

$$l = K\zeta\Big/\frac{d\zeta}{dy} = K\frac{dU}{dy}\Big/\frac{d^2U}{dy^2} \quad (46)$$

and
$$A = \text{constant } l^2\frac{dU}{dy}, \quad (47)$$

where $K$ is a constant and the zero suffix has been dropped as no longer necessary. The value of $A$ together with (44) leads immediately to the result that $u$ and $v$ are proportional to $l\,dU/dy$.

Two different methods of procedure are now possible: we may use the the preceding results to obtain either (a) a local value of $\tau$ or (b) a local value of $\partial\tau/\partial y$.

(a)
$$\tau = -\rho\,\overline{uv},$$

so that
$$\tau = \rho\frac{A^2}{l^2}\overline{\frac{\partial f}{\partial y'}\frac{\partial f}{\partial x'}} = \text{constant } \rho l^2\left(\frac{dU}{dy}\right)^2\overline{\frac{\partial f}{\partial y'}\frac{\partial f}{\partial x'}}. \quad (48)$$

Since the constant $K$ in $l$ is undetermined we may write

$$\tau = \rho l^2\left(\frac{dU}{dy}\right)^2, \quad (49)$$

where the unknown constant is supposed absorbed in $K$ in the formula for $l$. Equation (49) is identical with the equation (2) of the momentum transfer theory. (There appears to be a possible difference of sign, but from the derivation it is evident that it may be necessary to change the sign of $\tau$ in (49), so we can regard (2) and (49) as identical.) Here we have in addition a theoretical form for $l$, whereas in mixture length theories it is necessary to assume a form.

(b)
$$\frac{\partial\tau}{\partial y} = -\rho\frac{\partial}{\partial y}\overline{(uv)}$$

so that
$$\frac{\partial\tau}{\partial y} = \frac{\rho}{l}\frac{\partial}{\partial y'}\left(\frac{A^2}{l^2}\overline{\frac{\partial f}{\partial y'}\frac{\partial f}{\partial x'}}\right)$$

$$= \text{constant } \rho l\left(\frac{dU}{dy}\right)^2\frac{\partial}{\partial y'}\left(\overline{\frac{\partial f}{\partial y'}\frac{\partial f}{\partial x'}}\right). \quad (50)$$

Here again, since $K$ is unspecified, we may write the result as

$$\frac{\partial \tau}{\partial y} = \rho K l \left(\frac{dU}{dy}\right)^2, \qquad (51)$$

where the unknown constant is supposed absorbed in $K$ in the formula for $l$. From equation (46) we see that (51) is equivalent to

$$\frac{\partial \tau}{\partial y} = \rho l^2 \frac{dU}{dy} \frac{d^2U}{dy^2}, \qquad (52)$$

so that

$$-\frac{1}{\rho}\frac{\partial p}{\partial x}\left(=\frac{U_\tau^2}{h}\right) = l^2 \frac{dU}{dy}\frac{d^2U}{dy^2}. \qquad (53)$$

In this case we obtain the same equation as on the vorticity transfer theory (with the same proviso about sign as in the previous case. Cf. equations (32) and (53)).

If we consider the application of result (a) to the problem of flow along a plane wall under no pressure gradient, we should expect, since $y$ is the only length which appears in the problem, that the length $l$ found from (46) would be a multiple of $y$. If we substitute for $l$ from (46) in (49) and integrate, using the condition $dU/dy = \infty$ when $y = 0$, we find, in fact, precisely the same result (7) as by Prandtl's method,† and on substituting this result back again into (46) we find that $l = Ky$. There is no significant result to be obtained from (b).

If the viscosity were not neglected in (40), it would be impossible to satisfy the condition of similarity. It has in fact been pointed out in Chap. V, § 90 (p. 221) that for similar flow patterns the rate of dissipation of energy of the turbulent motion would be proportional to $\mu u^2 l^{-2}$, whereas the rate at which work is done by the Reynolds stresses per unit volume is proportional to $\rho u^3 l^{-1}$. ($u$ denotes $\sqrt{\overline{u^2}}$).

The similarity theory is the only theory so far proposed which yields a formula for the length involved; but the assumptions of the theory are still largely untested, and moreover, even in the cases of the fairly simple mean motions to which the theory has been applied, there are regions in the field of flow where the assumptions break down. For example, if $u$, $v$, $w$ are the turbulent velocity components, the assumption of similarity implies constant values of the ratios $u^2 : v^2 : w^2 : \overline{uv} : \overline{vw} : \overline{wu}$. From Chap. V, Fig. 47 (p. 194), it will be seen that for pressure flow between parallel planes $\overline{uv}/u^2$ is

† Cf. Kármán, loc. cit.; also Hydromechanische Probleme des Schiffsantriebs (Hamburg, 1932), pp. 63 et seq.; Journ. Aero. Sciences, 1 (1934), 8, 9; Proc. 4th Internat. Congress for Applied Mechanics, Cambridge, 1934, pp. 67 et seq.

fairly constant except near the walls and over the middle three-tenths of the distance between the walls. Near the walls, there is a marked departure from a constant value only when the viscous layer is approached. In the middle of the channel the ratio falls to zero, showing that there is no correlation between $u$ and $v$, a result to be expected from considerations of symmetry.

The restriction of the consideration of the turbulent mechanism at any point to the immediate neighbourhood of the point requires that $l$ should be small compared with a typical linear dimension of the system under examination—the half-width of a channel or the radius of a pipe, for example. In the cases considered in § 159 $l$ is, at best, only moderately small compared with the length involved, so that the results can, on this account, be only approximate. (A similar remark applies, however, to mixture length theories, since the mixture length is also assumed to be small.)

It has been pointed out in Chap. V, § 82 (p. 208), that the formula for $l$ can be found simply by dimensional arguments, without making the full assumption of similarity, by assuming that the value of $l$ at any point depends only on the distribution of mean velocity in the neighbourhood. Since a frame of reference moving with the mean velocity at the point may be taken, the value of $U$ itself cannot enter into the formula for $l$. The restriction to the immediate neighbourhood then implies that $l$ depends only on $dU/dy$ and $d^2U/dy^2$, and the formula follows. But the derivation in full from the principle of similarity is not without some interest, and a similar method can be applied to more complicated cases—to give formula (56) on p. 354, for example.

If the restriction to the immediate neighbourhood is removed, then $l$ may depend on $|U'/U''|$, $|U''/U'''|$, $|U'''/U''''|$, etc., where dashes denote differentiation with respect to $y$. The only case in which the simpler previous result would still be valid would be when the quotients were all proportional. This happens for flow in the absence of a pressure gradient, when $\tau$ is a constant and $U$ is given by (7). The turbulence flow patterns at different points may then be said to be strictly similar. Thus equation (7) when found by Kármán's method depends on fewer assumptions than the velocity distributions in § 159, for example. For any such case, the formula for $l$ will cease to be even a useful approximation near points where either $U' = 0$ or $U'' = 0$: the formula must therefore be expected to break down near the middle of a pipe or channel, whether we regard it as derived from the principle of similarity or simply from dimensional reasoning.

## 159. Velocity distributions for pressure flows between parallel walls and through circular pipes on the similarity theory (smooth or rough surfaces).

From (a) of the previous section (equations (46) and (49)) the velocity $U$ in a channel is obtained by integrating (27) ($\tau/y' = \rho U_\tau^2/h$) with use of the condition $dU/dy' = \infty$ at $y' = h$, where $y'$ is distance from the middle of the channel. The integral is

$$\frac{U_c - U}{U_\tau} = -\frac{1}{K_5}\left[\log_e\left(1 - \frac{y'^{\frac{1}{2}}}{h^{\frac{1}{2}}}\right) + \frac{y'^{\frac{1}{2}}}{h^{\frac{1}{2}}}\right] + b,\dagger \qquad (54)$$

where $K$ is given the suffix 5 to distinguish it from previous constants, and $b$ is an additive constant.

The comparison between experiment and this calculated result is shown in Fig. 99,‡ with $K_5$ ($= 0.295$) and $b$ ($= -0.172$) chosen to give agreement at $y'/h = 0.3$ and $0.7$; the agreement is good. (The experimental points are the same as in Figs. 95 and 97.) The value of $l/h$ is $2K_5[(y'/h)^{\frac{1}{2}} - y'/h]$, and a graph of $l/h$ is shown in the inset. The solution makes $U$ infinite at the walls.

Using (b) of the previous section (equations (46) and (53)), we can similarly obtain $U$ by integration with the condition $dU/dy' = \infty$ at $y' = h$. The solution is

$$\frac{U_c - U}{U_\tau} = \frac{2^{\frac{1}{2}}}{K_6}\left[1 - \left(1 - \frac{y'}{h}\right)^{\frac{1}{2}}\right] + b,\ddagger \qquad (55)$$

and is compared with experiment in Fig. 100 by choosing $K_6$ ($= 0.165$) and $b$ ($= -0.736$) to give agreement at $y'/h = 0.3$ and $0.7$; the agreement between the theoretical and experimental results is poor elsewhere. $l = 2K_6(h - y')$; $U$ is finite at the walls. The form of $l$ is the same as Prandtl's form, but in using Prandtl's form with the vorticity transfer theory we imposed the conditions $U = U_c$, $dU/dy' = 0$ at $y' = 0$, and the solution automatically made $dU/dy' = \infty$ at $y' = h$.

In order to calculate the velocity distribution for flow through a

† Kármán, *Göttinger Nachrichten, loc. cit.*
‡ Goldstein, *Proc. Roy. Soc.* A, **159** (1937), 480–483.

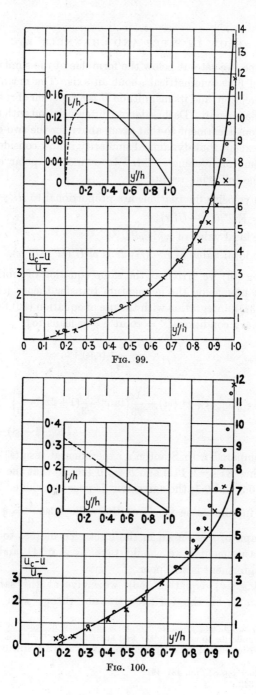

FIG. 99.

FIG. 100.

circular pipe, we set out below the formulae of the similarity theory for parallel flow symmetrical about an axis. The mean velocity is in the $x$-direction, and its magnitude $U$ is a function of $r$, the distance from the $x$-axis, only. Dashes denote differentiation with respect to $r$. An analogous treatment to that given above for the two-dimensional problem (except that dynamical similarity is now considered to hold in the frame of reference related to cylindrical polar coordinates) leads to the following results:[†]

   (i) the velocity fluctuations are proportional to $lU'$;

   (ii) $l = KU'/(U'' - U'/r)$;                                     (56)

   (iii) the local value of $|\tau|$ is $\rho l^2 U'^2$;

   (iv) the local value of $r^{-1}d(r\tau)/dr$ is $\rho l^2 U'(U''-U'/r)$.    (57)

By using (iii) we arrive at the same equation of motion for flow through a pipe as on the momentum transfer theory (cf. equations (30) and (31)); integration with the use of equation (56) for $l$ and with the boundary condition $U' = \infty$ at $r = a$ leads to[‡]

$$\frac{U_c - U}{U_\tau} = \frac{3}{K_7}f_1(r/a) + b,$$

where

$$f_1(x) = \tfrac{1}{6}\log_e(1 + x^{\frac{4}{3}} + x) + \frac{1}{\sqrt{3}}\left\{\tan^{-1}\frac{1}{\sqrt{3}}(1 + 2x^{\frac{1}{3}}) - \frac{\pi}{6}\right\}$$
$$- \tfrac{1}{3}\log_e(1 - x^{\frac{1}{3}}) - x^{\frac{1}{3}}.$$

$$(58)$$

(58) is compared with Stanton's experimental results‖ in Fig. 101, by choosing $K_7$ ($= 0\cdot171$) and $b$ ($= 0\cdot420$) so that the results agree at $r/a = 0\cdot3$ and $0\cdot7$; the agreement is poor elsewhere. The solution gives an infinite value for $U$ at the wall. Also $\dfrac{l}{a} = \tfrac{2}{3}K_7\left(\sqrt{\dfrac{a}{r}} - \dfrac{r}{a}\right)$, and a graph of $l$ is shown in the inset. To fit (58) to Nikuradse's 1933 measurements‖ we should take $K_7 = 0\cdot241$ and $b = 0\cdot578$; but the agreement is still poor.

By using (iv) (equation (56)) we arrive at the same equation of motion as on the vorticity transfer theory with symmetrical

---

[†] Goldstein, *op. cit.*, pp. 483–487. It is necessary in order to satisfy the condition for similarity to take only the first approximation for $U$, and not the second as in (43). Thus all terms of comparable magnitude have not, strictly speaking, been retained.

[‡] Goldstein, *op. cit.*, pp. 487–491.

‖ See p. 343.

turbulence (equation (38)); with $l$ as in (56) and with the boundary
condition $U' = \infty$ at $r = a$ we find on integration that†

$$\frac{U_c - U}{U_\tau} = \frac{\sqrt{3}}{K_8} \int_0^{r/a} \frac{\eta \, d\eta}{(1 - \eta^3)^{\frac{1}{2}}} + b, \qquad (59)$$

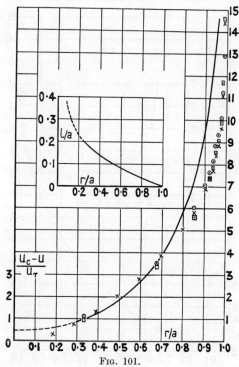

FIG. 101.

a result which is in excellent agreement with experiment when
$K_8$ (= 0·128) and $b$ (= 0·251) are chosen so that the theoretical
result agrees with experiment when $r/a = 0·3$ and $0·7$: the comparison is shown in Fig. 102. The solution (59) gives a finite value
for the velocity at the walls, and $l/a = \frac{2}{3}K_8(a^2/r^2 - r/a)$; a graph of
$l/a$ is shown in the inset. To fit (59) to Nikuradse's 1933 measurements we must take $K_8 = 0·181$, $b = 0·459$, and good agreement is
then obtained except near the wall.

The values found for $K$ in the two-dimensional and the three-
dimensional cases are again different.

† Goldstein, *op. cit.*, pp. 491, 492.

We may also note explicitly that the form of the theory which takes from the similarity hypothesis an expression for the rate at which momentum is communicated to unit volume, and which gives an equation of motion of the type of the vorticity transfer theory with symmetrical turbulence, leads to a result in excellent agreement

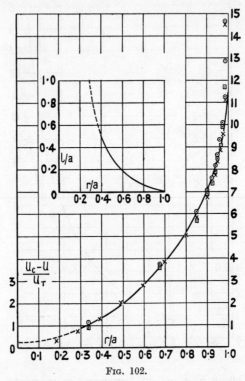

FIG. 102.

with observation for pipe flow, whereas the form which takes an expression for the shearing stress and gives an equation of motion of the type of the momentum transfer theory leads to a result in disagreement with observation; but that for pressure flow between parallel walls exactly the opposite is true.

## 160. Experimental determination of the mixture length.

Starting from the experimentally determined velocity distribution, and using the formula $\tau = \tau_0 r/a$, Nikuradse[†] has calculated

† Ver. deutsch. Ing., Forschungsheft 356 (1932), p. 21.

from equation (30) the value of the mixture length for smooth pipe flow
on the momentum transfer theory.  The results for various Reynolds

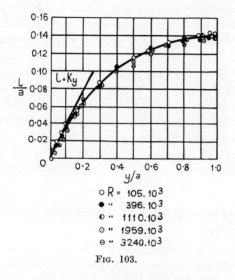

$$\circ\ R = 105.\ 10^3$$
$$\bullet\ ,,\qquad 396.\ 10^3$$
$$\circledcirc\ ,,\quad 1110.10^3$$
$$\odot\ ,,\quad 1959.10^3$$
$$\ominus\ ,,\quad 3240.10^3$$

FIG. 103.

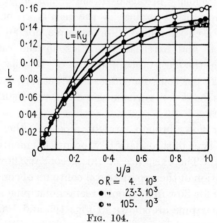
$$\circ\ R = 4.\quad 10^3$$
$$\bullet\ ,,\qquad 23\cdot3.10^3$$
$$\circledcirc\ ,,\quad 105.\quad 10^3$$

FIG. 104.

numbers are shown in Figs. 103 and 104 (wherein $y = a-r$ and denotes distance from the wall): there ceases to be any apparent scale
effect on $l$ for values of $R \geqslant 10^5$.  A curve similar to that in Fig. 103
has been given by Nikuradse† for pressure flow between parallel

† *Forschungsarbeiten des Ver. deutsch. Ing.*, No. 289 (1929), p. 42.

walls: curves for this case have also been given by Fritsch[†] and Dönch.[‡] A numerical differentiation of the experimental velocity distribution is necessary in order to obtain $l$; this process is usually inaccurate, since minor irregularities in the curve produce considerably greater irregularities in the value of the differential coefficient. The value of $l$ determined in this way cannot, therefore, be regarded as very reliable. Moreover, towards the middle of the pipe or channel it is necessary to calculate the ratio of two small quantities to determine $l$. On the other hand, if we start from either Prandtl's or Kármán's form for $l$, the velocity distribution is obtained by one or two integrations, so even if the assumed values are not quite right the calculated value of the velocity may not be far wrong. Thus we see why different values of $l$ may lead to values of $U$ very nearly the same.

### 161. Straight pipes of non-circular cross-section. Resistance. Secondary flow.

The resistance to turbulent flow has been measured in various smooth non-circular pipes: if four times the hydraulic mean depth and the mean velocity are used in the definition of $R$ it is found that when $\gamma$ is plotted against $R$ the results do not lie far from the results for a circular pipe. For example, for an annular pipe[||] having internal and external diameters in the ratio 0·833 the deviation from the resistance formula of Blasius (see equation (20)) is about $2\frac{1}{2}$ per cent. for values of $R$ between 2,000 and 8,000. Similar results hold for various shapes of rectangular[††] and triangular[‡‡] pipes.

In turbulent flow through straight pipes of other than circular cross-section a type of secondary flow occurs (as mentioned in Chap. II, §28 (p. 87)). The existence of this secondary flow is suggested by an examination of the experimental contours of constant velocity. The contours[||||] for flow in a smooth rectangular pipe and in an equilateral triangular pipe are shown in Figs 105 and 106, whereon the velocities are marked in metres per second: it will be noticed that

---

[†] *Zeitschr. f. angew. Math. u. Mech.* 8 (1928), 215.

[‡] *Forschungsarbeiten des Ver. deutsch. Ing.*, No. 282 (1926), p. 57.

[||] Winkel, *Zeitschr. f. angew. Math. u. Mech.* 3 (1923), 251–257.

[††] Schiller, *ibid.*, pp. 2–13; White and Davies, *Proc. Roy. Soc.* A, **119** (1928), 92–107; Nikuradse, *Ingenieur-Archiv*, 1 (1930), 306–332.

[‡‡] Schiller, *loc. cit.*; Nikuradse, *loc. cit.*

[||||] Nikuradse, *Forschungsarbeiten des Ver. deutsch. Ing.*, No. 281 (1926), pp. 13, 14.

instead of rapidly assuming an oval shape the contours near the walls
of the rectangular pipe, for example, show marked indentations near

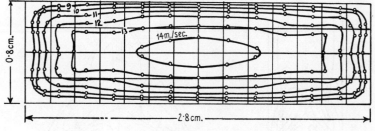

FIG. 105.

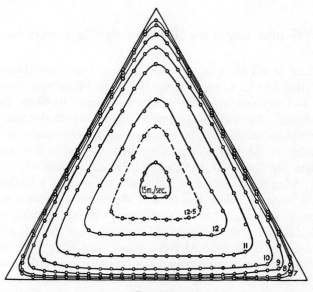

FIG. 106.

the corners. These indentations may be explained† if secondary
flows of the types shown in Figs. 107 and 108 exist. (The marked
curvature of the velocity contours implies a mean velocity component normal to the velocity contours from the concave to the

† This explanation was first given by Prandtl, *Verhandlungen des 2. internationalen
Kongresses für technische Mechanik, Zürich*, 1926, pp. 70–74.

convex side.) Direct experimental verification of the existence of secondary flows of the type shown has been obtained by a photographic method.†

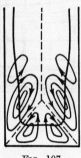

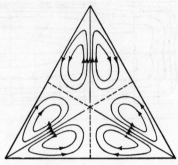

FIG. 107.                          FIG. 108.

## 162. The inlet length for turbulent flow in smooth circular pipes.

If care is taken to have the fluid free from disturbances at entry, the flow in a smooth pipe for some distance $x$ from the entry will be laminar even though turbulence develops farther downstream. The Reynolds number $Ux/\nu$ at which the transition to turbulence occurs may be expected to have the same order of magnitude as the Reynolds number for transition in flow along a flat plate (see Chap. VII, § 151, p. 325); i.e. $Ux/\nu$ will be of the order of $10^5$. After the transition to turbulence takes place, a further distance is required before the velocity distribution across a section takes its final form. When the conditions at the entry are disturbed the distance from the entry to the section where the velocity distribution first takes its final form is less than when there are no disturbances: the distance required is, therefore, dependent on the disturbances introduced at the entry. The only case where any attempt at calculation can be made is when the flow is fully turbulent at entry. Calculations from the momentum equation (of the type mentioned in Chap. VII, § 139 (p. 301), for laminar flow) have been made, using the $\frac{1}{7}$th power law for the velocity distribution in the boundary layer (equation (25), p. 340); the result obtained‡ for the inlet length is $0 \cdot 693 dR^{\frac{1}{4}}$, where $d$ is the pipe diameter and $R$ is

† Nikuradse, *Ingenieur-Archiv*, **1** (1930), 306–332.
‡ Latzko, *Zeitschr. f. angew. Math. u. Mech.* **1** (1921), 277–280.

$U_m d/\nu$. This result gives a shorter inlet length than any yet found in practice: the nearest result is 40 diameters found by Nikuradse[†] for disturbed entry conditions at a Reynolds number of $9 \times 10^5$, the theoretical result for this Reynolds number being 21 diameters.[‡] It is possible, however, that no experiments have yet approached fully turbulent conditions at entry.

## 163. Frictional intensity and resistance for turbulent flow along a smooth flat plate.

For flow in a smooth pipe the formulae

$$U/U_\tau = 8 \cdot 7(yU_\tau/\nu)^{\frac{1}{7}}$$

and 

$$\tau_0 = \rho U_\tau^2 = 0 \cdot 0225\rho U^2(\nu/Uy)^{\frac{1}{4}} \tag{60}$$

were deduced in § 155 (see eqns. (25) and (26)) for a certain range of Reynolds numbers from Blasius's empirical formula for the resistance. If we suppose that similar results hold throughout a boundary layer along a flat plate parallel to the steam, we may write

$$U/U_1 = (y/\delta)^{\frac{1}{7}} \tag{61}$$

and 

$$\tau_0/\rho U_1^2 = (U_\tau/U_1)^2 = 0 \cdot 0225(\nu/U_1\delta)^{\frac{1}{4}}, \tag{62}$$

where $\delta$ is the thickness of the boundary layer and $U_1$ is the velocity in the main stream at the edge of the boundary layer. These formulae are in moderate agreement with experiment for a certain range of Reynolds numbers.[||]

With the velocity distribution in (61)

$$\delta_1 = \int_0^\delta \left(1 - \frac{U}{U_1}\right) dy = \delta/8$$

and 

$$\vartheta = \int_0^\delta \left(1 - \frac{U}{U_1}\right)\frac{U}{U_1} dy = \tfrac{7}{72}\delta. \tag{63}$$

† *Ver. deutsch. Ing., Forschungsheft* No. 356 (1932).

‡ For further details and references see Schiller, *Handbuch der Experimentalphysik*, **4**, Part 4 (1932), 82–92.

|| Burgers (*Proc. 1st Internat. Congress for Applied Mechanics, Delft*, 1924, p. 121) states that (61) is in good agreement with experiment but that the coefficient 0·0225 in (62) is too large. According to Hansen (*Zeitschr. f. angew. Math. u. Mech.* **8** (1928), 185–199) the index $\frac{1}{7}$ in (61) is in good agreement for measurements near the plate, but for general agreement throughout the boundary layer a larger index is preferable. Dryden (*N.A.C.A. Report* No. 562 (1936), Figs. 18, 19) finds that at $U_1 x/\nu = 1 \cdot 6 \times 10^6$ and $2 \cdot 1 \times 10^6$ power law representations of the velocity distribution are unsatisfactory, but that a logarithmic formula represents the experimental results well as far as the data allow a comparison. See the footnote † on p. 363.

Since $U_1$ is constant, the momentum equation of the boundary layer (Chap. IV, equation (38), p. 133) is

$$d\vartheta/dx = \tau_0/(\rho U_1^2). \tag{64}$$

If we substitute for $\tau_0$ and $\vartheta$ from (62) and (63) and integrate, we find with the boundary condition $\delta = 0$ when $x = 0$† that

$$\delta = 0{\cdot}37(\nu/U_1 x)^{\frac{1}{5}}x. \tag{65}$$

The result obtained by substituting this value for $\delta$ in (62) is

$$\tau_0/\rho U_1^2 = 0{\cdot}0289(\nu/U_1 x)^{\frac{1}{5}}. \tag{66}$$

For a plate of length $l$ the frictional resistance coefficient $C_f$ $(= F/(\frac{1}{2}\rho U_1^2 l)$, where $F$ is the force per unit breadth on one side of the plate) is $0{\cdot}072R^{-\frac{1}{5}}$, where $R = U_1 l/\nu$.‡ When the experimental conditions ensure turbulence practically right from the leading edge, this formula is in good agreement with experiment up to $R = 3 \times 10^6$ if the coefficient $0{\cdot}072$, obtained by a rough estimation from pipe flow, is altered to $0{\cdot}074$ to obtain a better fit with the plate experiments. Correspondingly the coefficient $0{\cdot}0289$ in (66) would be altered to $0{\cdot}0296$. The effect of the laminar portion of the boundary layer near the leading edge can be allowed for, and good agreement obtained with experiment, by writing

$$C_f = 0{\cdot}074R^{-\frac{1}{5}} - AR^{-1},$$

where $A$ is a suitably chosen number.|| (See p. 366.)

Kármán has applied a slightly modified form of the logarithmic law (equation (8)) to the calculation of the skin-friction drag of a flat plate by means of the momentum integral equation.†† For the velocity distribution throughout the boundary layer he writes

$$\frac{U}{U_\tau} = \frac{1}{K}\left\{\log_e \frac{yU_\tau}{\nu} + h(y/\delta)\right\}, \tag{67}$$

where $h$ is an unknown function which may depend on the Reynolds

---

† We assume here that the flow is turbulent right from the leading edge.

‡ Kármán, *Zeitschr. f. angew. Math. u. Mech.* **1** (1921), 237–244.

|| Prandtl, *Ergebnisse der Aerodynamischen Versuchsanstalt zu Göttingen,* **3** (1927), 3–5. The comparisons were with experiments by Wieselsberger (*Göttinger Ergebnisse,* **1** (1925), 121–126) and Gebers (*Schiffbau,* **9** (1908), 435–452; 475–485) respectively, the value of $A$ which gives agreement with Gebers's results being 1,700. This value corresponds to a comparatively late transition to turbulence. (See p. 366.)

†† *Proc. 3rd Internat. Congress for Applied Mechanics, Stockholm,* 1930, **1**, 90–92; *Hydromechanische Probleme des Schiffsantriebs* (Hamburg, 1932), pp. 69–73. See also Prandtl, *Göttinger Ergebnisse,* **4** (1932), 18–29; *Aerodynamic Theory* (edited by Durand), **3** (Berlin, 1935), 145–154.

number: this function is introduced in order to improve the representation in the neighbourhood of the edge of the boundary layer. When $y = \delta$, $U = U_1$, so that

$$\frac{U_1}{U_\tau} = \frac{1}{K}\left(\log_e \frac{\delta U_\tau}{\nu} + h(1)\right). \tag{68}$$

Subtracting (68) from (67) we find that

$$U = U_1 - U_\tau f(y/\delta), \tag{69}$$

where    $f(y/\delta) = \dfrac{1}{K}\left[\log_e \dfrac{\delta}{y} + h(1) - h(y/\delta)\right].$†

We substitute from (69) into the momentum equation (64) and we find that

$$(U_\tau/U_1)^2 = \frac{d}{dx}\left[\frac{U_\tau}{U_1}\int_0^\delta f(y/\delta)\,dy - \frac{U_\tau^2}{U_1^2}\int_0^\delta [f(y/\delta)]^2\,dy\right],‡$$

i.e.    $$(U_\tau/U_1)^2 = \frac{d}{dx}\left[\frac{U_\tau}{U_1}c_1\delta - \frac{U_\tau^2}{U_1^2}c_2\delta\right], \tag{70}$$

where    $$c_1 = \int_0^1 f(\eta)\,d\eta, \qquad \eta = y/\delta,$$

and    $$c_2 = \int_0^1 [f(\eta)]^2\,d\eta.$$

Now from (68)    $$D\delta U_\tau/\nu = e^z, \tag{71}$$

where $\log D = h(1)$ and    $KU_1/U_\tau = z; \tag{72}$

and from (70)    $$\frac{DU_1 K^2}{\nu} = z^2 \frac{d}{dx}\left[\left(c_1 - \frac{Kc_2}{z}\right)e^z\right]. \tag{73}$$

Integrating (73) as it stands, we find that

$$\frac{DU_1 K^2}{\nu}x = z^2\left(c_1 - \frac{Kc_2}{z}\right)e^z - \int 2z\left(c_1 - \frac{Kc_2}{z}\right)e^z\,dz + \text{constant}$$

$$= z^2\left(c_1 - \frac{Kc_2}{z}\right)e^z - 2c_1(z-1)e^z + 2Kc_2 e^z + \text{constant}. \tag{74}$$

† Dryden (*N.A.C.A. Report* No. 562 (1936), Fig. 18) finds that the formula
$$U = U_1 - (U_\tau/K)\log_e(\delta/y)$$
represents well the not very extensive experimental data at $U_1 x/\nu = 1\cdot6 \times 10^6$ and $2\cdot1 \times 10^6$.

‡ It was pointed out when the logarithmic formula was derived that it is inapplicable for $yU_\tau/\nu < 30$. However, $\int U\,dy$ and $\int U^2\,dy$, when $U$ is given by (67), converge at $y = 0$, and the error introduced by retaining this form as far as the wall is small.

Thus

$$\frac{U_1 x}{\nu} = \frac{1}{K^2 D}[c_1 z^2 - (Kc_2 + 2c_1)z + 2(c_1 + Kc_2)]e^z - \frac{2}{K^2 D}(c_1 + Kc_2)$$
(75)

if $x = 0$ when $z = 0$.[†] For large values of $z$ an approximation is

$$\frac{U_1 x}{\nu} = \frac{1}{K^2 D}c_1 z^2 e^z,$$
(76)

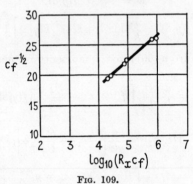

FIG. 109.

so that $\qquad \log_e \dfrac{U_1 x}{\nu} = \text{constant} + z + 2\log z,$

i.e. $\qquad\qquad c_f^{-\frac{1}{2}} = A' + B' \log_{10}(R_x c_f),$ (77)

where $c_f = U_\tau^2/(\frac{1}{2}U_1^2) = \tau_0/(\frac{1}{2}\rho U_1^2)$, $R_x = xU_1/\nu$, and $A'$ and $B'$ are constants.

In Fig. 109 $c_f^{-\frac{1}{2}}$ is plotted against $\log_{10}(R_x c_f)$, the results being obtained from measurements by Kempf.[‡] The theoretically predicted linear relation is in good agreement with experiment: the values of $A'$ and $B'$ found from the experiments are 1·7 and 4·15 respectively. The experiments are remarkable in that local values of the skin-friction coefficient $c_f$ were determined at points of a long plank by the direct measurement of the force on small movable plates.

If $C_f$ is the coefficient of mean friction (i.e. $F/(\frac{1}{2}\rho U_1^2 l)$, where $F$ is the frictional force per unit breadth on one side of a plate of length $l$) we can obtain a result for $C_f$ of precisely similar form to (77) by

[†] Here again we make the assumption that the flow is turbulent from the leading edge.

[‡] Kármán, *Journ. Aero. Sciences*, **1** (1934), 12; Kempf, *Werft, Reederei, Hafen*, **10** (1929), 234–239; 247–253.

starting from the integrated form of (70). Schoenherr[†] gives the result

$$C_f^{-\frac{1}{2}} = 4 \cdot 13 \log_{10}(R C_f) \qquad (R = l U_1/\nu),$$

the constants being chosen to give best agreement with experiment. For $R$ between $10^6$ and $10^9$ Prandtl[‡] finds that the formula

$$C_f = 0 \cdot 455 (\log_{10} R)^{-2 \cdot 58},$$

which is easier for the purposes of calculation, gives satisfactory values for $C_f$. A corresponding interpolation formula for $c_f$ is[||]

$$c_f = (2 \log_{10} R - 0 \cdot 65)^{-2 \cdot 3}.$$

Fig. 110, which is reproduced with some modifications from Schoenherr's paper,[††] shows the results of various experimental investigations[‡‡] of the frictional resistance of flat surfaces, together with the curves for $C_f$ represented by the formulae, quoted above, due to Schoenherr and to Prandtl. The curve corresponding to $C_f = 0 \cdot 074 R^{-\frac{1}{5}}$ (see p. 362), and Blasius's theoretical curve $C_f = 1 \cdot 328 R^{-\frac{1}{2}}$ for laminar flow, are also shown.

---

† Kármán, *loc. cit.*; Schoenherr, *Trans. Soc. Naval Architects and Marine Eng.* **40** (1932), 279–313.

‡ *Göttinger Ergebnisse*, **4** (1932), 27; *Aerodynamic Theory* (edited by Durand), **3** (1935), 153.

|| Schlichting, *Ingenieur-Archiv*, **7** (1936), 29.

†† *Loc. cit.*

‡‡ The key to the plotted points is as follows:

△ Froude, *British Association Report*, Brighton (1872), pp. 118–124.

⊗ Zahm, *Phil. Mag.* (6), **8** (1904), 58–66.

● Gebers, *Schiffbau*, **9** (1908), 435–452; 475–485.

+ Froude Tank, N.P.L., *Trans. N.E. Coast Inst. Engineers and Shipbuilders*, **32** (1915), 42–53; *Trans. Inst. Naval Architects*, **58** (1916), 65–73.

▲ Gibbons, *N.A.C.A. Report* No. 6, part 2 (1915); *N.A.C.A. First Annual Report* (1915), pp. 176–184.

⊘ Gebers, *Schiffbau*, **22** (1919), 687, 713, 738, 767, 791, 842, 899, 928.

⊖ Wieselsberger, *Göttinger Ergebnisse*, **1** (1925), 121–126.

⊕ Kempf and Kloess, *Werft, Reederei, Hafen*, **6** (1925), 435–443.

□ Kempf, *ibid.* **10** (1929), 234–239; 247–253.

○ Schoenherr, *loc. cit.*, 3-foot plane, smooth.

✦ *Ibid.*, 3-foot plane, leading edge roughened.

⊙ *Ibid.*, 6-foot plane, smooth.

○ *Ibid.*, 6-foot plane, artificial turbulence.

✕ Washington, unpublished (but see Schoenherr, *loc. cit.*)

The points attributed to Kempf were obtained from his values of the local coefficient $c_f$. For any length $x$,

$$F = \tfrac{1}{2}\rho U_1^2 x C_f = \int_0^x \tau_0 \, dx = \tfrac{1}{2}\rho U_1^2 \int_0^x c_f \, dx \, ;$$

hence $c_f = d(x C_f)/dx$. If $C_f^{-\frac{1}{2}} = 4 \cdot 13 \log_{10}(R_x C_f)$, as in Schoenherr's formula above, it follows after a little calculation that $c_f(0 \cdot 558 + 2 C_f^{-\frac{1}{2}}) = 0 \cdot 558 C_f$, and this relation was used to obtain values of $C$ from Kempf's values of $c_f$.

Near the leading edge of a plate at zero incidence the flow in the boundary layer is laminar; mathematically simple methods (which are at any rate approximately correct) for determining both the position of and the conditions at the point of transition to turbulence were discussed in Chap. VII, § 151. For a given value of $R_{xT}$ we can, therefore, determine the total skin-friction of a plate of length $l$, since we can determine the skin-friction in the laminar portion, the point of transition, and the skin-friction in the turbulent portion. We use Prandtl's condition that once transition has occurred the boundary layer behaves as though it had been turbulent right from the leading edge. Then an approximate correction may be found for the influence of the laminar portion by subtracting from $F$ (the frictional force per unit breadth) the turbulent frictional force for a length $x$ and adding the laminar frictional force for the same length, where the transition to turbulence is supposed to occur at the section $x$. The correction to $F$ is therefore $-\frac{1}{2}\rho U_1^2 x[C_{fT}-C_{fl}]$, where $C_{fT}$ and $C_{fl}$ are the turbulent and laminar coefficients of mean friction, and the square brackets indicate that values are to be taken at $R_{xT}$. The correction to $C_f$ is therefore $-(x/l)[C_{fT}-C_{fl}]$ or $-(R_{xT}/R)[C_{fT}-C_{fl}]$. We thus derive Prandtl's semi-empirical formula†

$$C_f = 0.455(\log_{10} R)^{-2.58} - A/R,$$

where $A$ depends on $R_{xT}$, being equal to $R_{xT}[C_{fT}-C_{fl}]$.

Three curves of $C_f$ against $R$ ($= lU_1/\nu$) according to this formula are drawn in Fig. 110; the different curves arise from choosing different values of $R_{xT}$. Prandtl chooses, as before (p. 362), a value 1,700 for $A$ to give agreement with Gebers's results for a smooth plate. The curve obtained by using this value of $A$ is drawn in Fig. 110; it corresponds to a value of $R_{xT}$ of about $5.3 \times 10^5$. The values of $A$ chosen for the other two curves shown in Fig. 110 are 600 and 300, which correspond to $R_{xT} = 1.9 \times 10^5$ and $10^5$ respectively.

Owing to our limited knowledge of the transition region it is usually considered essential to use large values of the Reynolds number $lU_1/\nu$ for measurements of turbulent skin-friction, in order that the laminar and transition regions should be as small as possible. Values of $R$ exceeding $10^6$ appear advisable.‡ The phenomena

---

† *Ergebnisse der Aerodynamischen Versuchsanstalt zu Göttingen*, **3** (1927), 3–5; **4** (1932), 27.

‡ See, for example, Bairstow, *Trans. Inst. Naval Architects*, **76** (1934), 329.

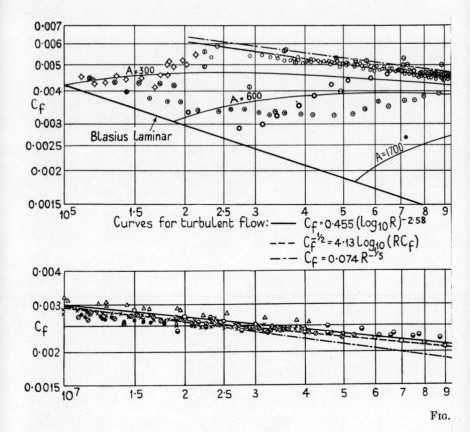

Curves for turbulent flow: —— $C_f = 0.455 (\log_{10} R)^{-2.58}$

$C_f^{-1/2} = 4.13 \log_{10}(RC_f)$

$C_f = 0.074 R^{-1/5}$

Fɪɢ.

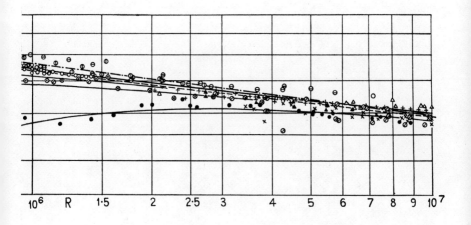

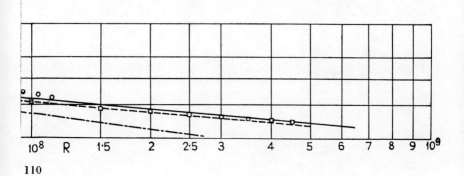

110

will be similar for shapes which have a slight curvature, such as ship models, so that in determining frictional resistance in model experiments the Reynolds number should be high enough to bring the experimental points outside the transition region if extrapolation to full scale is not to be difficult. It has been suggested† that fairly small models may be used if turbulence is induced by fixing, at the extreme forward end of the model, a transverse strip roughened with relatively coarse sand of uniform grain. The roughness itself causes an extra resistance, which can be eliminated by using two or more strips of different breadths and extrapolating the results to zero breadth.

### 164. The resisting moment on a smooth rotating disk.

The resisting moment on a smooth rotating disk has been considered by methods similar to those used in § 163.‡ The results are compared with experiment in Fig. 111, where $R = \omega a^2/\nu$, $\omega$ is the angular velocity, $a$ the radius, $M$ the resisting moment, and $C_M = M/(\frac{1}{2}\rho a^5 \omega^2)$. Theoretical and experimental results for laminar flow are also included. (See Chap. III, § 43, for the theoretical results.)

### 165. The calculation of the velocity distribution for flow along a flat plate on mixture length theories.

The methods we have used to obtain the frictional resistance in turbulent flow along a flat plate are based on a velocity distribution which is empirical for a large part of the boundary layer. It seems desirable to seek a more satisfactory theoretical form for this velocity distribution.

The measurements of Éliás‖ suggest that the velocity and temperature distributions for the flow in the boundary layer along a heated plate are (as near as can be ascertained from the somewhat scattered experimental values) of the same form. The momentum transfer theory predicts that these two distributions should be identical, whereas according to the vorticity transfer theory they are not

---

† See, for example, K. S. M. Davidson, *Journ. Applied Mechanics*, **3** (1936), A 41–A 46.

‡ Kármán (*Zeitschr. f. angew. Math. u. Mech.* **1** (1921), 247–250) uses the $\frac{1}{7}$th power velocity distribution. Goldstein (*Proc. Camb. Phil. Soc.* **31** (1935), 232–241) uses the logarithmic form and adjusts the constants to agree with experiment. See also Schultz-Grunow, *Zeitschr. f. angew. Math. u. Mech.* **15** (1935), 191–204.

‖ *Ibid.* **9** (1929), 434–453; **10** (1930), 1–14.

identical. This fact has been taken as an indication of the superiority of the momentum transfer theory for this particular motion. It has, however, been shown† that by a suitable choice

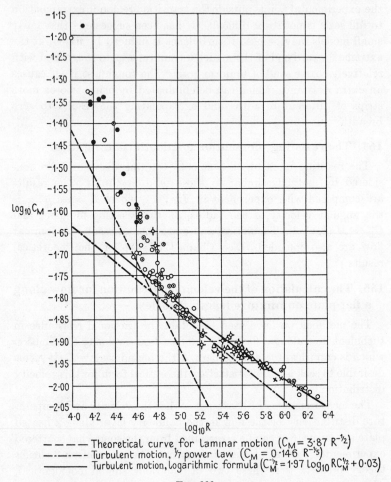

—————— Theoretical curve for laminar motion $(C_M = 3.87\ R^{-\frac{1}{2}})$
—·—·—·— Turbulent motion, $\frac{1}{7}$ power law $(C_M = 0.146\ R^{-\frac{1}{5}})$
——————— Turbulent motion, logarithmic formula $(C_M^{-\frac{1}{2}} = 1.97\ \log_{10} RC_M^{\frac{1}{2}} + 0.03)$

Fig. 111.

of certain constants the velocity distribution calculated by the vorticity transfer theory can be brought into excellent agreement with experiment except in the immediate neighbourhood of the wall, and that the corresponding temperature distribution (with a suitable

† Howarth, *Proc. Roy. Soc.* A, **154** (1936), 364–377.

choice of a constant) is very nearly the same as the velocity distribution, the difference being less than the scatter of the experimental points. The calculations, however, are not all that could be desired.

The equation of motion on the momentum or vorticity transfer theory is

$$U\frac{\partial U}{\partial x} + V\frac{\partial U}{\partial y} = \frac{\partial}{\partial y}\left\{l^2\left(\frac{\partial U}{\partial y}\right)^2\right\}, \tag{78}$$

or

$$U\frac{\partial U}{\partial x} + V\frac{\partial U}{\partial y} = l^2\frac{\partial U}{\partial y}\frac{\partial^2 U}{\partial y^2} \tag{79}$$

respectively, where $V$ is the velocity perpendicular to the plate. (Although the vorticity transfer theory does not give a velocity distribution in agreement with experiment in the immediate neighbourhood of a plane wall (cf. § 157, p. 344), it does not follow that the theory will not give satisfactory results over a large part of the turbulent boundary layer.)

The calculations are much simplified if it is assumed that the velocity distributions at different sections are similar: that is, that

$$\frac{U}{U_1} = f(\eta), \tag{80}$$

where $\eta = y/\delta$ and $\delta$ is the boundary layer thickness. For large Reynolds numbers this assumption is probably inaccurate: on the other hand, for the range of Reynolds numbers to which the published experimental data refer, the somewhat scattered experimental points satisfy the assumption of similarity within the experimental error.

The simplest assumption for $l$ would again be $l = Ky$, but this leads to a boundary layer thickness increasing linearly with the distance from the leading edge and to a constant skin-friction. It is, in fact, evident dimensionally that the skin-friction must be constant if we neglect viscosity, since the boundary conditions we apply, $U = U_1$ and $\partial U/\partial y = 0$ at $\eta = 1$, $V = 0$ and $\partial U/\partial y = \infty$ at $\eta = 0$, are also independent of the viscosity. In order to include the effect of viscosity we must either include the viscous term in the equation of motion near the wall and use the condition $\tau_0 = \mu(\partial U/\partial y)$ at the wall, or we must obtain a solution of the equations without viscosity which can be joined smoothly to the logarithmic velocity distribution (equation (8)) near the wall. Neither of these methods of attack is simple.  We can, however, obtain a solution if we modify the form of $l$ by writing

$$l^2 = ay^2\chi(x),\tag{81}$$

and then we find that we must take

$$\chi(x) = d\delta/dx\tag{82}$$

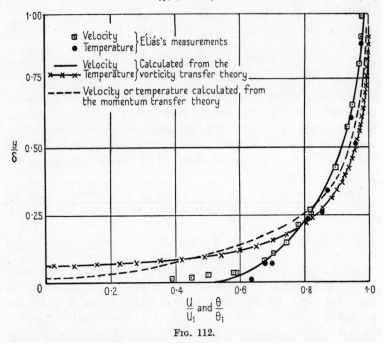

FIG. 112.

if the velocity distributions are to be similar, an unknown constant being included in $a$ in (81). The form of $\delta$ is not determined by the calculations, which give the velocity distribution as a function of $y/\delta$. Thus we can compare the calculated and experimental velocity distributions only by plotting $U/U_1$ against $y/\delta$. If $\delta$ is required it must be found from the experimental results.

With the modified form for $l$ in (81), the original partial differential equations reduce to ordinary third-order linear equations.  In either case the solution is of the form

$$U/U_1 = Af_1(a\eta)+Bf_2(a\eta),$$

the third independent solution being absent because if it were present it would be impossible to make $V$ vanish at the plate.  The boundary conditions $U/U_1 = 1$ and $\partial U/\partial y = 0$ at $\eta = 1$ serve to determine $A$ and $B$ in terms of $a$, which may in turn be found by making the observed and calculated curves of $U/U_1$ against $\eta$ pass through the same point when $\eta$ has a particular value.

Solutions for $U/U_1$ obtained in this way on both theories are shown in Fig. 112 and are compared with Éliás's measurements. The corresponding calculated temperature distributions are also shown.[†] The vorticity transfer theory seems, on the whole, to be in better agreement with experiment than the momentum transfer theory.

## 166. Flow in convergent and divergent channels.

Experiments have been carried out[‡] on pressure flow in smooth

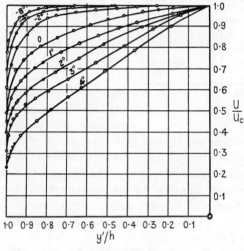

FIG. 113.

channels with various angles between the walls. Some of Nikuradse's results for the velocity distributions across various channels are reproduced in Fig. 113 (negative angles refer to channels converging in the

† In the original paper the formula $\delta = 0{\cdot}37x(xU_0/\nu)^{-\frac{1}{5}}$ was assumed in order to complete the solution. The comparison given in Fig. 112, however, in no way depends on this formula. The smallest value of $y$ at which the experimental value of $U$ could not be distinguished from $U_1$ was taken as the value of $\delta$ in determining $y/\delta$.

‡ *Forschungsarbeiten des Ver. deutsch. Ing.*, Nikuradse, No. 289 (1929); Dönch, No. 282 (1926); Kröner, No. 222 (1920). Reference may also be made to Gruschwitz, *Ingenieur-Archiv*, **2** (1931), 321–346; Demontis, *Publications scientifiques et techniques du ministère de l'air*, No. 87 (1936).

direction of flow; $U_c$ is the velocity in the middle of a channel; $h$ is the half-width and $y'$ the distance from the middle of the channel at any section). The symmetrical flow is stable when the channel is convergent or slightly divergent—even with a divergence of $4°$ there is no trace of instability. The velocity distributions for divergences of $5°$, $6°$, and $8°$, shown in Figs. 114, 115, and 116, are no longer symmetrical about the middle plane. No region of back

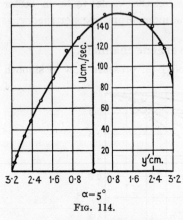

$\alpha = 5°$

FIG. 114.

flow at either wall is evident in Fig. 114; separation of the forward flow in the boundary layer along one of the walls is just beginning to occur at $5°$ (Nikuradse found that the first occurrence happened between $4·8°$ and $5·1°$, but could not be more exactly specified). At $6°$ the separation becomes apparent from the measurements taken: at which wall the separation occurs is a question settled by incidental changes in pressure, and once it occurs the pressure gradients are altered. For divergences of $5°$ and $6°$ there is no apparent tendency for separation to occur at the second wall, and the asymmetrical flow once established seems to be quite stable. At $8°$ a tendency appears for the asymmetrical distribution to switch over from wall to wall; each configuration is usually maintained for some time—long enough to allow measurements to be taken before a change occurs.

The existing methods of calculation of flow in turbulent boundary layers with pressure gradients are based largely on empirical relations obtained from experimental results for flow in convergent and divergent channels. It is therefore appropriate that at this stage we should discuss the various parameters involved.

For laminar flow along a flat plate in the absence of a pressure gradient

$$\tau_0/(\rho U_1^2) \propto R_\delta^{-1} \quad \text{or} \quad R_\delta \tau_0/(\rho U_1^2) = \text{constant},$$

where $R_\delta = \delta U_1/\nu$, $U_1$ is the velocity in the main stream at the edge of the boundary layer and $\delta$ is the boundary layer thickness. The parameter used for the discussion of a general problem of flow

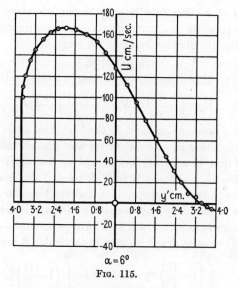

$\alpha = 6^\circ$

FIG. 115.

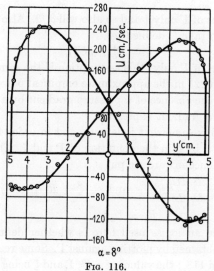

$\alpha = 8^\circ$

FIG. 116.

in a laminar boundary layer is, in the Kármán-Pohlhausen method,

$$\Lambda = \frac{dU_1}{dx}\frac{\delta^2}{\nu} = \left(\frac{1}{U_1}\frac{dU_1}{dx}\delta\right)R_\delta = \left(-\frac{1}{\rho U_1^2}\frac{dp}{dx}\delta\right)R_\delta,$$

and for laminar flow in the presence of a pressure gradient, $R_\delta \tau_0/(\rho U_1^2)$

is a function of $\Lambda$. For flow in a turbulent boundary layer in the absence of a pressure gradient, the empirical assumption of the $\frac{1}{7}$th power velocity distribution leads to the relation

$$\tau_0/(\rho U_1^2) \propto R_\delta^{-\frac{1}{4}}$$

(see eqn. (62)). It is a reasonable analogy to employ the parameter

$$\Gamma = \left( \frac{1}{U_1} \frac{dU_1}{dx} \delta \right) R_\delta^{\frac{1}{4}} = \left( -\frac{1}{\rho U_1^2} \frac{dp}{dx} \delta \right) R_\delta^{\frac{1}{4}}$$

for flow with a pressure gradient in the turbulent case: $\Gamma$ is positive for flow in a converging channel or for any accelerated flow, and is negative for flow in a diverging channel or for any retarded flow. When the flow in a channel is fully developed the boundary layer must be taken to reach to the middle, and the boundary layer thickness must be replaced by half the width of the channel.

If the analogy is to hold $\Gamma$ will be a form parameter,—i.e. the value of $\Gamma$ will fix the velocity graph, so that, in particular, $\delta_1$ and $\vartheta$ defined by

$$\delta_1 = \int\limits_0^\delta \left( 1 - \frac{U}{U_1} \right) dy \quad \text{and} \quad \vartheta = \int\limits_0^\delta \left( 1 - \frac{U}{U_1} \right) \frac{U}{U_1} dy$$

will be functions of $\Gamma$ only,[†] and so also will $\delta_1/\vartheta$. Also the analogous form for $\tau_0$ is to take $\zeta$, defined by

$$\zeta = R_\delta^{\frac{1}{4}} \tau_0/(\rho U_1^2),$$

to be a function of $\Gamma$ only: this assumption is the same as replacing the constant $0 \cdot 0225$ in the approximate empirical power-law formula (62) for flow in the absence of a pressure gradient by a function of $\Gamma$ when a pressure gradient is present.

The length $\delta$ is somewhat indefinite, and in order to be precise $\delta$ may be replaced by $\vartheta$ in the definition of $\Gamma$ and $\zeta$. This we suppose done, so that henceforward we put

$$\Gamma = \frac{\vartheta}{U_1} \frac{dU_1}{dx} R_\vartheta^{\frac{1}{4}} \quad \text{and} \quad \zeta = R_\vartheta^{\frac{1}{4}} \tau_0/(\rho U_1^2), \left.\right\} \tag{83}$$

where $\qquad\qquad\qquad R_\vartheta = U_1 \vartheta/\nu.$

A test of the analogies used here is whether definite curves for $\delta_1/\vartheta$ and $\zeta$ are obtained by plotting against $\Gamma$. Some results are shown in Figs. 117 and 118,[‡] the values of $\delta_1/\vartheta$, $\Gamma$ and $\zeta$ being obtained from

[†] This property cannot hold exactly (just as the corresponding property of $\Lambda$ is approximate) since the velocity distribution at any section depends on conditions upstream. Since $\delta_1$ and $\vartheta$ are integrals depending on the velocity distribution, any slight variations in the dependence of the velocity on $\Gamma$ will have less effect on $\delta_1$ and $\vartheta$,

[‡] These figures are taken from a paper by Howarth, *Proc. Roy. Soc.* A, **149** (1935), 558–586 (Figs. 10 and 11).

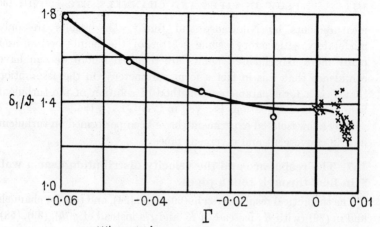

o Nikuradse's results, as corrected by Buri
x Buri's results

FIG. 117.

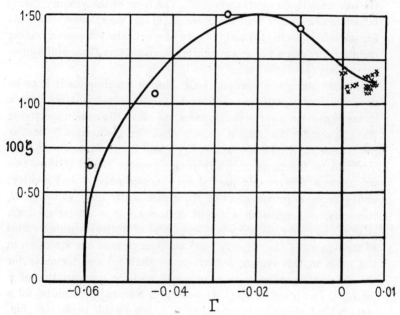

o Nikuradse's results, as corrected by Buri
x Buri's results

FIG. 118.

measurements by Nikuradse and Buri.[†] The results are only moderately satisfactory; some scatter of the points occurs, and many more experimental points are required before we can have confidence that $\Gamma$ is in fact a form parameter. On the assumption that $\Gamma$ is a form parameter a method of solution of the turbulent boundary layer equations is given in Chap. IX, § 194.

Only a few isolated experiments have been performed on turbulent flow in convergent and divergent pipes.[‡]

### 167. The resistance and the velocity distribution near a wall in flow through rough pipes.

The theoretical results given in (29), (36), (54), and (55) for channels and in (29) (with $K_2$ instead of $K_1$ and $r/a$ instead of $y'/h$), (39), (58), and (59) for pipes hold equally whether the walls are smooth or rough, but in neither instance do they hold right up to a wall: in the neighbourhood of a wall viscosity in the one case and roughness in the other are the important factors. The form of the velocity distribution near a smooth wall was given in (8) and the resistance formula for smooth pipes in (18) and (19); we now consider the corresponding results for rough walls and rough pipes. Equation (7) is still appropriate, but the modification given in (8) is not.

In considering even geometrically similar roughnesses it is to be expected that it will be necessary to take into account not only a linear dimension $\epsilon$ of the roughnesses but also a dimension specifying their spacing. We should expect that for geometrically similar roughnesses in a pipe or channel, $U_m/U_\tau$ and the resistance coefficient $\gamma$ would be functions of the Reynolds number $R$ and of these two parameters. For closely packed roughnesses only $R$ and $\epsilon$ enter; and $\gamma$ becomes independent of $R$ for sufficiently large values of $R$, the smallest permissible value of $R$ decreasing as $\epsilon/a$ (or $\epsilon/h$ for a channel) increases. When $\gamma$ is independent of $R$, the resistance varies as the square of the velocity; and we shall restrict our attention in the main to this region, for which we shall set out formulae for the distribution of velocity near a wall and for the variation of $\gamma$ (or $U_m/U_\tau$) with $\epsilon/a$ (or $\epsilon/h$). According to Nikuradse's results[||] for a certain kind of sand roughness, $\epsilon U_\tau/\nu \geqslant 100$ in this region (cf. Fig. 120). On the other hand, if $\epsilon U_\tau/\nu \leqslant 4$ the surface is to be counted

---

[†] *Zürich Dissertation*, 1931.          [‡] For references see p. 400.

[||] *Ver. deutsch. Ing., Forschungsheft* 361 (1933). The pipes were artificially roughened by means of uniform sand grains giving the values of $a/\epsilon$ shown in Fig. 119.

hydraulically smooth. (If the roughnesses do not project beyond the laminar sub-layer, an argument of the type given in Chap. VII, § 142 still applies; even if the roughnesses project into the layer in which the viscous stresses and the Reynolds stresses are of comparable magnitude, the result should not be much altered. The estimate thus found for the upper limit of $\epsilon U_\tau/\nu$ was 5·5: the experimental result is about 4.)

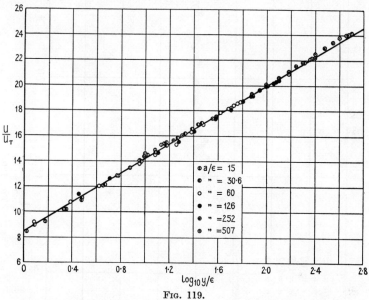

FIG. 119.

We now confine our attention to flows in which the skin-friction is proportional to the square of the velocity. Then in the neighbourhood of a wall $\epsilon$ is the important linear dimension, and we may write $U/U_\tau = g(y/\epsilon)$. In virtue of (7) it follows that

$$U/U_\tau = \text{constant} + K^{-1}\log_e y/\epsilon \qquad (84)$$

in the neighbourhood of the wall. Equation (84) is the form of the universal velocity distribution near a rough wall. Nikuradse's experimental results for pipe flow with a quadratic resistance law are shown in Fig. 119, together with the line

$$U/U_\tau = 8{\cdot}48 + 5{\cdot}75\log_{10} y/\epsilon. \qquad (85)$$

By combining (84) with (11) we find for flow in a pipe (in exactly the same way as we found (14)) that

$$U_c/U_\tau = \text{constant} + K^{-1}\log_e a/\epsilon,$$

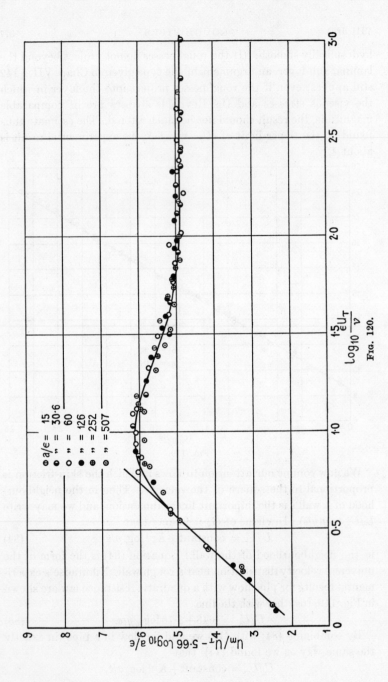

Fig. 120.

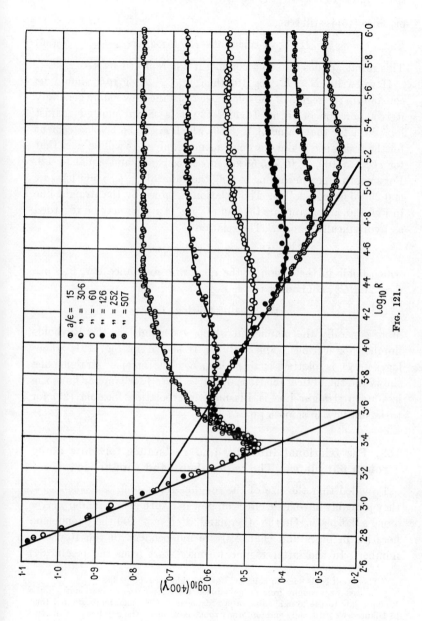

FIG. 121.

or, since (15) is still true,

$$U_m/U_\tau = \text{constant} + K^{-1}\log_e a/\epsilon.\dagger \qquad (86)$$

This result applies also to a channel if $a$ is replaced by $h$.

If we write $B = K^{-1}\log_e 10$, then $U_m/U_\tau - B\log_{10} a/\epsilon$ should be constant in the region of the quadratic resistance law. More generally it is a function of $\epsilon U_\tau/\nu$. In Fig. 120 it is shown plotted against $\log_{10} \epsilon U_\tau/\nu$ for that value of $B$ (5·66) which gives the best agreement for the resistance to flow through a smooth pipe. It will be seen that for $\epsilon U_\tau/\nu \geqslant 100$, $U_m/U_\tau - 5\cdot66\log_{10} a/\epsilon$ is constant and equal to 4·9. For smooth pipes, according to (18), the value of $U_m/U_\tau - 5\cdot66\log_{10} a/\epsilon$ is $0\cdot29 + 5\cdot66\log_{10} 2\epsilon U_\tau/\nu$. This relation is shown by the straight line in Fig. 120, and it appears that for $\epsilon U_\tau/\nu \leqslant 4$ a pipe may be regarded as hydraulically smooth. The relation

$$U_m/U_\tau = 4\cdot9 + 5\cdot66\log_{10} a/\epsilon,$$

which holds in the region of the quadratic resistance law, is, since $\gamma = 2U_\tau^2/U_m^2$, equivalent to

$$\gamma^{-\frac{1}{2}} = 3\cdot46 + 4\cdot00\log_{10} a/\epsilon.$$

More generally the manner in which $\gamma$ depends on the Reynolds number for various values of $a/\epsilon$ is shown in Fig. 121, where $\log_{10}(400\gamma)$ is plotted against $\log_{10} R$. The steeper straight line shows the theoretical relationship ($\gamma = 16R^{-1}$) for laminar flow; the less steep straight line is Blasius's interpolation formula (20) for turbulent flow in smooth pipes.‡

## 168. The frictional intensity and resistance for flow along rough flat plates. The 'equivalent sand roughness'.

Provided that the size of the roughnesses is such that $\epsilon U_\tau/\nu \leqslant 4$, they probably have little effect on flow in a turbulent boundary layer along a flat plate. On the other hand, if $\epsilon U_\tau/\nu \geqslant 100$ the phenomena here (as in pipes and channels) are independent of the Reynolds number. In the latter régime we may start from the result (84)

---

† This result is also due to Kármán. See the references on p. 350.

‡ In cast or wrought iron or galvanized steel pipes the transition from the 'smooth' law to the 'rough' (quadratic resistance) law is much more gradual than in Nikuradse's artificially and uniformly roughened pipes: the gap between the two has been bridged by using artificial non-uniform roughness. See Colebrook and White, *Proc. Roy. Soc.* A, **161** (1937), 367–381.

for the universal velocity distribution near a wall and then proceed as in § 163, but with $yU_\tau/\nu$ replaced by $y/\epsilon$. We thus obtain the approximate result

$$c_f^{-\frac{1}{2}} = A'' + B'' \log_{10}(c_f^{\frac{1}{2}} x/\epsilon), \tag{87}$$

where $c_f = \tau_0/(\frac{1}{2}\rho U_1^2)$ as on p. 364. Kármán[†] gives $A'' = 5 \cdot 8$ and $B'' = 4 \cdot 15$ as the result of calculation based on data from rough pipes and channels. A corresponding formula to (87) holds for $C_f$.

More elaborate calculations have been carried out by Prandtl and Schlichting,[‡] who also took into account the region in which the skin-friction is not proportional to the square of the velocity. They took from pipe experiments the velocity distribution near a wall, and for purposes of calculation replaced it by the distributions $U/U_\tau = A + 5 \cdot 75 \log_{10} y/\epsilon$, where

$$\left.\begin{aligned}
A &= 5 \cdot 5 + 5 \cdot 75 \log_{10} \epsilon U_\tau/\nu &(\epsilon U_\tau/\nu \leqslant 7 \cdot 08), \\
&= 9 \cdot 58 &(7 \cdot 08 \leqslant \epsilon U_\tau/\nu \leqslant 14 \cdot 1), \\
&= 11 \cdot 50 - 1 \cdot 62 \log_{10} \epsilon U_\tau/\nu &(14 \cdot 1 \leqslant \epsilon U_\tau/\nu \leqslant 70 \cdot 8), \\
&= 8 \cdot 48 &(70 \cdot 8 \leqslant \epsilon U_\tau/\nu),
\end{aligned}\right\} \tag{88}$$

the two middle equations replacing the transition from the 'hydraulically smooth' to the 'completely rough' régime. (It will be seen that the values of $\epsilon U_\tau/\nu$ at the limits of this transition region, as taken by Prandtl and Schlichting, are rather different from those (4 and 100) mentioned above.) If $\delta$ is the boundary layer thickness and $U_1$ the velocity at the edge of the boundary layer, then (88) holds when $U = U_1$ and $y = \delta$. Thus we obtain a relation between $U_1/U_\tau$, $\epsilon U_\tau/\nu$, and $\delta/\epsilon$. The momentum integral with the velocity distribution given by (88) provides a further relation between these quantities and $x$. Elimination of $\delta$ between these two relations leads to the determination of $U_\tau/U_1$ as a function of $x/\epsilon$ and $xU_1/\nu$. Reference may be made to the original paper for further details.[||]

According to Schlichting,[††] in the 'completely rough' régime (skin-friction proportional to the square of the velocity), the results

† *Journ. Aero. Sciences*, **1** (1934), 18.

‡ *Werft, Reederei, Hafen*, **14** (1934), 1–4.

|| See also Prandtl, *Aerodynamic Theory* (edited by Durand), **3** (1935), 153, 154.

†† *Ingenieur-Archiv*, **7** (1936), 29.

of these calculations are well represented by the interpolation formulae

$$c_f = (2\cdot87 + 1\cdot58\log_{10} x/\epsilon)^{-2\cdot5},$$
$$C_f = (1\cdot89 + 1\cdot62\log_{10} x/\epsilon)^{-2\cdot5},$$

where $c_f$, $C_f$ are the coefficients of local skin-friction and of mean frictional intensity for a length $x$, respectively, as defined on p. 364.

Schlichting has also carried out experiments on roughnesses of different kinds in the region in which the resistance is independent of the Reynolds number, and introduces a length $\epsilon_s$, called the 'equivalent sand roughness', which is such that a uniform roughness formed by sand grains of this size and of the type used by Nikuradse in his pipe experiments would produce the same resistance as that actually found.[†] If $\epsilon_s$ is known for any type of roughness, then the formulae and calculations for rough pipes, channels, and plates set out above may be applied with $\epsilon_s$ in place of $\epsilon$. Schlichting also investigated the influence of the density of the roughnesses: it appears that the greatest resistance does not always correspond with the greatest possible density.

## 169. Turbulent flow under pressure in curved channels.

Consider the flow under pressure in a channel whose walls are concentric cylinders, the pressure gradient being parallel to the walls and perpendicular to their common axis. In the fully developed region (where the velocity distribution at any one section is identical with that at any other) the angular momentum of a portion of the fluid about the axis of the cylinders is constant, and the sum of the moments of the pressures and shearing stresses acting on the portion of fluid must vanish. Hence

$$\frac{\partial}{\partial r}(\tau r^2) = \frac{\partial p}{\partial \theta} r, \tag{89}$$

where $r$ is distance measured from the axis, $\theta$ is angular distance round the axis, and $\tau$ is the shearing stress $\overline{p_{r\theta}}$. On the momentum transfer theory (a fluid element being supposed to keep unaltered its moment of momentum about the common axis of the cylinders)

$$v_\theta = -L_r\left(\frac{dU}{dr} + \frac{U}{r}\right) \quad \text{and} \quad \tau = -\rho\overline{v_r v_\theta} = \rho\overline{L_r v_r}\left(\frac{dU}{dr} + \frac{U}{r}\right), \tag{90}$$

where $L_r$ and $v_r$ are the components of the mixture length vector

[†] *Ingenieur-Archiv*, 7 (1936), 1–34. See also Colebrook and White, *loc. cit.*

and the turbulent velocity in the direction of $r$ increasing, $v_\theta$ is the component of the turbulent velocity in the direction of $\theta$ increasing, and $U$ is the mean velocity in the direction of motion, i.e. in the direction of $\theta$ increasing. The two-dimensional (or modified) vorticity transfer theory leads to the equation

$$\frac{1}{\rho}\frac{1}{r}\frac{\partial p}{\partial \theta} = \overline{L_r v_r}\frac{d}{dr}\left(\frac{dU}{dr}+\frac{U}{r}\right), \qquad (91)$$

since the mean vorticity is $dU/dr+U/r$. The first difficulty is to find a satisfactory assumption for $\overline{L_r v_r}$. If we proceed as in two-dimensional straight flow we should put $v_r \propto v_\theta$ and

$$\overline{L_r v_r} = l^2\left(\frac{dU}{dr}+\frac{U}{r}\right), \qquad (92)$$

but this assumption is even more doubtful than in the case of straight flow. Until a form for $\overline{L_r v_r}$ is chosen we cannot even deduce the form of $l$ from experimental velocity distributions. Further, the fact that the flow is curved would in any case necessitate some modification of the forms used for $l$ in straight flow.

By integration of (89), we find that

$$\tau = \frac{1}{2}\frac{\partial p}{\partial \theta}\left[1-\left(\frac{r_m}{r}\right)^2\right], \qquad (93)$$

where $r_m$ is the value of $r$ for which $\tau$ vanishes. Now according to (90) $\tau$ vanishes when $dU/dr+U/r = 0$, whatever assumption we make about $\overline{L_r v_r}$. It can, however, be shown experimentally[†] that $\tau$ does not vanish either when $dU/dr+U/r = 0$ or when $dU/dr-U/r = 0$. The experimental results are shown in Fig. 122, in which $U/U_m$ and $\tau/\tau_i$ are plotted against distance from the centre of curvature of the walls, where $U_m$ is the average of $U$ over a cross-section and $\tau_i$ is the wall friction at the inner wall. The points where $dU/dr = \pm U/r$, found by drawing the curves $U/r = $ constant and $rU = $ constant and choosing the constants so that the curves are tangential to the observed curve of $U$, are shown by arrows.

For flow in a straight channel

$$(U_c-U)/\sqrt{(\tau_0/\rho)}$$

is a universal function of $y/h$, where $h$ is the half-width (see equation (11)). Alternatively $\tau_0/\rho$ may be replaced by $-(h/\rho)\,\partial p/\partial x$. For a

† Wattendorf, *Proc. Roy. Soc.* A, **148** (1935), 565–598.

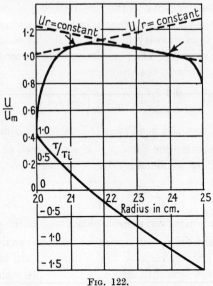

Fig. 122.

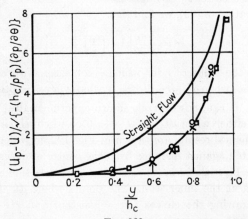

Fig. 123.

possible extension to curved channels we may take as reference velocity the potential velocity $U_p$ given by $U_p r = (Ur)_{\max}$. The point where the actual velocity coincides with $U_p$ may be taken as origin, thus dividing the channel into two unequal parts. Denoting by $h_c$ the width of one of these, Wattendorf plots

$$(U_p - U)/\sqrt{\{-(h_c/\rho r_p)(\partial p/\partial \theta)\}}$$

VIII. 169]   COAXIAL CYLINDERS: INNER CYLINDER ROTATING    385

against $y/h_c$, where $r_p$ is the radius of the circle at which $U_p = U$. For each channel two curves are obtained, one for each part of the channel: these are shown in Fig. 123 for Wattendorf's two channels, and are compared with the corresponding result for straight flow. The squares and triangles refer to the inner portions of two different channels: the circles and crosses to the outer portions. The four curves for curved flow coincide but are quite different from that for straight flow: this clearly indicates the need for experiments with channels of weaker curvature. (The inner and outer radii in Wattendorf's experiments were 45 and 50 cm. in one experiment and 20 and 25 cm. in another.)

## 170. Turbulent flow between coaxial rotating cylinders.

(a) Taylor† has considered both theoretically and experimentally turbulent motion between two rotating cylinders under no pressure gradient, and Wattendorf (loc. cit.) has also made measurements of such a motion. On the vorticity transfer theory (equation (91)) the mean vorticity is constant throughout the fluid whatever be the form of $\overline{L_r v_r}$, and the velocity $U = Ar + B/r$, where $A$ and $B$ are constants. From the measurements of Taylor and Wattendorf with the *inner cylinder rotating* and the outer one at rest it is apparent that $rU$ is constant over about 80 per cent. of the region between the cylinders (but not near to the walls). Thus the solution above is satisfactory for 80 per cent. of the channel if $A$ vanishes. On the momentum transfer theory (equation (90)) $d(rU)/dr$ is a non-zero constant (a multiple, in fact, of the couple applied to the cylinders); and this is in definite contradiction to the experimental result $rU = $ constant. It can, however, be shown to be in better agreement with experiment than the result $rU = $ constant near the walls. Fig. 124 shows the observed velocity distribution between two rotating cylinders together with the observed velocity distributions for pressure flow in curved and straight channels, the measurements being taken by Wattendorf (loc. cit.). The squares refer to the measurements in flow between rotating cylinders, the circles to pressure flow in a straight channel, and the crosses and triangles to pressure flow in curved channels. (For the flow between rotating cylinders, $(Ur)_{max}$ must be taken to denote the maximum value of $Ur$ when the region near the inner cylinder is excluded.)

† *Proc. Roy. Soc.* A, **151** (1935), 494–512; **157** (1936), 546–578.

It should be added that on the momentum transfer theory the distribution of temperature in the fluid, when the cylinders are maintained at different temperatures, should be identical with the distribution of $Ur$, but that, when the rotating inner cylinder was maintained at a higher temperature than the outer one, Taylor

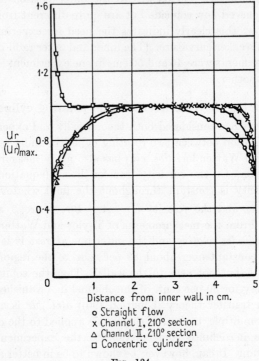

o Straight flow
x Channel I, 210° section
△ Channel II, 210° section
□ Concentric cylinders

FIG. 124.

found a continuous decrease of temperature outwards, the variation being 22 per cent. over the region where the variation in $Ur$ is less than 0·4 per cent. (See Chap. XV, § 283.)

(b) Fig. 125 shows the distribution of velocity in water contained between cylinders of radii $R_1 = 4·05$ cm. and $R_2 = 3·13$ cm. when only the *outer cylinder rotates*. $U$ is the velocity at a distance $r$ from the common axis, and $N$ is the number of revolutions per sec. Measurements were made at $N = 33·3, 30·0, 26·7$ and $23·3$.

Unlike most cases of turbulent motion past solid surfaces, this is a case in which the slope of the velocity distribution curve at the outer cylinder is easily measurable, so the tangential stress $\tau_0$ at the outer cylinder can be calculated from the formula

$$\frac{\tau_0}{\rho U_1^2} = \frac{\mu}{4\pi^2 \rho R_1^2 N^2}\left(\frac{dU}{dr} - \frac{U}{r}\right)_{r=R_1},$$

where $U_1$ is the velocity of the outer cylinder, and is equal to $2\pi R_1 N$.

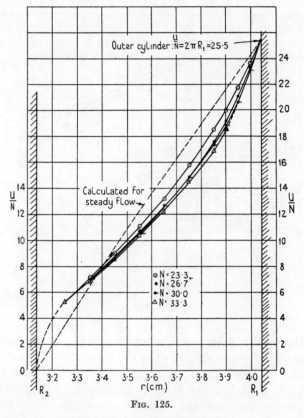

FIG. 125.

The results obtained by applying this formula to the measurements shown in Fig. 125 are given in the following table.

| $N$ | $\dfrac{\tau_0}{\rho U_1^2}$ | $\dfrac{tU_1}{\nu}$ |
|---|---|---|
| 23·3 | $2·4 \times 10^{-5}$ | $5·4 \times 10^4$ |
| 26·7 | $2·6 \times 10^{-5}$ | $6·2 \times 10^4$ |
| 30·0 | $2·6 \times 10^{-5}$ | $7·0 \times 10^4$ |
| 33·3 | $2·5 \times 10^{-5}$ | $7·7 \times 10^4$ |

For comparison with the results of direct torque measurements described below, the values of $tU_1/\nu$, where $t$ ($= R_1 - R_2$) is the thickness of the annulus between the two cylinders, are given in the table in addition to those of $\tau_0/\rho U_1^2$, and are marked in Fig. 126 below ($t/R_1 = 0.227$).

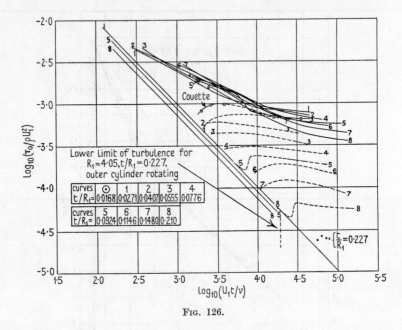

FIG. 126.

(c) The results of torque measurements made by balancing the torque acting on one cylinder when the other rotates are shown in Fig. 126. In that figure $\log_{10}\tau_0/\rho U_1^2$ is shown plotted against $\log_{10}(U_1 t/\nu)$, where $U_1$ is the velocity of the outer cylinder relative to the inner one. Separate curves are given for each value of $t/R_1$, the full lines representing values when the inner cylinder rotates and the dotted lines when the outer cylinder rotates. The straight lines at 45° to the axes are theoretical results for non-turbulent flow. The effect of the stability which characterizes the motion when the outer cylinder rotates is to reduce the friction. The instability caused by the curvature when the inner cylinder rotates does not seem to affect the resistance appreciably.

(d) The critical speed at which vortices begin to be produced when

the inner cylinder rotates has been calculated. When $t/R_1$ is small
it can be expressed by the equation†

$$2[\log_{10} U_1 t/\nu]_{\text{crit}} + \log_{10} t/R_1 = 3\cdot232.$$

This relationship, which represents well the observed critical speeds

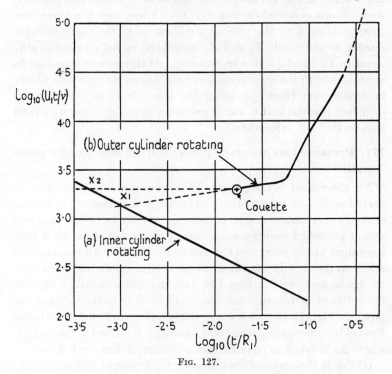

Fig. 127.

when the inner cylinder is rotating, is shown as a straight line in
Fig. 127, where a curve representing the observations of the speeds
at which turbulence sets in when the outer cylinder is rotating is
also shown.

The effect of rotation on the critical speed is very great. When
$\log_{10} t/R_1 = -1$, so that $t = 0\cdot1R_1$, it will be seen that the critical
speed with the outer cylinder rotating is about 60 times as great as
when the inner cylinder rotates. When $t/R_1 = \frac{1}{30}$ this ratio is about
10. In the case of the smallest value of $t/R_1$ for which results have

† Taylor, *Proc. Roy. Soc.* A, **157** (1936), 558. The relation is deduced from the
results of the calculations referred to in Chap. V, § 75.

been published—namely those of Couette[†] for which $t/R_1 = 0.0168$—the critical speed with the outer cylinder rotating is still about 6 times as great as when the inner cylinder rotates. To estimate the ratio $t/R_1$ at which rotation will have no effect on the critical speed the upper curve in Fig. 127 may be produced backwards in two plausible ways, shown as dotted lines in Fig. 127. These meet the lower line, corresponding with the critical condition when the inner cylinder rotates, at the points $X_1$ and $X_2$ corresponding with values of $t/R_1$ equal to $10^{-3}$ and $\frac{1}{2} \times 10^{-3}$ respectively. If therefore it is desired to approximate to the hydrodynamical conditions between two planes in relative movement by using the space between two rotating cylinders, it seems that it will be necessary to use cylinders with radii so large that $t/R_1$ is less than $10^{-3}$.[‡]

### 171. Pressure loss in curved pipes. Curved flow along a plane wall.

No theoretical investigations have been published on flow in curved pipes, and the experimental investigations cannot yet be considered as systematic except perhaps for 90° bends. Very often two sets of published results are apparently in conflict. It is clear that conditions at the entry and the exit to the bend have a considerable effect on the flow in the bend. Some typical results obtained with 90° bends are shown in Figs. 128, 129, and 130, in which $L$ denotes the radius of curvature of the axis of the bend and $a$ the radius of the pipe. In Fig. 128 various ways of plotting the results for a particular Reynolds number are shown. The variable $\xi$ is defined as $\Delta p/(4\rho U_m^2)$, where $\Delta p$ is taken to have various meanings, defined as follows:

(1) $\Delta p$ is the pressure drop in a straight pipe of the same axial length as the bend.

(2) $\Delta p$ is the excess pressure drop in an otherwise straight pipe containing the bend over that in a straight pipe of the same axial length.

(3) $\Delta p$ is the total pressure drop due to the bend, i.e. (1) plus (2).

(4) $\Delta p$ is the total pressure drop in a fixed axial length of the pipe, defined as the curved axial length when $L/a = 40$. For other values of the ratio $L/a$ this fixed length will include a portion of straight pipe at each end of the bend.

[†] *Ann. de Chimie et de Physique* (6) **21** (1890), 433–510, especially pp. 457–465.

[‡] Figures 125, 126, 127 are due to Taylor, *Proc. Roy. Soc.* A, **157** (1936), 546–578.

(5) $\Delta p$ is the total pressure drop between two fixed points joined by different arrangements of two straights and one 90° bend. The two fixed points are one on each axis of two straight lengths of pipe perpendicular to each other, and the points are at the ends of a bend with $L/a = 40$. This method of plotting shows the effect of joining up two perpendicular lengths of straight pipe by bends of different radius.

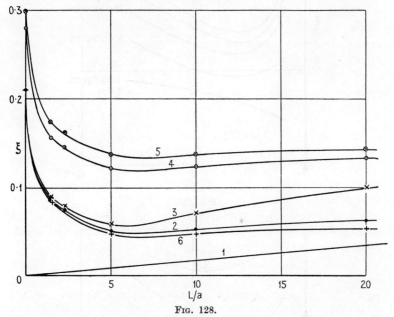

FIG. 128.

(6) $\Delta p$ is the excess pressure drop in the pipe line over that in a straight length equal to the sum of the two straights up to the point of intersection.

The experiments from which these results are taken were performed by Davis† using bends of malleable, cast and wrought iron and long straight inlet and exit pipes of lap-welded iron. Care was taken to have these straight pipes sufficiently long to include the whole loss. The internal diameter of the pipe was $2\frac{1}{16}$ inches,‡ and on an estimation of $\nu$ the Reynolds number was about 183,000.

In Fig. 129 $\xi$, defined as in (2) above, is plotted against the

† *Trans. Amer. Soc. Civil Engineers*, **62** (1909), 97–112.

‡ There is included here one result which was obtained by using a tee bend with one side plugged and with an internal diameter of $2\frac{5}{8}$ inches, the diameter of the straight being $2\frac{1}{16}$ inches as before.

Reynolds number for a particular pipe; the various curves arise from the different distances downstream at which the pressure loss was measured, the distances being specified in the figure. In Fig. 130

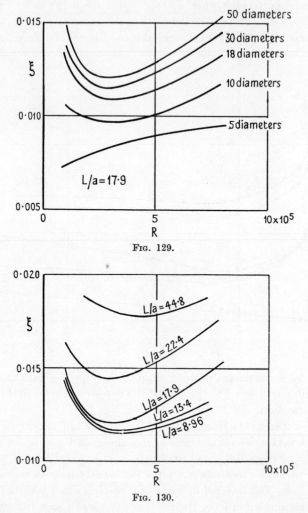

FIG. 129.

FIG. 130.

$\xi$ has the same meaning; the pressure loss is measured 50 diameters downstream from the bend for various pipes. The measurements were taken by Gregorig.[†]

† *Zürich Dissertation*, 1933.

Experiments† have been carried out on curved flow along a plane wall, the streamlines of the flow in the main stream outside the boundary layer being parallel to the boundary.

## 172. Velocity fluctuations in turbulent flow through a pipe.

Various methods of measuring velocity fluctuations in turbulent flow were described in Chap. VI,—namely use of the hot wire anemometer (§ 119), the 'Schlieren' method with a kinematograph camera (§ 134), and use of the ultramicroscope (§ 136). The last method measures observed maxima, whilst the others, by taking records of the fluctuations, can be used to determine either the observed maxima or the root-mean-square values of the turbulent velocities. The distribution of the turbulent velocities has been shown experimentally to follow the error law.‡

The maxima measured by the ultramicroscope method seem to be roughly three times the root-mean-square values as found by the other methods. Typical results for the maxima $u_1$, $v_1$, $w_1$, of the velocity fluctuations at any point in a circular pipe of radius $a$, in the axial, radial, and circumferential directions, are shown in Fig. 131,‖ where $u_1/U$, $v_1/U$, $w_1/U$ ($U$ being the mean velocity at the point) are plotted against $r/a$.

Fage and Townend†† have determined $u_1$, $v_1$, $w_1$ for flow along a square pipe ($u_1$ now denoting the maximum fluctuation along the pipe, $v_1$ the maximum fluctuation normal to one of the walls, and $w_1$ the maximum component perpendicular to $u_1$ and $v_1$): they conclude that near the wall $u_1/U$ tends to a constant value, $v_1/U$ tends to zero, and $w_1/U$ increases to a maximum value. (The measurements in Fig. 131 for a circular pipe would have to be carried nearer the wall before we could be sure whether or not corresponding results are true.) The results indicate that the fluctuating pressure gradients parallel to the axis of the pipe are negligible compared with the transverse gradients in a thin layer near the wall.‡‡ In the laminar sub-layer, apart from pressure gradients and inertia, $U+u_1$ and $w_1$ increase linearly with the distance $y$ measured normally from the wall; further (since $U$ also increases approximately linearly

† Gruschwitz, *Ingenieur-Archiv*, **6** (1935), 355–365.
‡ Simmons and Salter, *Proc. Roy. Soc.* A, **145** (1934), pp. 212–234; Townend, *ibid.*, pp. 180–211.
‖ Fage, *Phil. Mag.* (7), **21** (1936), 100, Fig. 13.
†† *Proc. Roy. Soc.* A, **135** (1932), 657–677.
‡‡ The argument is due to Taylor, *ibid.* **135** (1932), 678–684.

with $y$) we should expect $u_1/U$ and $w_1/U$ to remain finite at the wall and $d(u_1/U)/dy$ and $d(w_1/U)/dy$ to be zero there. The presence of local turbulent pressure gradients parallel to the axis of the tube probably produces increasing values of $u_1/U$ as the surface is

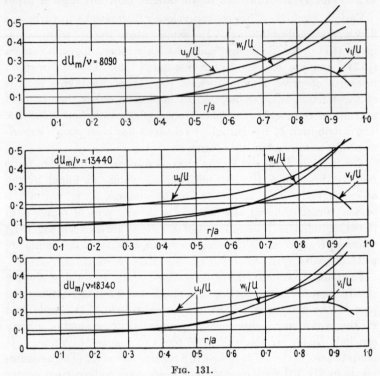

FIG. 131.

approached, and similarly for local transverse pressure gradients and $w_1/U$. Thus the fact that $d(w_1/U)/dy$ is much greater than $d(u_1/U)/dy$ suggests that the transverse turbulent pressure gradients are much greater than the gradients parallel to the axis, and to that extent justifies the assumption of the momentum transfer theory that pressure gradients parallel to the axis of the pipe are negligible. Dimensional analysis suggests that this remark will be true for values of $y$ less than $\alpha\nu/U_\tau$, where $\alpha$ is some constant; and we can make an estimate of $\alpha$ from Fage and Townend's earlier measurements. They found that $d(u_1/U)/dy \ll d(w_1/U)/dy$ for $y \leqslant 0.2s$, where $s$ is the half-width of the pipe. The Reynolds number formed from the hydraulic mean depth ($\frac{1}{2}s$) and the mean velocity was 1,280; on the assumption that the resistance coefficient is the same as that for a circular pipe (again at a Reynolds number formed from the hydraulic mean depth and the mean velocity), we find that $U_\tau$ is $0{\cdot}07U_m$ approximately. Thus $\alpha \doteq 36$, and the region where the results of Fage and Townend show that the fluctuating pressure gradient parallel to the pipe axis is negligible is little more than the viscous layer.

### 173. The distribution of the dissipation of energy in a channel over its cross-section.†

For pressure flow in a channel Wattendorf and Kuethe‡ have measured the root-mean-square value, $u$, of the component velocity fluctuation in the direction of the mean flow. Prandtl and Reichardt|| have made correlation measurements from which the length $\lambda$ (Chap. V, § 91 (p. 224)) can be determined, so the rate of dissipation $\overline{W}$ of the energy of turbulent flow can be calculated from the formula

$$\overline{W} = 15\mu u^2/\lambda^2 \tag{94}$$

(see Chap. V, equation (81) (p. 224)). Denote by $y'$ distance measured from the central plane. The rate at which the fluid between $y' = 0$ and $y' = Y$ does work on the fluid between $y' = Y$ and $y' = h$ (where $h$ is the half-breadth) is $U(Y)\tau = U(Y)\tau_0 Y/h$, where $U(Y)$ denotes the mean velocity at $y' = Y$. The rate at which the pressure gradient does work on the fluid between $y' = 0$ and $y' = Y$ is

$$- \int_0^Y \frac{\partial p}{\partial x} U \, dy' = \frac{\tau_0}{h} \int_0^Y U \, dy'.$$

The difference between these results, which is

$$\frac{\tau_0}{h}\left[ \int_0^Y U \, dy' - U(Y)Y \right] = -\frac{\tau_0}{h} \int_0^Y y' \frac{dU}{dy'} \, dy',$$

must be equal to the rate of dissipation of energy by viscosity acting on the mean flow together with the rate of transformation of the energy of the mean flow into energy of turbulent flow. Thus the total

† Taylor, *Proc. Roy. Soc.* A, **151** (1935), 455–464.
‡ *Physics*, **5** (1934), 153–164.
|| *Deutsche Forschung*, Part **21** (1934), 110–121.

rate of transformation of energy of mean flow into energy of turbulent flow between $y' = 0$ and $y' = Y$ is

$$-\frac{\tau_0}{h} \int_0^Y y' \frac{dU}{dy'}\, dy' - \int_0^Y \mu \left(\frac{dU}{dy'}\right)^2 dy',$$

and hence the rate of transformation per unit volume is

$$-\frac{\tau_0}{h} y' \frac{dU}{dy'} - \mu \left(\frac{dU}{dy'}\right)^2. \quad (95)$$

Except in the regions near the walls we may write this result as

$$-\frac{\tau_0}{h} y' \frac{dU}{dy'} \quad (96)$$

approximately. This result for the rate of degradation of mean energy into turbulent energy is compared with the rate of dissipation of energy given by (94) in Fig. 132. It will be noticed that the rate of dissipation is greater than the rate of degradation between $y' = 0$ and $y'/h = 0.6$, and less for greater values of $y'/h$. Equation (94) is based on the assumption of isotropy, and since the flow is no longer isotropic for $y'/h > 0.7$ the conclusion as regards dis-

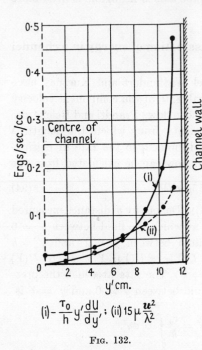

$$(i) - \frac{\tau_0}{h} y' \frac{dU}{dy'}; \quad (ii)\ 15\,\mu \frac{u^2}{\lambda^2}$$

Fig. 132.

sipation in the region $1 > y'/h > 0.7$ should be treated with reserve. The result for the degradation will also be affected by viscous action in the immediate neighbourhood of the walls.

## 174. Measurements of correlation between the longitudinal turbulent velocity components at the axis and elsewhere in flow through a pipe.

The correlation, $R$, between the longitudinal turbulent component of velocity $u_c$ at the centre of a circular pipe and $u$ at radius $r$ has been measured in a pipe $2\frac{7}{8}$ inches diameter. The results are shown in

Fig. 133. The experimental points obtained by two methods which were described in Chap. VI, § 120 are distinguished in Fig. 133. They fall on one curve.

$R$ falls from its initial value 1·0 at $r = 0$ to zero at $r = 0.6$ inch.

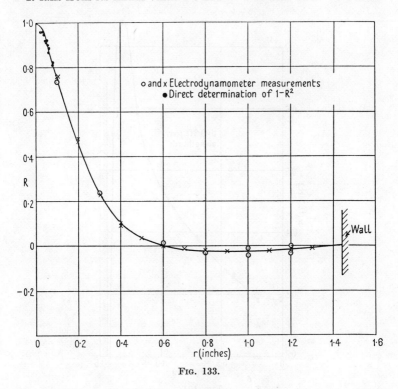

o and x Electrodynamometer measurements
● Direct determination of $1-R^2$

Wall

$r$(inches)

Fig. 133.

In the outer part of the pipe, between $r = 0.6$ inch and $r = 1.44$ inches, $R$ is negative. It can in fact be proved that $R$ must be negative in part of the section if the mean flow across any section does not vary with time, for in that case

$$\int_0^a (U+u)r\,dr = \text{constant} = \int_0^a Ur\,dr, \qquad (97)$$

where $U$ is the time-mean velocity and $a$ the radius of the pipe. Hence at each instant

$$\int_0^a ur\,dr = 0, \qquad (98)$$

and therefore $\qquad u_c \int_0^a ur\,dr = 0$

or $\qquad\qquad \int_0^a u_c ur\,dr = 0,$ (99)

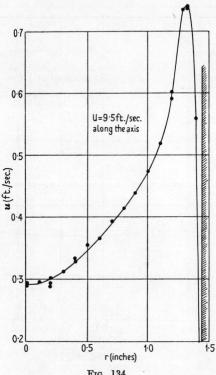

U = 9·5 ft./sec.
along the axis

$u$ (ft./sec.)

r (inches)

Fig. 134.

since at any instant $u_c$ is a constant in the integration. This equation is true at each instant of time, and therefore

$$\int_0^a \overline{u_c u}\,r\,dr = 0. \qquad (100)$$

Since $\overline{u_c u} = R u_\mathrm{c} u$, where $u_\mathrm{c} = \sqrt{(\overline{u_c^2})}$ and $u = \sqrt{(\overline{u^2})}$, it follows that

$$u_\mathrm{c} \int_0^a R u r\,dr = 0,$$

so $\qquad\qquad \int_0^a R u r\,dr = 0.$ (101)

The values of $u$ were measured by means of the electrodynamometer at the same time as those of $R$. The results are shown in Fig. 134. The corresponding values of $rRu$ are shown in Fig. 135, with $r$ and $u$ in the same units as in Fig. 134. By measuring the areas under

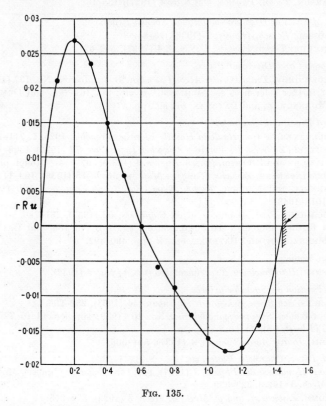

FIG. 135.

the positive and negative parts of the curve in Fig. 135 it is found that (101) is verified.†

## ADDITIONAL REFERENCES

*Treatises and Collected Works, etc.*

SCHILLER, *Handbuch der Experimentalphysik,* **4,** part 4 (1932).

STANTON, *Friction* (London, 1923), pp. 48–86.

DRYDEN, MURNAGHAN, and BATEMAN, *Hydrodynamics* (Bulletin No. 84 of the National Research Council, Washington, 1932), pp. 438–492.

† Taylor, *Proc. Roy. Soc.* A, **157** (1936), 537–546.

KÁRMÁN, *Journ. Aero. Sciences*, 1 (1934), 1–20, and *Proc. 4th Internat. Congress for Applied Mechanics, Cambridge*, 1934, pp. 54–91, gives bibliographies.

*Derivation of Logarithmic Law for Velocity Distribution.*
IZAKSON, *Techn. Physics, U.S.S.R.* 4 (1937), 155–162.

*Flow of Gases at High Speeds through Smooth Pipes.*
FRÖSSEL, *Forsch. Ingwes.* 7 (1936), 75–84.
LELCHUK, *Techn. Physics, U.S.S.R.* 4 (1937), 592–621.

*Convergent and Divergent Pipes.*
VEDERNIKOV, *Trans. Centr. Aero-Hydrodyn. Inst., Moscow*, No. 137 (1926).
For further references see SCHILLER, *op. cit.*, pp. 161–175, and DRYDEN, MURNAGHAN, and BATEMAN, *op. cit.*, pp. 476, 477.

*Curved Pipes and Channels.*
BOUCHAYER, *3ᵉ Congrès de la Houille Blanche, Grenoble*, 1925, 1, 711–727.
KEUTNER, *Zeitschr. des Vereines deutscher Ingenieure*, 77 (1933), 1205–1209.
FRITZSCHE and RICHTER, *Forsch. Ingwes.* 4 (1933), 307–314. (Rough pipes.)
WIESELSBERGER, *Zeitschr. f. angew. Math. u. Mech.* 15 (1935), 109–111.
YARNELL and NAGLER, *Trans. Amer. Soc. Civil Engineers*, 100 (1935), 1018–1043.
MOCKMORE, *Proc. Amer. Soc. Civil Engineers*, 63 (1937), 251–286.
For further references see SCHILLER, *op. cit.*, pp. 175–185, and DRYDEN, MURNAGHAN, and· BATEMAN, *op. cit.*, pp. 490–492.

*Open Channels.*
EISNER, *Handbuch der Experimentalphysik*, 4, part 4 (1932).

*Skin-Friction on Smooth Surfaces.*
GEBERS, *Schiffbau, Schiffahrt, Hafenbau*, 34 (1933), 252–256.
NORDSTRÖM, *Skeppsbyggnadskonst*, No. 10 (1933; supplement to *Teknisk Tidskrift*, 63), 71–86.
ZAHM, *Journ. Aero. Sciences*, 4 (1937), 504–506.

*Rough Pipes, Channels, Plates, etc.*
HOPF, *Handbuch der Physik*, 7 (1927), 147, 148; *Zeitschr. f. angew. Math. u. Mech.* 3 (1923), 329–339.
FROMM, *Zeitschr. f. angew. Math. u. Mech.* 3 (1923), 339–358.
STREETER, *Proc. Amer. Soc. Civil Engineers*, 61 (1935), 163–186.
TRIPP, *Journ. Aero. Sciences*, 4 (1936), 10, 11.
SKOGLUND, *Journ. Aero. Sciences*, 4 (1936), 28, 29.
LOYTZANSKY, *Trans. Centr. Aero-Hydrodyn. Inst., Moscow*, No. 250 (1936)·
FEDIAEVSKY, *ibid.*
DE MARCHI, *Energia Elettr.* 13 (1936), 421–454.
COLEBROOK and WHITE, *Journ. Inst. Civil Engineers* (1937), 99–118.
KEMPF, *Jahrb. Schiffbautechn. Ges.* 38 (1937), 159–176; *Engineering*, 143 (1937), 417–419.

*Velocity Fluctuations in Pipes and Channels.*
REICHARDT, *Zeitschr. f. angew. Math. u. Mech.* 13 (1933), 177–180.
WATTENDORF, *Journ. Aero. Sciences*, 3 (1936), 200–202.

# FLOW ROUND SYMMETRICAL CYLINDERS. DRAG

## 175. Introduction.

IN this chapter we consider flow past long symmetrical cylindrical bodies in a stream parallel to a plane of symmetry. The dependence of some general flow characteristics on the thickness parameter is first considered, with special reference to flow past symmetrical aerofoils. The characteristics of flow past a circular cylinder are then described. We shall deal exclusively with the drag and the pressures experienced by a body, and with the flow in the immediate neighbourhood of the boundary, reserving for Chap. XIII consideration of the flow in the wake.

## 176. The effect of changes of thickness on the drag of a symmetrical aerofoil.

The dependence of drag on the thickness parameter, which is here taken to be the ratio of the maximum thickness ($T$) to the chord length ($c$), is best illustrated by comparative measurements on sections belonging to the same family. The family of symmetrical sections shown in Fig. 136† was derived from a circle by conformal transformations of the generalized Joukowski type.‡ The data to be analysed were obtained at the N.P.L. from measurements on large models spanning a 7-foot tunnel.† The flow past each model was very closely two-dimensional.

A curve showing the change in $C_D$ (drag per unit length$/(\frac{1}{2}\rho U_0^2 c)$), where $U_0$ is the undisturbed stream velocity) with $T/c$, at a Reynolds number $U_0 c/\nu$ equal to $4 \times 10^5$, is reproduced in Fig. 137. $C_D$ increases with $T/c$, at first slowly and then more rapidly. The curve must become much steeper at values of $T/c$ greater than those shown, since there is reason to believe that when $T/c = 1$, $C_D$ has a value differing only slightly from its value for a circular cylinder. This value (for the same Reynolds number) is 0·32, so the rise in $C_D$ over the range $0·4 < T/c < 1·0$ must be from 0·033 to about ten times this value.‖

---

† Fage, Falkner, and Walker, *A.R.C. Reports and Memoranda*, No. 1241 (1929).
‡ Glauert, *ibid.*, No. 911 (1924).
‖ The drag per unit length of a very long flat plate normal to the stream and of

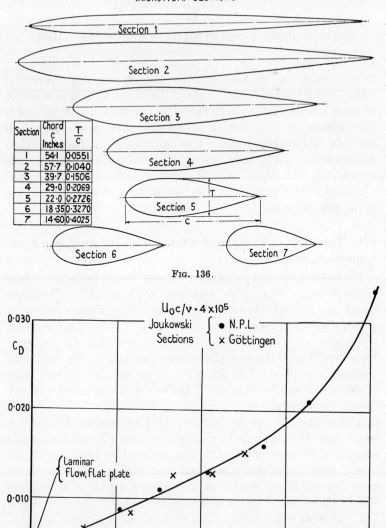

JOUKOWSKI SECTIONS

| Section | Chord c Inches | $\frac{T}{c}$ |
|---------|------|--------|
| 1 | 54·1 | 0·0551 |
| 2 | 57·7 | 0·1040 |
| 3 | 39·7 | 0·1506 |
| 4 | 29·0 | 0·2069 |
| 5 | 22·0 | 0·2726 |
| 6 | 18·35 | 0·3270 |
| 7 | 14·60 | 0·4025 |

FIG. 136.

$U_0 c / \nu \cdot 4 \times 10^5$

Joukowski Sections { ● N.P.L.   × Göttingen

FIG. 137.

Included in Fig. 137 are results obtained at Göttingen† on a closely similar series of symmetrical sections; these results will be seen to lie closely on the curve obtained at the N.P.L. Fig. 137 also shows that the value of $C_D$ at $T/c = 0$, obtained by extrapolation, is very nearly the same as the value for laminar flow past a flat plate at the same Reynolds number.

It is often necessary in practice to fair a bluff cylindrical obstacle. If the cross-stream breadth of the obstacle is denoted by $T$, the fair-

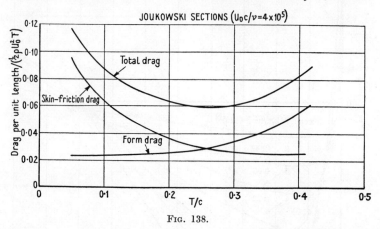

FIG. 138.

ing which will have the lowest drag is the one for which the drag coefficient obtained by dividing the drag per unit length by $\frac{1}{2}\rho U_0^2 T$ is a minimum. This drag coefficient (at $U_0 c/\nu = 4 \times 10^5$) for the sections of Fig. 136 is plotted against $T/c$ in Fig. 138. The section which has the lowest drag coefficient is one whose chord length is about four times the thickness. If a greater chord length is taken the increase in the skin-friction drag more than compensates for the slight drop in the form drag, whereas if a smaller chord length is taken the drop in the skin-friction drag is smaller than the rise in the form drag.

## 177. Symmetrical aerofoils. Normal pressure.

The experiments on the symmetrical sections of Fig. 136 included measurements of the distribution of normal pressure. The values

breadth $b$ is about $0.92\rho U_0^2 b$—i.e. at a Reynolds number of $4 \times 10^5$, about 6 times the drag of a circular cylinder having a diameter equal to the width of the plate.

† *Ergebnisse der Aerodynamischen Versuchsanstalt zu Göttingen*, **1** (1925), 50–112.

obtained, expressed in the non-dimensional form $(p-p_0)/(\frac{1}{2}\rho U_0^2)$, where $p$ is the pressure on the surface and $p_0$ is the pressure in the undisturbed stream, are plotted in Fig. 139† on a common base repre-

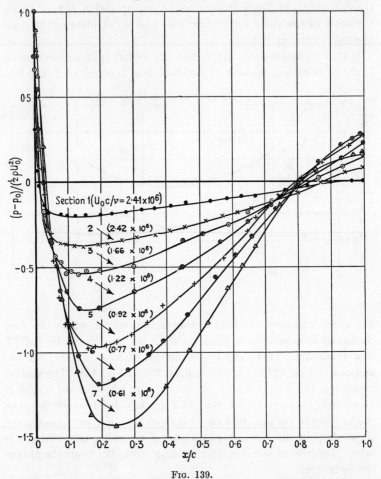

Section 1 $(U_0 c/\nu = 2 \cdot 41 \times 10^6)$

2 $(2 \cdot 42 \times 10^6)$

3 $(1 \cdot 66 \times 10^6)$

4 $(1 \cdot 22 \times 10^6)$

5 $(0 \cdot 92 \times 10^6)$

6 $(0 \cdot 77 \times 10^6)$

7 $(0 \cdot 61 \times 10^6)$

$x/c$

$(p-p_0)/(\frac{1}{2}\rho U_0^2)$

Fig. 139.

senting the chord of a model, the abscissae being distances along the middle line of a section expressed as fractions of the chord length. The pressure distribution undergoes a marked change as the thickness parameter increases: whereas the maximum value of $-(p-p_0)/(\frac{1}{2}\rho U_0^2)$ is only 0·20 for section 1, for section 7 it is 1·44.

† See footnote † on p. 401.

There is also a progressive shift towards the tail of the position at which the maximum negative pressure occurs.

Since these sections are of the standard Joukowski type, it is possible to determine by direct calculation the pressure distributions which they would experience in a stream of inviscid fluid on the Kutta-Joukowski theory. The theoretical distributions for sections 3 and 7 (after a small correction has been applied for interference

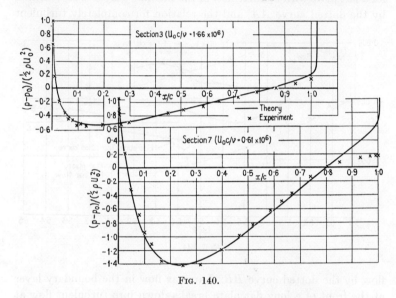

FIG. 140.

arising from constraint of the tunnel walls) are shown in Fig. 140, which also includes results obtained from measurements on these sections in a 7-foot wind tunnel (shown by crosses). The actual pressure distributions in air are very nearly the same, except at the tail, as the theoretical distributions in an inviscid fluid: the discrepancy at the tail (due to the presence of a wake) occurs earlier, and is more marked, on the thick than on the thin section. It should be mentioned that for convenience of manufacture the models used were made from the theoretical shapes with short lengths of the sharp tails cut off.† The comparisons are, however, not vitiated, since they are made on a base representing the chord of the model.

† The lengths cut off were $0·02c$ (section 3) and $0·005c$ (section 7).

### 178. Symmetrical aerofoils.  Skin-friction drag and form drag.

Values of the frictional coefficient $C_f$ for each of the sections of Fig. 136 can be obtained by dividing the skin-friction drag per unit span by $\frac{1}{2}\rho U_0^2 \times$ the periphery of the section;† these values are plotted against $\log R$ in Fig. 141, where $R$ is $U_0 c/\nu$. The relation for a flat plate with laminar flow in the boundary layer is represented by the dotted curve $AA$, and the relation for completely turbulent

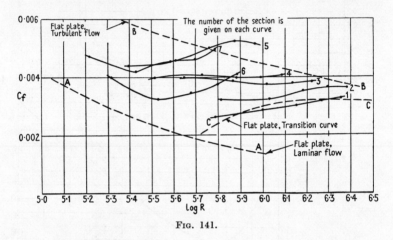

FIG. 141.

flow by the dotted curve $BB$. Laminar flow in the boundary layer at the front of a long flat plate breaks down into turbulent flow at some distance from the leading edge; and the position of the transitional region depends not only on the speed but also on the turbulence in the general stream (cf. Chap. VII, § 151). The value of $C_f$ for a plate with laminar flow over the forward part and turbulent flow over the after part is therefore intermediate between the values for completely laminar and completely turbulent flow; the greater the length of the plate the closer does the value approach that for turbulent flow. The particular transition curve $C_f = 0.074R^{-\frac{1}{5}} - 1700R^{-1}$‡ is shown as a dotted line $CC$ in Fig. 141.

---

† The skin-friction drag was obtained by subtracting the form drag (deduced from the measured normal pressure distribution) from the measured total drag.

The periphery of a thin section is roughly equal to twice the chord, so $C_D$ (skin-friction) $= 2C_f$ approximately.

‡ See Chap. VIII, § 163, p. 362.

An interesting feature of Fig. 141 is that the transition curves for the Joukowski sections lie almost entirely within the space enclosed by the three dotted lines for the flat plate. The curves for the two thinnest sections, 1 and 2, follow closely the general trend of the transition curve $CC$; with increasing Reynolds numbers they approach the turbulent curve for a flat plate. It is to be expected from tests on thin sections ($T/c < 0.08$) in the Compressed Air Tunnel at the N.P.L.,† and from the work of B. M. Jones,‡ that at Reynolds numbers well above those covered in the experiments the values of $C_f$ would fall quite near the turbulent curve for a flat plate. But although the average intensity of skin-friction on a thin section is not appreciably different from that for a flat plate at the same Reynolds number, it does not follow that the distribution on the surface is the same. The intensity of friction at any point of a surface depends both on the speed outside the boundary layer and on the pressure gradient along the surface: in the case of a flat plate this speed may be taken as constant and the pressure gradient as zero, whereas both the speed and the pressure gradient vary along a curved surface.

Another feature of Fig. 141 is that the curves for the thick sections (5, 6, and 7) approach the turbulent curve for a plate in an appreciably lower range of Reynolds numbers than those for the thin sections. It is not surprising that the curves for a thick section should show different tendencies from those for a thin section, since the flow experiences important changes as the thickness is increased: there is a change in the distribution of normal pressure (see Fig. 139), which is associated with a change in the distribution of skin-friction. Moreover the curvature of the surface becomes greater with increasing thickness, and eventually a shape is obtained for which the boundary layer leaves the surface and a dead-air region is formed at the back. In these circumstances it is not to be expected that the average frictional intensity on the surface should be the same as for a flat plate, and the comparison ceases to have any real meaning.

Since the drag of a flat plate arises entirely from the frictional stresses on its surface, and the drag of a circular cylinder (except at

† Relf, *Journ. Roy. Aero. Soc.* **39** (1935), 1–30.
‡ *Ibid.* **33** (1929), 357–385; also *A.R.C. Reports and Memoranda*, No 1115 (1928) and No. 1199 (1929).

low Reynolds numbers) almost entirely from normal pressures, we should expect the contribution of the frictional stresses to the drag of a body to become proportionately smaller the greater the thickness parameter. Fig. 142 shows that this is the case with the Joukowski sections: whereas the frictional drag of the thinnest section $(T/c = 0.055)$ is about 85 per cent. of the total drag, that of the thickest section $(T/c = 0.403)$ is only about 35 per cent. (The

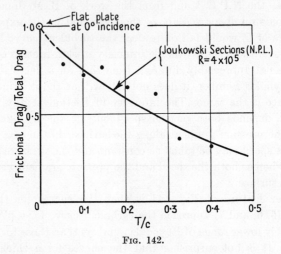

FIG. 142.

frictional drag of a circular cylinder, at the same Reynolds number, is only about 3 per cent. of the total drag.)

The relative contributions of the normal pressures and surface stresses to the total drag of a body depend greatly on its shape. When the body has a low drag coefficient separation of the boundary layer from the surface occurs late (if at all), and the frictional drag forms a large part of the total drag. On the other hand, for a very bluff body the drag is due almost entirely to normal pressures.

It is necessary to emphasize that the shapes of curves such as those in Figs. 137, 138 and 142 depend both on the family of sections and on the Reynolds number chosen. The diagrams serve, however, to show that in general the drag coefficient, $D/(\frac{1}{2}\rho U_0^2 c)$, of a cylindrical body increases with the thickness parameter; that a large part of the drag of a thin section arises from skin-friction; and that form drag makes a large contribution to the drag of a thick section.

## 179. The intensity of skin-friction at a symmetrical aerofoil.

A complete analysis of the stresses acting on a body involves the distributions both of normal pressure and of tangential or frictional stress. Direct measurement of frictional intensity is very difficult, and it is found more convenient to use an indirect method which involves prediction of the frictional intensity, $\tau_0$, from the relation

$$\tau_0 = \mu \left( \frac{\partial U}{\partial y} \right)_{y=0},$$

where $U$ is the tangential velocity at a distance $y$ from the surface. Even then a reliable measure of $\tau_0$ can be obtained only when the velocity gradient is known from measurements taken very close to the surface, and for this purpose surface tubes of the type designed by Stanton (see Chap. VI, § 126) may be employed.

A model has to be specially designed to accommodate these surface tubes, and so far very little work has been undertaken with them. The distribution of $\tau_0/(\frac{1}{2}\rho U_0^2)$ around the surface of Joukowski section No. 3 has been determined by Fage and Falkner,[†] using the particular form of surface tube shown in Fig. 74, Chap. VI, p. 279. The model, whose chord was 39·7 inches, was mounted with very small clearances between the walls of a 7-foot tunnel, and observations were confined to the section midway between the walls, where the flow was closely two-dimensional. To obtain a smooth surface the middle part of the model was formed from a hollow gun-metal casting accurately milled to shape and polished.

Measurements of velocity at distances between 0·002 and 0·003 inch from the surface were made with the small surface tubes. At the Reynolds numbers of the measurements the velocity gradients at the surface could be predicted on the assumption that the velocity increased linearly from zero at the surface to the value measured at these small distances. The intensity of skin-friction ($\tau_0$) was then obtained by multiplying the velocity gradient by the viscosity of the air. The results obtained at wind speeds of 60 and 80 feet per second ($R = U_0 c/\nu = 1·25 \times 10^6$ and $1·66 \times 10^6$) are reproduced in Fig. 143. (Strictly speaking, these results relate to the upper surface of the model when the plane of symmetry is inclined at the small angle

† *Proc. Roy. Soc.* A, **129** (1930), 378–410.

0·18° to the direction of the undisturbed stream; but it is improbable that the results for 0° incidence are noticeably different.) The intensity of friction rises to a maximum value near the nose, then falls, and afterwards rises to a second and greater maximum value. Measurements described below show that a transition from laminar to turbulent flow takes place in the boundary layer between the two maxima: the first peak value is therefore associated with laminar flow and the second with turbulent flow. Beyond the second peak

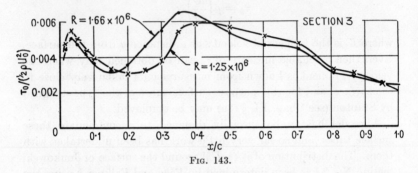

Fig. 143.

the frictional intensity falls steadily as the tail is approached. The value is still fairly high at the tail itself, whence it would appear that the boundary layer remains in contact with the surface right to the tail.

The total drag predicted from measurements of total-head losses in the wake was found to be $0·0054\rho U_0^2 c$ per unit length of span. The form drag predicted from the normal pressures was about $0·0014\rho U_0^2 c$, so the frictional drag obtained by subtracting this value of the form drag from the total drag was $0·0040\rho U_0^2 c$. The value of the frictional drag obtained from the measurements of frictional intensity with the surface tubes was $0·0043\rho U_0^2 c$.

## 180. The boundary layer thickness and the velocity distribution in the boundary layer at a symmetrical aerofoil.

The experiments on the model with the symmetrical section No. 3 also included measurements of the distribution of total head, taken with small exploring tubes across several sections of the boundary layer. Near the surface there is a loss of total head: with increasing distance from the surface the total head increases and eventually reaches a value nearly the same as that in the undisturbed stream. From the explorations of total head a rough measure was obtained of the thickness $\delta$ of the boundary layer at several positions on the aerofoil section: the results are shown in Fig. 144. The layer thickens rapidly in the region $0.25c < x < 0.30c$, where $x$ is distance from the nose parallel to the chord: this rapid thickening, which is a common

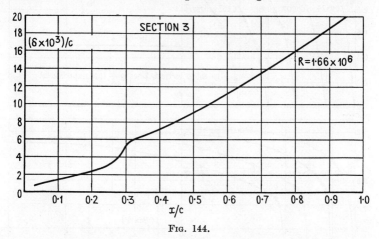

Fig. 144.

feature of flow in a boundary layer, is associated with a transition from laminar to turbulent flow. There is no indication in Fig. 144 of the rapid thickening of the boundary layer near the tail which would occur if the layer separated from the surface. The thickness of the laminar layer over the front of the model was found to decrease with speed, and the transition occurred earlier as the speed was increased.

The velocity distribution in the boundary layer can be predicted directly from explorations of total head, if it is assumed that the static pressure is constant through the layer and equal to the value measured on the surface. The distributions obtained (i.e. $U/U_0$ as a function of $y/c$, where $U$ is the velocity in the layer at a normal distance $y$ from the surface and $U_0$ is the velocity of the undisturbed stream) are shown in Fig. 145. The change in the character of the velocity distribution in the layer with increasing values of $x/c$ is perhaps more clearly shown by the curves in Fig. 146 relating $U/U_1$ with $y/\delta$, where $U_1$ is the velocity just outside the layer. The curves for the front part of the layer have shapes commonly associated with laminar flow, whereas those for the rear part are typical of turbulent flow.

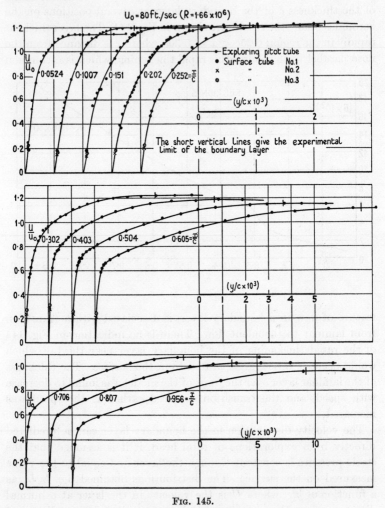

Fig. 145.

## 181. The effect of shape on drag and on scale effect.

The drag of a smooth aerofoil section depends on its shape as well as on the thickness parameter. The measurements which have been made to determine the effect of shape on drag are very numerous.

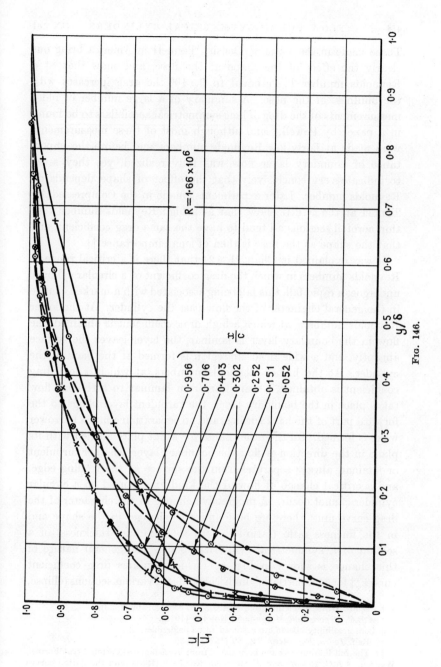

$$\frac{U}{U_1}$$

$$\frac{x}{c}$$

0·956
0·706
0·403
0·302
0·252
0·151
0·052

$R = 1·66 \times 10^{6}$

FIG. 146.

Those made in the Variable Density Tunnel[†] in America bring out clearly the effect of the shape at the nose, and show that at a Reynolds number $U_0 c/\nu$ equal to $3 \times 10^6$ the drag increases with the bluffness at the nose. A summary of a large number of older measurements of the drag of long symmetrical aerofoils is to be found in a paper by Powell;[‡] and although most of these measurements were made at fairly low Reynolds numbers, and before the importance of boundary layer flow was fully realized, yet they serve to indicate very conclusively that the effect of shape depends on Reynolds number. Later experiments[||] made in the Compressed Air Tunnel at the N.P.L. show that at a high Reynolds number very thin aerofoil sections all tend to have the same drag coefficient, and that the shape at the nose is then of small importance.[††]

It was explained in Chap. II, § 24 that there is a critical range of Reynolds numbers in which the drag coefficient of a circular cylinder undergoes a rapid fall, this fall being associated with a marked change in the general character of the flow past the cylinder. At the lower Reynolds numbers, at which a high drag coefficient is obtained, the flow in the boundary layer is laminar, the layer leaves the surface abruptly, and a wide dead-air region is formed at the back of the cylinder. At the higher Reynolds numbers, at which a low drag coefficient is obtained, a transition from laminar to turbulent flow takes place in the boundary layer, the turbulent layer clings to the forward part of the back surface, and on separation a much narrower wake is formed. On the other hand, for a flat plate aligned with its plane in the direction of flow the boundary layer, whether turbulent or laminar, always separates from the surface at the trailing edge, and a critical change of flow of the kind experienced on a circular cylinder cannot occur. A progressive change in the character of the drag curve must therefore be found with changes in the shape and in the fineness ratio (ratio of chord to maximum thickness) of a section of a symmetrical cylindrical body. The general nature of this change is shown in Fig. 147, which contains drag coefficient curves[‡‡] for symmetrical cylindrical bodies of various sections (ellipse,

† Jacobs, Ward, and Pinkerton, *N.A.C.A. Report* No. 460 (1933).

‡ *A.R.C. Reports and Memoranda*, No. 416 (1918).

|| Compressibility effects are assumed to be negligible.

†† Relf, *Journ. Roy. Aero. Soc.* **39** (1935), 1–30.

‡‡ The full line curves have been taken from 'Resistance of Certain Strut Forms', Warden, *A.R.C. Reports and Memoranda*, No. 1599 (1934), and the dotted curves

lenticular form, strut, oval, and aerofoil) and various fineness ratios. The drag coefficient plotted is the drag per unit length divided by $\frac{1}{2}\rho U_0^2 T$, where $T$ is the maximum thickness. The most noticeable feature is the difference between the shapes of the curves for the stream-line sections (Gloster, Oval, R.A.F. 30) and those for the elliptic sections. Whereas all the curves for the stream-line sections

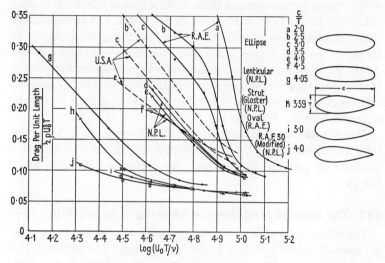

FIG. 147.

lie within a comparatively narrow band and have roughly the same shape, a shape characterized by a gradual fall of drag coefficient with increasing $U_0 T/\nu$, the curves for the elliptic sections show a rapid fall of drag coefficient with increasing $U_0 T/\nu$ over the range tested. The curves for the elliptic cylinders were obtained from tests made in three aerodynamic laboratories—R.A.E.; Navy Aerodynamic Laboratory, Washington; and N.P.L.: for fineness ratios 2·5 and 3·0 the curves obtained at the R.A.E. are displaced appreciably from those obtained at Washington. The differences can be attributed to differences in the steadiness of the air-streams in which the models were tested. Nevertheless, the series of curves show that for fineness

from 'Forces on Elliptic Cylinders in a Uniform Air-Stream', Zahm, Smith, and Louden, *N.A.C.A. Report* No. 289 (1928). Reference may also be made to 'Wind Tunnel Tests of Seven Struts', Hartshorn, *A.R.C. Reports and Memoranda*, No. 1327 (1930); 'Aerodynamic Theory and Test of Strut Forms', Smith, *N.A.C.A. Report* No. 311 (Part I) (1929) and No. 335 (Part II) (1929).

ratios between 2·0 and 3·5 the fall in the drag coefficient with increasing Reynolds numbers becomes less rapid as the fineness ratio is increased. Increasing the fineness ratio from 3·5 to 4·5 has comparatively little effect on drag, although the general tendency seems to be for the drag to decrease as the fineness ratio is increased.

The effect of a change of shape at the same fineness ratio (about 4·0) is shown by the curves $e$ (Ellipse, R.A.E. and U.S.A.), $g$ (Lenticular, N.P.L.), and $j$ (R.A.F. 30 (modified), N.P.L.). As the sectional shape changes from elliptic to stream-line the critical region of Reynolds numbers occurs earlier, and eventually tends to disappear. At the highest Reynolds numbers reached the drag coefficient of a bluff elliptic section is still greater than that of a stream-line section: this result is largely due to the fact that the form drag of a body increases with its bluffness.

The skin-friction drag coefficient for the Gloster strut, based on the total surface area, at $U_0 T/\nu = 1·06 \times 10^5$ (i.e. at $U_0 c/\nu = 3·6 \times 10^5$) is about 30 per cent. higher than the corresponding value for a flat plate.

## 182. The effect of roughness on the drag of an aerofoil.

The influence of surface roughness on the drag of an aerofoil section is especially marked at high Reynolds numbers. The nature of the effect is shown by tests† in the Compressed Air Tunnel on a symmetrical section N.A.C.A. 0012 (8-in. chord) with a smooth surface, and with the surface roughened by painting it with a paste made by mixing carborundum powder with a suitable lacquer. Two grades of carborundum were used; the average size of the particles of the first grade ($FF$) was about 0·001 inch and that of the second ($FFF$) about 0·0004 inch. Curves obtained from these tests are shown in Fig. 148 ($R = U_0 c/\nu$). It will be seen that roughening causes the drag curve to separate from that for the smooth surface at a Reynolds number which is lower for the greater roughness, and that subsequently the drag coefficient rises to a value greater than that for the smooth surface by an amount which is approximately proportional to the size of the particles constituting the roughness.

The drag curve for turbulent flow past a smooth flat plate is included in Fig. 148.

† Relf, *Journ. Inst. Civil Engineers,* **3** (1936), 534, 535; *Aircraft Engineering,* **8** (1936), 166.

The Reynolds numbers at which the curves for the rough surfaces depart from that for the smooth surface are in the neighbourhood of the values predicted for a flat plate by Prandtl and Schlichting.† The initial rate of increase of drag coefficient is appreciably greater than that predicted for the flat plate, but the increase at a Reynolds

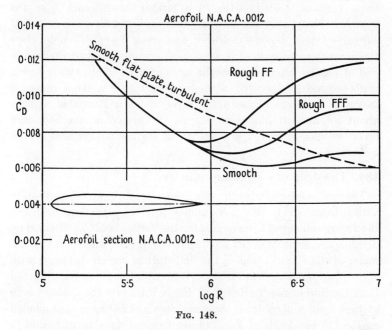

FIG. 148.

number approaching $10^7$ (Reynolds number of flight) is of about the same order of magnitude.  It is not to be expected that the results for a symmetrical section would agree quantitatively with those predicted for a flat plate, since the effect of excrescences on boundary layer flow will be different on curved and flat surfaces.  Nevertheless, the qualitative correspondence of the results strengthens the evidence that comparatively small degrees of roughness are of great practical importance at high Reynolds numbers.

## 183. Flow past a circular cylinder.

The motion around a long circular cylinder immersed in a fluid stream is especially interesting for the variety of changes which

† *Werft, Reederei, Hafen,* **15** (1934), 1–4.

occur with an increase in the Reynolds number. At a low Reynolds number the effects of viscosity are sensible at large distances from the cylinder; in particular the fluid at the back is appreciably retarded. At higher Reynolds numbers two symmetrical standing vortices are formed at the back.† With increasing Reynolds numbers these vortices stretch farther and farther downstream from the cylinder. Eventually the standing vortices are drawn out to a considerable length, become distorted, and break down. Then develops the characteristic state of flow in which vortices are shed alternately and at regular intervals from the sides of the cylinder, with vortex trails behind:‡ this type of flow persists over a large range of Reynolds numbers. Eventually, at or above a Reynolds number about $10^5$, another important change of flow occurs: the boundary layer becomes turbulent, and leaves the surface farther back on the cylinder.||

## 184. The drag of a circular cylinder.

The flow past a cylinder changes from a steady to an unsteady state with an increase in the Reynolds number, and the mean drag coefficient shows large changes. Curves showing the variation of $C_D$ (obtained by dividing the mean drag per unit length by $\frac{1}{2}\rho U_0^2 d$, where $d$ is the diameter of the cylinder and $U_0$ the undisturbed stream velocity) with $R \, (= U_0 d/\nu)$ are shown in Fig. 149.†† The full-line curve was obtained from measurements by Relf‡‡ at the N.P.L., and the dotted curve at high values of $R$ from measurements at Göttingen.|||| Included on the left-hand side of Fig. 149 are two curves, due to Lamb††† and to Bairstow‡‡‡ respectively, representing theoretical predictions of $C_D$ at low values of $R$. At very low values of $R$ the drag is approximately proportional to $U_0$, so $C_D$ falls rapidly with an increase in $R$.

---

† See Chap. II, § 20, p. 64, Figs. 26 and 27. Photographs of the standing vortices have been obtained by Camichel, *Engineering*, **123** (1927), 27–30; Thom, *Proc. Roy. Soc.* A, **141** (1933), 651–669. See Fage, *ibid.* **144** (1934), 381–386, for photographs of changes from flow of the standing-vortex type to flow of the unsteady type.

‡ Chap. I, § 10; Chap. II, § 20; Chap. XIII, § 242.

|| See Chap. II, § 24, p. 72.

†† For the curve of $Nd/U_0$ in Fig. 149, see § 186 below.

‡‡ *A.R.C. Reports and Memoranda*, No. 102 (1914). See also Eisner, *Mitteilungen der Preussischen Versuchsanstalt für Wasserbau und Schiffbau*, No. 4 (Berlin, 1929).

|||| *Ergebnisse der Aerodynamischen Versuchsanstalt zu Göttingen*, **2** (1923), 23.

††† *Phil. Mag.* (6), **21** (1911), 112–121; *Hydrodynamics* (1932), pp. 615, 616.

‡‡‡ Bairstow, Cave, and Lang, *Phil. Trans.* A, **223** (1923), 383–432. See also Southwell and Squire, *ibid.* **232** (1933–4), 27–64.

The value of $R$ at which the unsteady régime commences is uncertain, but is probably about 50.† As $R$ increases $C_D$ falls steadily to the minimum value 0·95 at $R = 1800$, and then rises to about 1·2 at $R = 3 \times 10^4$. A marked fall from 1·2 to about 0·36 occurs at about $R = 10^5$. Fig. 149 shows that this fall occurs earlier in the N.P.L. experiments than in those at Göttingen: this is due to the fact that the turbulence is greater in the N.P.L. tunnel than in the Göttingen tunnel.

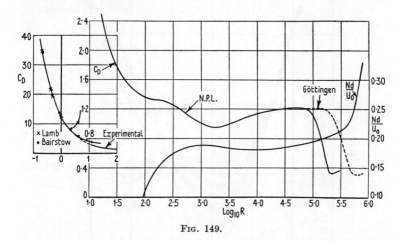

FIG. 149.

## 185. Changes in the pressure distribution at a circular cylinder, and in the drag coefficient, for a motion started from rest. Oscillating lift.

In a motion which is started from rest the flow pattern, initially the potential flow pattern, changes to the pattern corresponding to the final régime. The changes in the pressure distribution during the initial period have been found by Schwabe for an impulsive start with a constant velocity $U_0$.‡ When the régime is that in which two symmetrical vortices are formed behind the cylinder the flow closes up again behind the vortices, so there is a stagnation point behind

---

† The value depends on the steadiness of the general stream: standing vortices can be obtained at higher Reynolds number if the flow in the general stream is very steady. See also Chap. XIII, § 240, p. 552.

‡ *Ingenieur-Archiv*, **6** (1935), 34–50. The pressure was found from the velocity, and the velocity was found from photographs of the flow. The paper includes photographs of the development of the standing vortices.

them; this rear stagnation point, which is on the cylinder for the
potential flow, moves downstream as the pattern changes. Schwabe's
results are shown in Fig. 150 as curves of $(p-p_0)/(\frac{1}{2}\rho U_0^2)$ plotted
against the angular distance, $\theta$, from the forward stagnation point
for various values of $D/a$, where $D$ is the distance of the stagnation

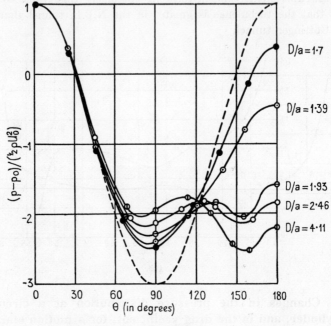

FIG. 150.

point in the rear from the centre of the cylinder, and $a$ is the
radius. The dotted line represents the result for potential flow.
The Reynolds number $U_0 d/\nu$ was 560, where $d$ is the diameter of the
cylinder.

The rise in the drag coefficient (drag per unit length divided by
$\frac{1}{2}\rho U_0^2 d$) from the beginning of the motion, before eddies become
detached from the cylinder, is shown in Fig. 151, where $C_D$ is plotted
against $U_0 t/a$ for a Reynolds number of 580, $t$ being the time. The drag
coefficient rises to 2·07 at $U_0 t/a = 9$; its value under steady conditions
is just over 1, so the consumption of energy in the development of
the eddies is large enough to cause the drag coefficient to rise initially
to nearly double the value it has under steady conditions.

When asymmetry has made its appearance the alternating eddy formation causes an alternating lift on the cylinder, and from an instantaneous pressure distribution Schwabe found at a Reynolds number of 735 a lift coefficient of 0·45 at a certain stage. The drag coefficient was then 1·09, so the lift was more than 40 per cent. of the drag.

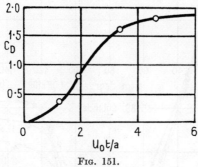

FIG. 151.

## 186. The frequency with which vortices are shed from a circular cylinder.

The motion in the wake when vortices are shed alternately from the opposite sides of a cylinder is discussed in Chapter XIII; but, the vortices being shed fairly regularly, a brief reference should be made here to an interesting connexion between the drag coefficient and the frequency ($N$) with which the vortices are shed. A curve† relating $Nd/U_0$ with $R$ is included with the drag curves in Fig. 149, and shows that a drop in $C_D$ is accompanied by a rise in vortex frequency. This is very apparent for the ranges $10^2 < R < 10^3$ and $10^5 < R < 10^6$. The latter range is that in which the large drop in $C_D$ occurs, and in it no definite periods can be measured; the frequencies of the most prominent disturbances were, however, recorded, and there was no doubt that they increased rapidly with $R$. For the range $10^2 < R < 10^5$ the flow in the wake was periodic; above $R = 10^5$ the flow was aperiodic.

## 187. The normal pressure at a circular cylinder.

Curves representative of the distribution of normal pressure on a cylinder are shown in Fig. 152.‡ These curves were obtained from

† Relf and Simmons, *A.R.C. Reports and Memoranda*, No. 917 (1924).
‡ Fage and Falkner, *ibid.*, No. 1369 (1931).

observations taken at the median sections of two long cylinders, of diameters 2·93 inches and 5·89 inches, which were mounted in the normal stream of a 4-foot wind tunnel (N.P.L. type), and also in the

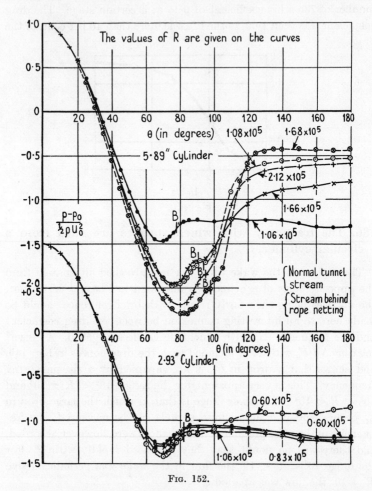

Fig. 152.

disturbed stream behind a rope netting.† The curves give the variation of the pressure coefficient $(p-p_0)/(\frac{1}{2}\rho U_0^2)$ with $\theta$, where $p$ is the normal pressure at an angular distance $\theta$ from the forward stagnation

† For further examples of experimental normal pressure curves see Chap. I, Fig. 4, p. 24, which shows measurements by Flachsbart at $R = 6\cdot7 \times 10^5$ and $1\cdot86 \times 10^5$.

point, and $p_0$ is the static pressure in the general wind stream. No correction has been made for the interference of the tunnel walls. The curves for the 2·93-inch cylinder were taken over a range of Reynolds numbers from $0·60 \times 10^5$ to $1·06 \times 10^5$. The pressure coefficient falls from a maximum value $1·0$ at $\theta = 0°$ to a minimum value at about $\theta = 70°$, then rises slightly, and from about $80°$ to $180°$ remains fairly constant. The curves for the 5·89-inch cylinder show the characteristic changes in pressure distribution which occur in the critical range of Reynolds numbers ($10^5 < R < 2·5 \times 10^5$): the coefficient of minimum pressure falls steadily and the point at which it occurs moves farther around the cylinder as $R$ increases, whilst the pressure coefficient at the back of the cylinder rises. Another important feature (to which reference will be made later) is the marked inflexion in each curve at the point marked $B$ just beyond the region of minimum pressure.

Curves† showing the normal pressure distribution at lower Reynolds numbers are reproduced in Fig. 153. These correspond to the range (roughly $3·5 < \log R < 4·5$) in which $C_D$ increases as $R$ increases (Fig. 149), and show the progressive drop in pressure at the rear of the cylinder. In this range of $R$ laminar boundary layers separate from the surface and are continued as free vortex-layers in the main body of the fluid, the flow in which becomes turbulent downstream of the position of separation from the solid surface.‡ As the Reynolds number increases the transition to turbulence in the free vortex-layers approaches the cylinder surface; the motion in the 'dead-water' region behind the cylinder is thereby modified more and more, with the result that the pressure in the rear decreases more than in proportion to the square of the speed.‖ Increased turbulence in the main stream causes the transition to turbulence in

† Linke, *Physik. Zeitschr.* **32** (1931), 900–914; Schiller and Linke, *Zeitschr. f. Flugtechn. u. Motorluftschiffahrt*, **24** (1933), 193–198.

‡ For further details see Chap. XIII, § 241.

‖ The same phenomenon has been observed by Flachsbart (*Zeitschr. f. angew. Math. u. Mech.* **15** (1935), 32–37) for a flat plate normal to the stream. No pressure measurements were taken; but it was found that the drag coefficient, which was about 1·96 at $R = 6,000$ (where $R$ is the Reynolds number referred to the breadth of the plate), remains constant as $R$ decreases to about 4,000. For values of $R$ below 3,000 there is a marked drop, down to 1·67 at $R = 1,600$, after which, with further decrease of $R$, the drag coefficient rises. The movement away from the plate, as $R$ decreases from 3,000 to 1,600, of the transition to turbulence in the vortex-layers springing from the edges of the plate was clearly shown by smoke photographs. (Below 1,600 the transition to turbulence takes place at a very considerable distance from the plate.) See also Chap. XIII, § 241.

the free vortex-layers to occur nearer the cylinder, and the rise in $C_D$ then occurs for a range of smaller values of $R$. The transition to turbulence may also be artificially hastened by fixing wires parallel to a cylinder generator in the boundary layer near or before the position

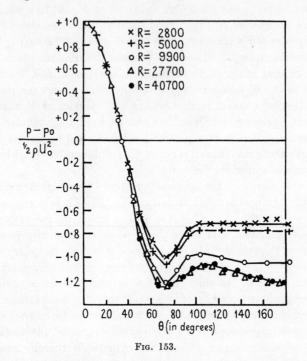

Fig. 153.

of separation. Suitably chosen wires in suitable positions cause the rise in $C_D$ to occur sharply between about $R = 2{,}000$ and $R = 5{,}000$ (and also to proceed rather farther than without wires—i.e. somewhat above the value $1\cdot2$ shown in Fig. 149).[†]

## 188. The skin-friction drag and the form drag of a circular cylinder below the critical range of Reynolds numbers.

Fig. 154[‡] shows the relative contributions to the total drag of a circular cylinder made by skin-friction and by normal pressure for values of $R$ less than about $2 \times 10^4$. Except at $R = 10$ and 20 the

† Schiller and Linke, loc. cit.
‡ Thom, A.R.C. Reports and Memoranda, No. 1194 (1929).

form drag was predicted from measurements of normal pressure, and the skin-friction drag was taken as the difference between the total drag and the form drag. The skin-friction and form drags at $R = 10$ and 20 were determined by Thom from a numerical solution of the equations for the flow of a viscous fluid around a cylinder. The figure shows that at $R = 10$ the contribution of the skin-friction

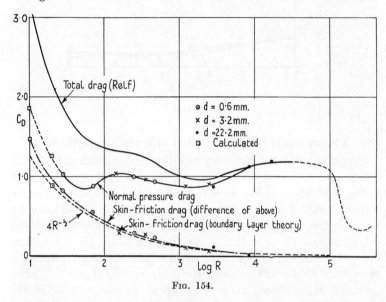

FIG. 154.

amounts to about 43 per cent. of the total drag; the contribution then falls with increasing $R$, and becomes quite small at $R = 10^4$. A broken line in the diagrams shows that the skin-friction drag over the range $30 < R < 10^4$ is closely given by the relation $C_D = 4R^{-\frac{1}{2}}$. This relation is due to Thom,† who calculated the skin-friction up to 60° from the forward stagnation point by using his approximate solution in closed form of the boundary layer equations (see Chap. IV, § 64), and took values between 60° and 90° from experiment, thus deducing $3 \cdot 84R^{-\frac{1}{2}}$ as the skin-friction drag coefficient for the front half of the cylinder. With a small addition for the contribution of the rear half, Thom gave $4R^{-\frac{1}{2}}$ as probably a close estimate. The relation has been verified up to $R = 4 \times 10^4$ by Schiller and Linke‡ within the accuracy obtained by experiment. (See Fig. 155. The

† *Ibid.*, No. 1176 (1928).          ‡ *Loc. cit.* in footnote † on p. 423.

skin-friction drag was again obtained as the difference of the total and normal pressure drags.)

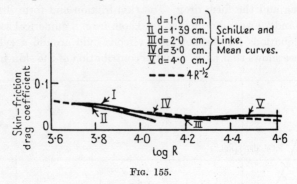

FIG. 155.

### 189. The intensity of skin-friction at a circular cylinder.

Curves showing the distribution of the frictional intensity ($\tau_0$) on the smooth surface of the 5·89-inch cylinder of § 187 are reproduced in Fig. 156; these curves were plotted from observations taken by Fage and Falkner† with small surface tubes. The frictional intensity increases from zero at $\theta = 0°$ to a maximum near $\theta = 60°$: beyond this position the type of distribution depends on the Reynolds number and on the turbulence in the stream. The intensity measured in the normal stream at the lowest Reynolds number ($1·06 \times 10^5$), after attaining its maximum value, falls rapidly to zero, whereas in the normal stream at the higher Reynolds numbers, and in the disturbed stream behind the rope netting, the intensity falls less rapidly to a minimum value and then rises to a second maximum before falling to zero. The intensity falls to zero at the position on the cylinder where the boundary layer separates from the surface; the minimum values correspond with transitions from laminar to turbulent flow in the boundary layer. The curves of Fig. 156 for $R = 1·66 \times 10^5$ and $1·68 \times 10^5$ show that the addition of artificial turbulence into the tunnel stream produces a rearward movement of the position of the minimum.‡

† *Loc. cit.* in the footnote on p. 422.

‡ This result is surprising, since on flat plates and aerofoils increased turbulence causes the region of minimum frictional intensity (i.e. the transition to turbulence in the boundary layer) to occur earlier. The different behaviour on the circular cylinder presumably arises from changes in the normal pressure distribution caused by the addition of turbulence in the critical region. (Figs. 152, 157.)

The distributions of frictional intensity in the range $70° < \theta < 140°$ are shown in conjunction with those of normal pressure in Fig. 157. The inflexion in each pressure curve (marked $B$), just beyond the region of maximum negative pressure, occurs where the frictional intensity is a minimum: a region of fairly constant pressure at the back begins (at $C$) where the friction falls to zero—that is, immediately after the boundary layer has left the surface of the cylinder.

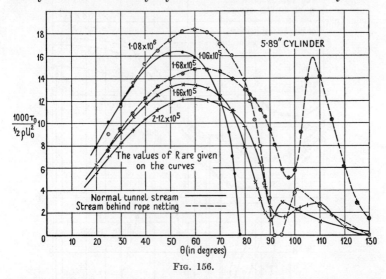

FIG. 156.

Values of the form drag, skin-friction drag, and total drag of the 5·89-inch cylinder (predicted from the curves of normal pressure and of frictional intensity given in Figs. 152 and 156 respectively) are given in Table 14. (The Reynolds numbers here are higher than those

TABLE 14

*Results for 5·89-inch cylinder in 4-foot wind tunnel*

| | | $C_D$ (tunnel) | | | |
|---|---|---|---|---|---|
| $R$ | | *From normal pressure* | *From skin-friction* | *Total* | $C_D$. *Infinite stream* |
| Standard stream of tunnel | $2·12 \times 10^5$ | 0·524 | 0·010 | 0·534 | 0·504 |
| | 1·66 | 0·782 | 0·010 | 0·792 | 0·746 |
| | 1·06 | 1·236 | 0·010 | 1·246 | 1·174 |
| Stream behind rope netting | 1·68 | 0·328 | 0·018 | 0·346 | 0·326 |
| | 1·08 | 0·470 | 0·016 | 0·486 | 0·458 |

considered in § 188.) The contribution of skin-friction to the drag of the cylinder is small: thus at the lowest Reynolds number ($1·06 \times 10^5$: standard stream) the contribution amounts to only 0·8

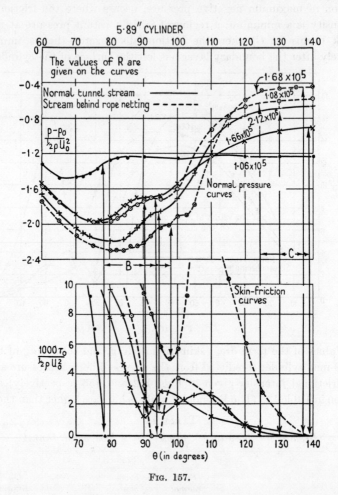

FIG. 157.

per cent. of the total drag, and at the highest Reynolds number ($2·12 \times 10^5$—i.e. above the critical range) to about 2·0 per cent.

The last column of the table gives the total drag coefficient in an infinite stream, obtained by correcting approximately for the interference of the tunnel walls.

## 190. The velocity distribution in the boundary layer at a circular cylinder.

The velocity distribution within the boundary layer can be determined from measurements of total head and of normal pressure, on the assumption that the static pressure across each section of the layer is constant and equal to the pressure on the surface. Velocity curves† obtained in this manner for an 8·9-inch cylinder mounted

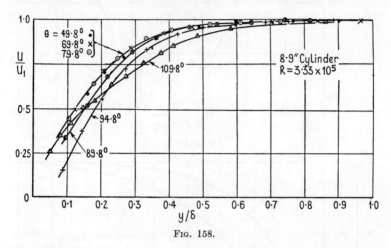

FIG. 158.

between the walls of a 4-foot wind tunnel (N.P.L. type) are shown in Fig. 158. The normal pressure curve (Fig. 159) has a marked inflexion at $\theta = 96\cdot4°$, and the earlier discussion leads us to believe that the transition from laminar to turbulent flow occurs near this region. The velocity curves of Fig. 158 relate $U/U_1$ and $y/\delta$, where $\delta$ is the thickness of the layer and $U_1$ the velocity just outside the layer, and they show the change in the character of the velocity distribution which occurs between $\theta = 94\cdot8°$ and $\theta = 109\cdot8°$. Evidence that the flow in the boundary layer is laminar over the front part of the cylinder is given in Fig. 160, where the measured velocity distribution is seen to be in close agreement with that predicted by theory. This comparison has been restricted to the outer part of the layer, since no observations were taken nearer the surface because the exploring tube, although small, was large compared with the thickness of the layer.

† Fage, *A.R.C. Reports and Memoranda*, No. 1179 (1929).

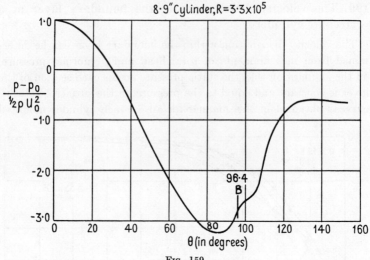

FIG. 159.

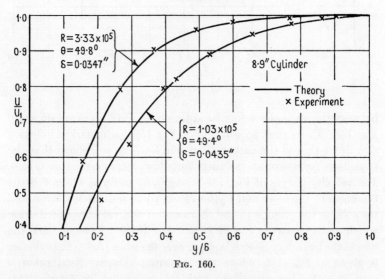

FIG. 160.

## 191. The effect of turbulence in the stream on the drag of a circular cylinder in the critical region.

We have already seen that the large fall in $C_D$ which occurs in the critical range of $R$ (i.e., for a tunnel of the N.P.L. type, between $R = 8 \times 10^4$ and $R = 2 \times 10^5$ approximately) is intimately related with changes in the flow in the boundary layer over the front of the cylinder, and in particular with the transition from laminar to turbulent flow in the layer. We should expect the flow characteristics in this range of $R$ to be influenced both by turbulence in the general stream and by surface roughness. The curves in Fig. 161 have been plotted from measurements of drag on 6·09-inch and 2·375-inch cylinders mounted in the turbulent stream behind a rope netting of

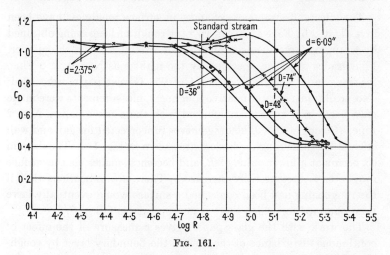

FIG. 161.

square mesh:† the diameter of the rope was 0·25 inch, and the horizontal and vertical spacing of the ropes was 1·5 inches. The symbol $D$ in Fig. 161 represents the distance of the axis of the cylinder behind the netting. The cylinders extended between the walls of a 4-foot wind tunnel (N.P.L. type), and the results in Fig. 161 are uncorrected for wall or gap interference effects. Although the resemblance between the shapes of the curves obtained in streams with artificial turbulence and in the normal stream of the tunnel is noteworthy, the outstanding feature is the lateral shift of the 'turbulent' curve in the direction of decreasing Reynolds numbers as the distance from the screen decreases—that is, as the turbulence increases. Advantage is taken of this phenomenon in comparing the turbulence in wind tunnels of different designs.‡ (A comparison of this kind is to be found

† Fage and Warsap, *A.R.C. Reports and Memoranda*, No. 1283 (1930).
‡ Dryden and Heald, *N.A.C.A. Report* No. 231 (1925). The body usually used is a sphere. See Chap. XI, § 219.

in Fig. 149: it will be seen that the critical range occurs earlier in the N.P.L. curve than in the Göttingen curve, from which we infer that the flow in the Göttingen tunnel is less turbulent than in the N.P.L. tunnel.)

## 192. The effects of roughness and of concentrated excrescences on the drag of a circular cylinder in the critical region.

The effects of systematic changes in surface roughness have been investigated by Fage and Warsap,† the rough surfaces being obtained by wrapping large sheets of glass paper around a 6·09-inch cylinder. Five grades of paper, specified by the makers as Nos. 0, 1, 2 Fine, 2 Strong, and 3, and also a sheet of 058 Garnet paper were used; also, to obtain a greater relative roughness, measurements were made on a 2·375-inch cylinder roughened with the coarser papers Nos. 2 Fine, 2 Strong, and 3. The drag curves (uncorrected for gap and wall effects) for these rough cylinders are shown in Fig. 162. The drop in $C_D$ occurs at a lower value of $R$, and becomes smaller, as the surface roughness is progressively increased. There are indications that if the roughening had been continued a surface would eventually have been obtained for which there is no appreciable fall in $C_D$.

The work with the glass papers gives a measure of the effect of continuous disturbance of the flow in the boundary layer by roughness arising from innumerable small excrescences of unknown size and shape; but it gives no information on the effect of a localized roughness. It can be shown that the flow becomes particularly sensitive to surface roughness as the transition region is approached. Measured values of the drag of the 6·09-inch cylinder, when wires of various diameters were placed along generators at $\theta = \pm 65°$, are plotted in Fig. 163: these positions of the wires were forward of the transition region. The selected diameters ($\epsilon$) of the wires were 0·001, 0·002, 0·005, 0·010, and 0·020 inch respectively. The corresponding values of $\epsilon/\delta$, where $\delta$ is the thickness of the boundary layer at $\theta = 65°$, may be taken as 0·03, 0·06, 0·14, 0·29, and 0·57, approximately, so that each of the wires was totally immersed in the boundary layer. It will be seen that the shape of a drag curve undergoes a regular change as the diameter of the wires is increased: even the finest wires (of diameter 0·001 inch), which project into the fluid

† Loc. cit. on p. 431.

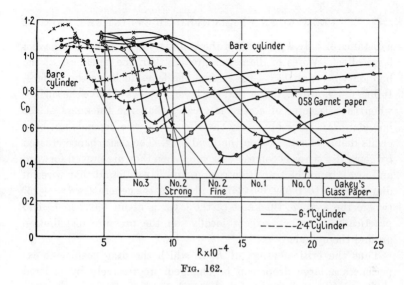

Fig. 162.

EFFECT ON $C_D$ OF WIRES AT $\theta = \pm 65°$
6·09" CYLINDER

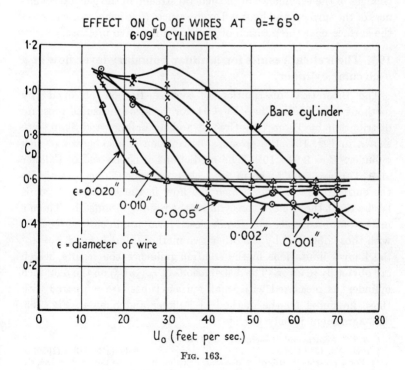

Fig. 163.

a distance of only 3 per cent. of the boundary layer thickness, have a large effect. Wires of diameters greater than the thickness of the boundary layer at $\theta = \pm 65°$ caused the layer to leave the surface. It does not, however, follow that the layer will always leave the surface when the diameter of a wire is greater than the thickness of the boundary layer, for obviously the effect of a wire depends not only on its diameter but also on the local flow. The disturbance created by generator wires becomes less severe when they are moved forward and away from the transition region, and it was found that wires of diameters 0·016, 0·032, 0·048 inch, placed at the generators $\theta = \pm 25°$ (where the average thickness of the layer is about 0·032 inch) had practically no effect, except locally, on the pressure distribution around the cylinder.

Thus the critical range of $R$, in which the drag coefficient experiences a large drop, can be changed progressively by ordered changes in the turbulence of the general stream, in the general roughness of the surface, and by small excrescences (fine generator wires) on the surface near the position of maximum negative pressure.

## 193. Theoretical results for laminar boundary layer flow at a circular cylinder.

The theoretical distribution of surface friction obtained (by methods described in Chap. IV, § 64) for the same measured pressure distribution by Thom,[†] by Green,[‡] and by Falkner and Skan,[||] are shown in Fig. 164: the pressure distribution is also shown (6-inch cylinder: $R = 0·94 \times 10^5$). The predictions of Green and of Falkner and Skan are seen to be in very close agreement over practically the entire range of $\theta$ for which the flow in the boundary layer is laminar. At the front of the cylinder the results obtained by Thom's method (approximate solution in closed form) are in close agreement with those obtained by the other two methods, but beyond $\theta = 45°$ the known limitations of the solution influence the results, which are obviously in error. The distributions of $\tau_0/(\frac{1}{2}\rho U_0^2)$ on the 5·89-inch cylinder, as measured with small surface tubes, are compared with those predicted by the method of Falkner and Skan in Fig. 165. The agreement is only fair.[††]

† A.R.C. Reports and Memoranda, No. 1176 (1928).
‡ Ibid., No. 1313 (1930).                    || Ibid., No. 1314 (1930).
†† For a pressure distribution at a circular cylinder, such as that in Fig. 164, the method of expansion in series described in Chap. IV, § 58 is probably the best method.

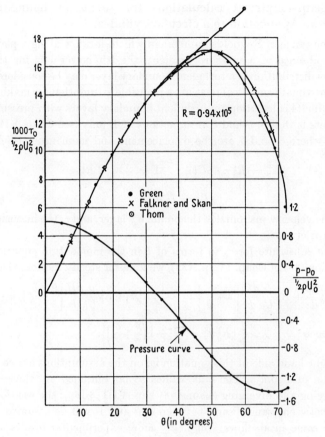

FIG. 164.

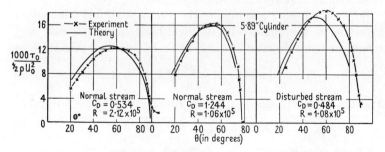

FIG. 165.

### 194. Semi-empirical calculations for turbulent boundary layers. Application to a circular cylinder.

In the range of Reynolds numbers in which, for flow along a plate in the absence of a pressure gradient, the $\frac{1}{7}$th power law for the velocity distribution in a turbulent boundary layer may be considered an approximate empirical relation, a plausible assumption for making approximate calculations in turbulent boundary layers with pressure gradients is that $\delta_1/\vartheta$ and $\zeta$ are functions of $\Gamma$ only (see Chap. VIII, § 166), where $\delta_1$ and $\vartheta$ are the displacement and momentum thicknesses,

$$\zeta = R_\vartheta^{\frac{1}{4}}\tau_0/(\rho U_1^2), \qquad \Gamma = \frac{\vartheta}{U_1}\frac{dU_1}{dx}R_\vartheta^{\frac{1}{4}}, \tag{1}$$

$$R\vartheta = U_1\vartheta/\nu, \tag{2}$$

$U_1$ is the velocity just outside the boundary layer, and $\tau_0$ the intensity of skin-friction.

If we substitute for $\tau_0$ in terms of $\zeta$ in the momentum equation (equation (38) of Chap. IV, p. 133), we find, for steady motion, that

$$\zeta = R_\vartheta^{\frac{1}{4}}\frac{d\vartheta}{dx} + R_\vartheta^{\frac{1}{4}}\frac{\vartheta}{U_1}\frac{dU_1}{dx}\left[2 + \frac{\delta_1}{\vartheta}\right] = \frac{4}{5}\frac{d}{dx}[R_\vartheta^{\frac{1}{4}}\vartheta] + R_\vartheta^{\frac{1}{4}}\frac{\vartheta}{U_1}\frac{dU_1}{dx}\left[\frac{9}{5} + \frac{\delta_1}{\vartheta}\right], \tag{3}$$

and hence

$$\frac{d}{dx}[R_\vartheta^{\frac{1}{4}}\vartheta] = \frac{5}{4}\left\{\zeta - \Gamma\left(\frac{9}{5} + \frac{\delta_1}{\vartheta}\right)\right\}. \tag{4}$$

The right-hand side of this equation is, on the assumptions above, a function of $\Gamma$ only. The only values so far obtained are those in converging and diverging channels (Chap. VIII, § 166): the relationships between $\zeta$ and $\Gamma$ and between $\delta_1/\vartheta$ and $\Gamma$ will be assumed to be the same in all flows; but many more experimental results are necessary to determine to what extent and in what circumstances these assumptions are approximately correct.

Buri† found that, according to the results of his and Nikuradse's experiments, the function on the right-hand side of (4) was, very roughly, a linear function, $A - B\Gamma$, of $\Gamma$. Equation (4) then becomes

$$\frac{d}{dx}[R_\vartheta^{\frac{1}{4}}\vartheta] + B\left(\frac{1}{U_1}\frac{dU_1}{dx}\right)(R_\vartheta^{\frac{1}{4}}\vartheta) = A. \tag{5}$$

This is a linear equation of the first order for $R_\vartheta^{\frac{1}{4}}\vartheta$, whose integral is

$$U_1^B R_\vartheta^{\frac{1}{4}}\vartheta = A\int U_1^B\,dx + \text{constant}. \tag{6}$$

† *Zürich Dissertation*, 1931. See also Prandtl, *Aerodynamic Theory* (edited by Durand), **3** (Berlin, 1935), 155–160.

The values of $A$ and $B$ from Nikuradse's experiments may be taken as 0·017 and 5 approximately; the values from Buri's experiments, in which the flow started practically from rest, are 0·015 and 4 approximately.

An alternative, and rather less crude, method of solution has been used by Howarth,[†] who solves (3) numerically after reducing it to non-dimensional form. If $U_0$ is a typical velocity, $d$ a typical length, and

$$R = \frac{U_0 d}{\nu}, \qquad U_1' = \frac{U_1}{U_0}, \qquad x' = \frac{x}{d}, \qquad \chi = R^{\frac{1}{4}}\left(\frac{\vartheta}{d}\right)^{\frac{5}{4}}, \qquad (7)$$

then

$$\Gamma = \frac{\chi}{U_1'^{\frac{1}{4}}}\frac{dU_1'}{dx'}; \qquad (8)$$

and after some reduction (3) becomes

$$\frac{d\chi}{dx'} + \frac{5}{4}\frac{\chi}{U_1'}\frac{dU_1'}{dx'}\left[2 + \frac{\delta_1}{\vartheta}\right] = \frac{5}{4}\frac{\zeta}{U_1'^{\frac{1}{4}}}. \qquad (9)$$

$U_1'$ is a known function of $x'$, and $\zeta$ and $\delta_1/\vartheta$ are treated as known functions of $\Gamma$ (see Chap. VIII, Figs. 117 and 118), so they are known functions of $\chi$ and $x'$. Hence (9) may be solved numerically or graphically if the initial value of $\chi$ (or $\vartheta$) is known. If the boundary layer along the surface of an obstacle in a stream is considered to be turbulent right from the forward stagnation point, then, since $U_1' = 0$ at the forward stagnation point, it follows from (9) that we must take $\chi = 0$ there if $\chi$ is not to be infinite. Equation (9) can be solved with this initial value for $\chi$.

When allowance is made for a laminar portion of the boundary layer, we must first make some assumption concerning the transition to turbulence. In the present state of our knowledge it is usual to assume a sudden transition (see Chap. VII, § 151). We may then make the assumption that once transition has taken place the conditions in the boundary layer are the same as if it had been turbulent right from the forward stagnation point: this is the easiest assumption to work with. An alternative assumption is that $\vartheta$ is continuous at transition.

Once $\chi$ is known both $\Gamma$ and $\vartheta$ are known, and hence also $\zeta$ and $\delta_1$. The values of $\zeta$ give the skin-friction; separation occurs when $\zeta = 0$. According to the curve that Buri drew through his experimental results, the value of $\Gamma$ when $\zeta = 0$ is $-0·07$; Howarth's curve makes

† *Proc. Roy. Soc.* A, **149** (1935), 558–586, especially 574–579.

the corresponding value of $\Gamma$ equal to $-0.06$. (According to Prandtl, in more general circumstances $\Gamma$ may vary within the range $-0.05$ to $-0.09$ at separation.)

It is to be remarked that if separation occurs for any fixed value of $\Gamma$, then from (8) the condition for separation in terms of $\chi$ is independent of the Reynolds number $R$ if $U_1'$ does not vary with $R$. Since this is true also for equation (9) and the boundary condition $\chi = 0$ at $x' = 0$, it follows that for a boundary layer turbulent right from the forward stagnation point there is no scale effect on the position of separation except in so far as $U_1'$ alters. When there is practically no variation in $U_1'$ (i.e. in the non-dimensional pressure distribution) with Reynolds number, a scale effect on the position of separation arises only for boundary layers which are partly laminar and partly turbulent, as a result of the conditions for transition to turbulence.

Calculations have been carried out by Howarth† for the case of the pressure distribution at a circular cylinder measured by Fage and Falkner at a Reynolds number of $2.12 \times 10^5$ and reproduced in Fig. 152 on p. 422. The transition to turbulence was assumed to take place at the point of inflexion in the graph of $U_1$. With $\vartheta$ continuous at transition and with the graph of $U_1$ as originally calculated, no separation was found. (Calculations for a completely turbulent boundary layer gave separation at $\theta = 111°$.) With a small change in the graph of $U_1$ between $\theta = 100°$ and $\theta = 120°$—a small change well within the experimental error—separation was found at $116°$; and so long as the small change is sufficient to produce separation, the position of separation is insensitive to the actual change made.‡ Experimentally it is difficult to estimate with exactness the position of separation; it appears to be between $122°$ and $130°$. In view of the crudeness of the assumptions in the calculation, the uncertainty of the position at which the transition to turbulence should be taken, and the experimental difficulties in locating separation, the result of the calculation may be considered satisfactory.

Further applications of this method of calculation (to boundary layers at the surfaces of bodies experiencing lift) will be described

---

† *Proc. Camb. Phil. Soc.* **31** (1935), 585–588.

‡ Calculations made in the same way for the pressure distribution measured by Flachsbart at a Reynolds number of $6.7 \times 10^5$ (Chap. I, Fig. 4, p. 24, full line curve) gave separation at $\theta = 117°$. No experimental information concerning the position of separation in these experiments is recorded.

in Chap. X, § 213, together with an alternative method due to Gruschwitz[†] which has been studied mainly in connexion with boundary layers at the surfaces of aerofoils also experiencing lift.

## 195. Cylinders of finite span. The effect of aspect ratio on drag.

In this chapter we have so far been concerned exclusively with two-dimensional flow. When the span is finite a departure of the flow from the two-dimensional pattern occurs towards the ends. The effect of this change of flow on the drag coefficient of two bluff forms,—viz. a circular cylinder and a flat plate normal to the stream, is shown in Table 15. The drag coefficient, obtained by dividing the

TABLE 15[‡]

| Circular cylinder $R = 88{,}000$ | | Flat plate normal to stream.[\|] $R = 68{,}000$ to $170{,}000$ | |
| --- | --- | --- | --- |
| $\dfrac{Length}{Diameter}$ | $\dfrac{C_D}{C_D \text{ for infinite span}}$ | $\dfrac{Length}{Breadth}$ | $\dfrac{C_D}{C_D \text{ for infinite span}}$ |
| ∞ | 1 | ∞ | 1 |
| 40 | 0·82 | 17·8 | 0·70 |
| 20 | 0·76 | 12·0 | 0·64 |
| 10 | 0·68 | 10·0 | 0·64 |
| 5 | 0·62 | 8·0 | 0·63 |
| 2·96 | 0·62 | 4·0 | 0·59 |
| 1·98 | 0·57 | 2·0 | 0·57 |
| 1·0 | 0·53 | 1·0 | 0·55 |

drag by the product of $\frac{1}{2}\rho U_0^2$, the span and the cross-stream breadth, decreases with the span: the drag for a length equal to the cross-stream breadth is a little greater than one-half of that of the same length of an infinite span.

The effect of the ends on profile drag (non-lifting bodies) depends not only on the ratio of length to breadth, but also on the sectional form. On low-drag forms the effect is considerably smaller than on bluff forms.

[†] *Ingenieur-Archiv*, **2** (1931), 321–346; *Zeitschr. f. Flugtechn. u. Motorluftschiffahrt*, **23** (1932), 308–312.

[‡] *Ergebnisse der Aerodynamischen Versuchsanstalt zu Göttingen*, **2** (1923), 28 and 34.

[\|] Schubauer and Dryden (*N.A.C.A. Report* No. 546 (1935)), experimenting on a flat plate of aspect ratio 6 normal to the stream, found that the drag coefficient increases with increasing turbulence in the main stream. If $u$ is the root-mean-square value of the longitudinal velocity fluctuation in a stream of mean velocity $U_0$, then $C_D$ was 1·326 for $u/U_0 = 0$·03, whereas the extrapolated value for $u/U_0$ zero was $C_D = 1$·246. Change of Reynolds number (except at low values) has no effect on $C_D$.

## ADDITIONAL REFERENCES

*Treatise.*

MUTTRAY, *Handbuch der Experimentalphysik*, **4**, part 2 (Leipzig, 1932), 233–336. (This article deals also with bodies of revolution.)

*Flow in a separating boundary layer at an elliptic cylinder.*

SCHUBAUER, *N.A.C.A. Report* No. 527 (1935).

*Pressure distribution at, and drag on, a strut form.*

ERMISCH, *Abhandl. Aero. Inst. Aachen*, No. 6 (1927), pp. 43–46.

*Effect of shape on drag and on scale effect in the critical region.*

HOPF, *Handbuch der Physik* (Berlin, 1927), p. 169.

*Motion of an elliptic cylinder.*

RICHARDS, *Phil. Trans.* A, **233** (1934), 279–301.

*Flow past cylinders at low Reynolds numbers.*

HOMANN, *Zeitschr. f. angew. Math. u. Mech.* **16** (1936), 153–164.

*Drag of rough and smooth cylinders at high Reynolds numbers.*

ACKERET, *Schweiz. Bauztg.* **108** (1936), 25–36.

*Separation of the boundary layer on a cylinder and on an aerofoil.*

GREEN, *A.R.C. Reports and Memoranda*, No. 1396 (1931).

*Laminar boundary layer calculations at the surface of a circular cylinder.*

HOWARTH, *A.R.C. Reports and Memoranda*, No. 1632 (1935).

*Turbulence in boundary layers and its effect on the drag of aerofoils.*

JACOBS, *Journ. Soc. Automotive Engineers*, **41** (1937), 468–472.

*Calculation of skin-friction drag and of form drag.*

An account of a method of calculating the profile drag of an aerofoil (i.e. the skin-friction drag plus the form drag) by integration of the momentum equation of the boundary layer will be published in the *A.R.C. Reports and Memoranda* during 1938 in a paper by YOUNG and SQUIRE.

Further information concerning the drag of aerofoils can also be found in Chap. X and in several of the papers in the list of references at the end of that chapter.

# X

# FLOW PAST ASYMMETRICAL CYLINDERS.
# AEROFOILS. LIFT

## 196. Introduction.

In the present chapter we are concerned with asymmetrical flow past long cylindrical bodies: the asymmetry may arise from the shape of a body and its inclination to the free stream, or from inclination alone if its section is symmetrical. In either case the aerofoil is the most important body to be considered.

Many investigations have been made to determine the best wing shape to meet particular requirements in aeroplane design. The findings of such investigations, although of great practical value, do not concern us here: our aim is to consider aerofoil flow in its broadest aspect, and particularly those features which are associated with the boundary layer. Some consideration will be given to the dependence of the performance of an aerofoil on its sectional shape: further detailed information of this kind may be found in the publications of aerodynamical laboratories.[†]

The lift and drag coefficients of an aerofoil are usually obtained from wind-tunnel tests on a model of finite aspect ratio (usually 6 to 1) and rectangular plan form. Unless otherwise stated, we are here concerned with two-dimensional flow (infinite span): coefficients for this type of flow can be predicted by the vortex theory of aerofoils from those measured on an aerofoil of finite span.[‡]

---

[†] The results of performance tests on 859 aerofoils, collected from the published reports of aerodynamical laboratories in England, U.S.A., France, Germany, Italy, and Belgium, have been published by the National Advisory Committee for Aeronautics (U.S.A.) in 'Aerodynamic Characteristics of Airfoils', Reports No. 93 (1920); No. 124 (1921); No. 182 (1923); No. 244 (1926); No. 286 (1928); No. 315 (1929). See also 'The Characteristics of 78 Related Airfoil Sections from Tests in the Variable-Density Tunnel', Jacobs, Ward, and Pinkerton, *N.A.C.A. Report* No. 460 (1933); 'Caratteristiche Aerodinamiche di Ali', *Supplemento ai Rendiconti Tecnici*, Parts 1–11 (1925–6).

[‡] For an explanation of the vortex theory and of its many practical applications (e.g., the prediction of the effect of a change of aspect ratio on lift and drag, the interference of tunnel boundaries on lift and drag, the interference between neighbouring aerofoils), see Glauert, *Aerofoil and Airscrew Theory* (Cambridge, 1926); Kármán and Burgers, *Aerodynamic Theory* (edited by W. F. Durand), **2** (Berlin, 1935).

## 197. The lift of aerofoils, and scale effect.

Curves of $C_L$ against $\alpha$, obtained from tests[†] in the C.A.T.[‡] on the aerofoil section R.A.F. 34, are shown in Fig. 166: $C_L$ is the lift per unit span divided by $\frac{1}{2}\rho U_0^2 c$ (where $c$ is the chord length and $U_0$ the undisturbed velocity of the stream) and $\alpha$ is the inclination of the chord of the section to the direction of the free stream. For each value of the Reynolds number $R$ (equal to $U_0 c/\nu$) the lift coefficient

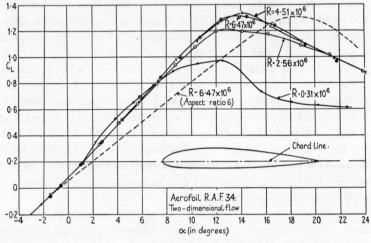

FIG. 166.

increases at first almost linearly with incidence: shortly after departure from the linear relation it reaches a maximum value at a certain critical incidence, called the stalling angle. For the aerofoil to which Fig. 166 relates the maximum value of $C_L$ and the stalling angle increase with $R$.[‖] Over the linear part of the $(C_L, \alpha)$ curve the slope is about 5·7 per radian. (The theoretical slope for thin aerofoils is $2\pi$ per radian:[††] the experimental slope is less because of the departure of the flow from the ideal form. See Chap. II, § 25, pp. 76, 77.)

The highest value of $R$ that can be reached with an aerofoil of 6 inches chord in any existing atmospheric tunnel of the ordinary type

† Relf, Jones, and Bell, *A.R.C. Reports and Memoranda*, No. 1706 (1936).

‡ The Compressed Air Tunnel at the National Physical Laboratory, Teddington.

‖ For $R = 6·47 \times 10^6$ the curve in Fig. 166 should be corrected near the maximum so that $C_{L\,max} = 1·395$ at $\alpha = 14·5°$.

†† See Glauert, *Aerofoil and Airscrew Theory* (1926), p. 86.

is less than $10^6$, whereas the value reached by the wing of an aeroplane in flight may exceed $10^7$. The curves of Fig. 166 therefore indicate clearly that the maximum value of $C_L$ measured in an atmospheric

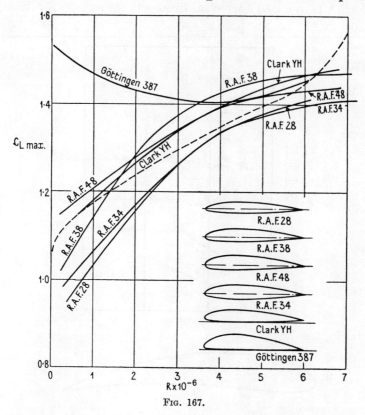

Fig. 167.

tunnel may differ appreciably from that experienced in flight on a wing of the same section.

In Fig. 167 the variation with $R$ of the maximum value of $C_L$ is shown for several sections.[†] The maximum lift of every section except Göttingen 387 rises with the Reynolds number; that of Göttingen 387 falls as the Reynolds number increases up to about $4 \times 10^6$, and then slowly rises.[‡] The Reynolds number for the wing of an aeroplane

[†] For the meaning of the broken line in Fig. 167 see § 213, p. 486.

[‡] Both the maximum thickness and the centre line camber of Göttingen 387 are appreciably greater than those of the other aerofoils. For further discussion of the influence of shape on scale effect see § 207.

of average size when landing is about $3\cdot5\times10^6$: at this value every section of Fig. 167 has a maximum lift coefficient about $1\cdot3$ or $1\cdot4$. It should be added that the maximum lift coefficients measured in the C.A.T. on a model aeroplane fitted with the sections R.A.F. 28,

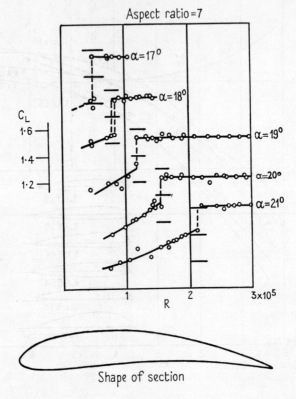

Shape of section

FIG. 168.

34, and 38 successively were practically the same as those measured on the aeroplane in flight at the same Reynolds numbers.†

When the span of an aerofoil is finite, trailing vortices spring from the trailing edge, and the downwash due to these vortices reduces

† Relf, *Journ. Roy. Aero. Soc.* **39** (1935), 1–30. Since the publication of this paper good agreement has also been obtained for R.A.F. 48 and Göttingen 387. See also Stack, 'Tests in the Variable-Density Tunnel to investigate the Effects of Scale and Turbulence on Airfoil Characteristics', *N.A.C.A. Technical Note* No. 364 (1931); Jacobs, 'The Aerodynamic Characteristics of eight very Thick Airfoils from Tests in the Variable-Density Wind Tunnel', *N.A.C.A. Report* No. 391 (1931).

both the mean incidence (relative to the local flow) and the mean lift coefficient. Below the stalling angle the lift coefficient for an aspect ratio 6 : 1 is about 75 per cent. of that for infinite aspect ratio at the same incidence to the free stream. (Compare the broken and full-line curves in Fig. 166.) The reduction in lift depends not only on the aspect ratio but on the slope of the $(C_L, \alpha)$ curve; its amount can be predicted with good accuracy by the vortex theory of aerofoils.

Scale effect on $C_L$ at Reynolds numbers within the range $5 \times 10^4$ to $3 \times 10^5$ is shown in Fig. 168. The results, due to Krasilschikov and Volkov,[†] refer to angles of incidence near the stalling angle and to a rectangular aerofoil of aspect ratio 7 : 1. Since the aspect ratio is finite the effective incidence alters along the span; the conditions at different sections are different. The results are interesting in that they show a discontinuity in $C_L$ at each incidence for a certain Reynolds number. At lower Reynolds numbers, over a certain portion of the middle part of the span, away from the wing tips, it is probable that the boundary layer is laminar up to the position where the forward flow separates from the surface; at higher Reynolds numbers, a transition to turbulence occurs, and there is no laminar separation. (Cf. § 213 below.)

## 198. The effect of turbulence in the stream on the maximum lift of aerofoils.

The maximum lift coefficient of a section depends not only on the shape and on the Reynolds number, but also on the turbulence in the free stream. This is clearly shown by the curves of Figs. 169 a and b, where $C_{L\,\mathrm{max}}$ is plotted against $R$ for two aerofoils (R.A.F. 28 and Göttingen 387) tested in the C.A.T. behind turbulence screens.[‡] Two screens were used, constructed from vertical steel strips $\frac{1}{4}$ inch wide and spaced 3 inches and 1·5 inches apart respectively: the aerofoils were placed in the disturbed stream 21 inches (2·6 chord lengths) behind the screen. The increased turbulence produces an increase in maximum lift in both cases; otherwise the effect of the increased turbulence on the two aerofoils is markedly different.

Similar experiments[||] have been made in the Guggenheim Aeronautical Laboratory at the California Institute of Technology on a

---

† *Trans. Centr. Aero-Hydrodyn. Inst.*, *Moscow*, No. 254 (1936).

‡ Relf, *loc. cit.*

|| 'A Theoretical Investigation of the Maximum Lift Coefficient', Kármán and Millikan, *Journ. Applied Mechanics*, **2** (1935), 21–27.

model of the N.A.C.A. 2412 aerofoil. To obtain several different degrees of turbulence a wire grid or screen was placed at various distances upstream of the model; and at each position of the grid measurements of $C_{L\max}$ were taken over a wide range of Reynolds numbers. The curves obtained are reproduced in Fig. 170; each

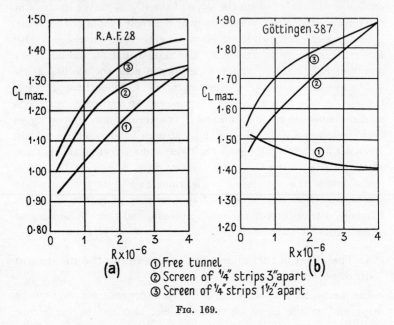

FIG. 169.

curve relates to a fixed grid position and therefore to a given degree of turbulence, the degree of turbulence at the model decreasing as the distance behind the screen increases. The values of $R_c$ are the Reynolds numbers at which the drag coefficient of a sphere at the position of the model relative to the grid would be 0·3 (see Chap. XI, § 219, p. 499). The curves show that $C_{L\max}$ increases as the degree of turbulence increases.

The maximum lift coefficient depends also on surface roughness,† especially on the upper surface near the leading edge. At present the effects on $C_{L\max}$ of Reynolds number, turbulence in the general stream, and surface roughness can be found only by experiment: the tendency is for $C_{L\max}$ to increase with $R$ and with turbulence,

† The effect of roughness is considered in more detail in § 211.

and to decrease with surface roughness, but there are exceptions (e.g. Göttingen 387, Fig. 167). The phenomena are complicated and further research is needed.

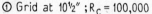

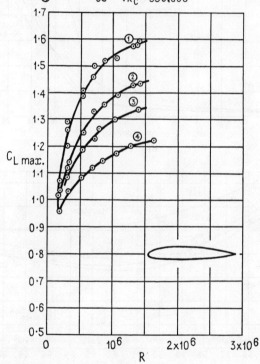

FIG. 170.

## 199. The drag of aerofoils, and scale effect.

Curves of profile drag coefficient $C_D$ against incidence at various values of $R$ are shown in Fig. 171. The drag coefficient does not change appreciably with incidence over the range $-2° < \alpha < 6°$; but above this range a marked rise, at first slow and then rapid, occurs.

The variation of $C_D$ with $R$ at constant values of $\alpha$ is shown in

Fig. 172, which also includes drag curves (dotted lines) for laminar and completely turbulent flows along a smooth flat plate. Over the middle of the range of values of log $R$ in Fig. 172 the curve for

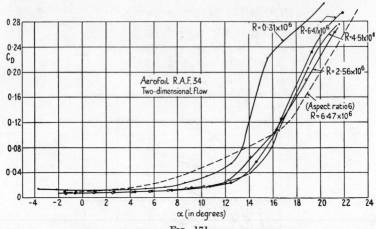

FIG. 171.

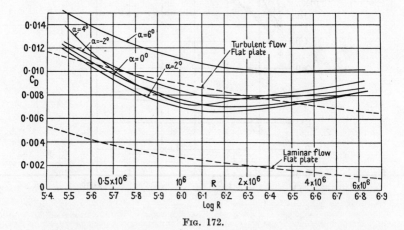

FIG. 172.

minimum drag ($\alpha = 2°$) lies below the turbulent curve for a flat plate: at higher values of $R$ the two curves have crossed one another. This feature is common to all thin aerofoil sections.

Profile drag is compounded of two parts, a form drag associated with normal pressures, and a frictional drag arising from the frictional stresses on the surface. At low Reynolds numbers an appreciable

part of the frictional drag is associated with laminar flow in the
boundary layer; but this part becomes progressively smaller as the
Reynolds number increases, and eventually nearly all the frictional
drag is associated with turbulent flow. The nature of the results
illustrated in Fig. 172 suggests that at low Reynolds numbers there is

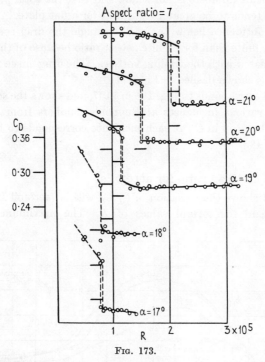

FIG. 173.

a relatively large form drag, which makes the minimum drag coefficient
for a thin aerofoil lie near the turbulent curve for a flat plate rather
than near the laminar curve. With increasing Reynolds numbers
the frictional drag coefficient rises as a larger part of the boundary
layer becomes turbulent, but the form drag coefficient decreases.
At first the second effect is the greater and the total drag coefficient
falls; but later the form drag becomes small, and the drag curve
rises to the turbulent curve for a flat plate. At the highest Reynolds
number of Fig. 172, the form drag for low incidences is small, the
flow in the boundary layer is turbulent except for a small region
near the leading edge, and practically the whole of the drag arises

from skin-friction. These conditions hold for all thin aerofoils of good shape (thickness $< 0.14$ chord).†

The evidence obtained from thick aerofoil sections is not so conclusive, especially when the camber of the centre line is large. Present indications are that the form drag coefficient of a thick section at high Reynolds numbers is not small, and that the total profile drag coefficient tends to be greater than that for a flat plate.

At any incidence below the stalling angle the drag coefficient is greater for finite than for infinite aspect ratio because of the induced drag associated with the trailing vortices. The drag curve for aspect ratio 6 : 1 is shown dashed in Fig. 171.‡

Fig. 173 corresponds to Fig. 168 in § 197, and shows the scale effect on $C_D$ at various incidences at Reynolds numbers from $5 \times 10^4$ to $3 \times 10^5$. The drop in $C_D$ at each incidence corresponds to the rise in $C_L$ in Fig. 168.

## 200. The lift/drag ratio for aerofoils.

Fig. 174 shows the variation of $C_L/C_D$ with $\alpha$ (aerofoil R.A.F. 34, infinite span) for several values of $R$. The maximum value of

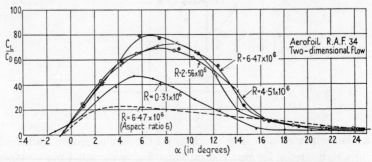

Fig. 174.

$C_L/C_D$ increases with increasing values of the Reynolds number. At $R = 6.47 \times 10^6$, $\alpha = 7.0°$, the lift on the aerofoil is about 80 times the drag. The maximum value of $C_L/C_D$ for an aerofoil of moderate aspect ratio is considerably smaller, mainly because of the increase

† In Figs. 171 and 172 the increase in the profile drag coefficient as the incidence departs from the value for minimum drag is due largely to the progressive increase in form drag.

‡ For the effect on drag of surface roughness see § 182 and § 211.

in drag arising from the trailing vortices. The dotted curve shows
that for an aspect ratio 6:1 the maximum value of $C_L/C_D$ is about
21·5 and occurs at an incidence of about 4·5°.

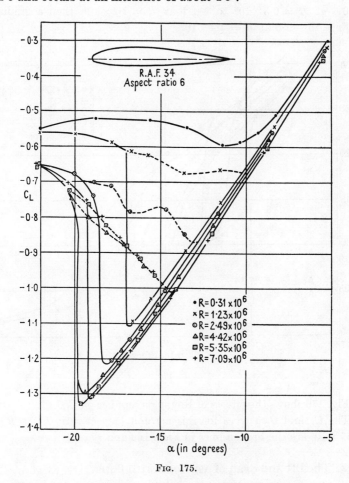

FIG. 175.

## 201. The lift and drag of aerofoils at negative incidences, and scale effect.

Fig. 175 shows the effects of a change of the Reynolds number on
the lift coefficient of an aerofoil (R.A.F. 34, aspect ratio 6) at negative incidences.† Except at the lowest Reynolds number, double

† Williams and Brown, *A.R.C. Reports and Memoranda*, No. 1772 (1937).

readings were obtained at the stall. The values when $-\alpha$ increased are shown by the full lines, those when $-\alpha$ decreased by the broken lines. There is a large change in negative maximum $C_L$ from $-0.6$ at $R = 0.3 \times 10^6$ to $-1.1$ at $R = 1.2 \times 10^6$, then a gradual change to $-1.3$ at $R = 4.4 \times 10^6$.

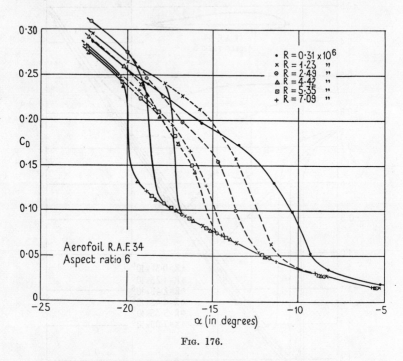

FIG. 176.

Fig. 176 shows the effects of Reynolds number on $C_D$.

The $C_L$ and $C_D$ curves for this aerofoil (aspect ratio 6) at $R = 6.47 \times 10^6$ are the broken lines in Figs. 166 and 171.

## 202. The lift and drag of aerofoils with flaps.

Fig. 177† shows the effect on $C_L$ of a trailing-edge flap (width $= 0.15c$) normal to the under surface of the aerofoil R.A.F. 34 (aspect ratio 6) at $R = 4.32 \times 10^6$. The flap increases the lift coefficient of the aerofoil at all incidences; but there is no appreciable change in the stalling angle. The air flow breaks away from the sharp edge of the

† Williams and Brown, *loc. cit.*

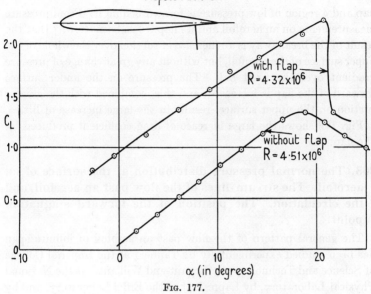

R.A.F. 34 with and without 15% split flap.
Aspect ratio 6

Fig. 177.

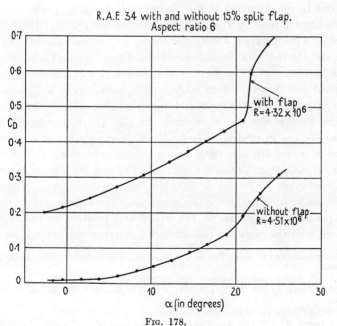

R.A.F. 34 with and without 15% split flap.
Aspect ratio 6

Fig. 178.

flap and a region of low pressure is formed behind it. From pressure measurements on an aerofoil and its flap it has been shown† that the result of the breakaway is an increase of suction over the whole of the upper surface of the aerofoil, but without any great change of pressure gradient over the front half. The pressure on the under surface forward of the flap is increased, and this, combined with the greater suction on the upper surface, results in the large increase of lift.

Fig. 178 shows the large increase of drag coefficient produced by the flap.

## 203. The normal pressure distribution at the surface of an aerofoil. The stream-lines of the flow past an aerofoil, and the circulation. The position of the forward stagnation point.

The general pattern of the flow past an aerofoil of infinite span has been studied experimentally by Tanner‡ at the Imperial College of Science and Technology, by Bryant and Williams‖ at the National Physical Laboratory, by Lapresle†† at the Eiffel Laboratory, and by van der Hegge Zijnen‡‡ at the Technical University, Delft.

The aerofoil used by Tanner had a symmetrical section (R.A.F. 30, chord 10·5 inches, span 4 feet): the model was mounted between the floor and the roof of a rectangular tunnel 4 feet high and 5 feet wide, at an incidence of 7° to the direction of the stream; measurements of speed and direction were made over a wide field in the plane of symmetry normal to the span; and the distribution of normal pressure on the surface was also measured. In Fig. 179 the experimental values (points in circles) of $(p-p_0)/(\frac{1}{2}\rho U_0^2)$ are plotted against $x/c$, where $x$ is the distance from the nose measured along the chord; both the maximum pressure coefficient (1·0) and the minimum pressure coefficient (−3·0) occur near the nose, the maximum pressure on the under surface and the minimum pressure on the upper surface. The pressure is negative on the upper and positive on the under surface, and the negative pressure on the upper surface makes an

---

† Gruschwitz and Schrenk, *Zeitschr. f. Flugtechn. u. Motorluftschiffahrt*, **23** (1932), 599, 600.

‡ *A.R.C. Reports and Memoranda*, No. 1353 (1931).

‖ *Phil. Trans.* A, **225** (1925), 199–237; *A.R.C. Reports and Memoranda*, No. 989 (1926).

†† *Service Technique et Industriel de l'Aéronautique*, No. 43 (1927).

‡‡ *Rapport A.* 129 *van den Rijks-Studiedienst voor de Luchtvaart, Amsterdam* (1926).

appreciably greater contribution to the lift than the positive pressure
on the under surface.

The full-line curve of Fig. 179 shows the distribution of normal

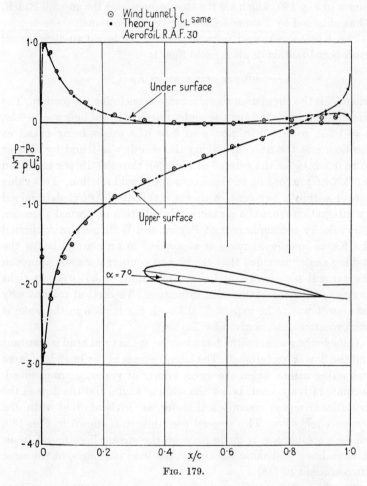

FIG. 179.

pressure for an inviscid flow with a circulation giving the same lift,
according to the Kutta-Joukowski formula, as that measured in the
wind tunnel: the experimental values lie close to this theoretical
curve except for a short part of the tail.

The circulation around the aerofoil gives rise to a region of high
velocity, with a crowding together of the stream-lines, above the

aerofoil, and a region of low velocity, with a wider spacing of the stream-lines, below, together with an up-flow at the nose and a down-flow at the tail. These features are illustrated by the broken curves in Fig. 180, which are the stream-lines past the aerofoil R.A.F. 30 as obtained by Tanner from velocity observations.[†]

The Kutta-Joukowski formula for the lift in a two-dimensional irrotational motion of an inviscid fluid is

$$\text{lift per unit span} = K\rho U_0,$$

where $K$ is the circulation round a contour enclosing the section. The circulation round a rectangular contour defined by the lines $x = -0.5c$, $x = 1.25c$, $y = -0.5c$, and $y = 0.5c$ (the origin being taken at the nose and the axis of $x$ along the chord) was found by Tanner to be $0.352cU_0$ for the aerofoil of Fig. 180; thus the lift per unit span is $0.352\rho cU_0^2$ according to the Kutta-Joukowski relation. This value agreed within 2 per cent. with the value ($0.359\rho cU_0^2$) determined by integration from the measured distribution of normal pressure. The velocity measurements of Bryant and Williams also confirmed the Kutta-Joukowski relation as applied to an aerofoil below the stalling angle, provided that the chosen contour does not approach the aerofoil too closely, and that the trailing wake is cut at right angles to the general direction of motion. Theoretical reasons why this result was to be expected, although the motion in the wake is not irrotational, were given by Taylor.[‡]

Outside the wake and the boundary layer the total head is constant and the flow is irrotational. The lateral extent of the boundary layer and wake, within which the direct effects of viscosity are confined, is comparatively small below the stalling angle, and the flow in the irrotational region resembles that for an inviscid fluid with the correct circulation. This general resemblance is shown in Fig. 180, where a comparison is made between the stream-lines found in air (broken lines) and those for an inviscid fluid (full lines) at the same lift coefficient (0.718).

Near the surface of an aerofoil there are, however, important differences between the flow of a real fluid and that of an inviscid fluid. In addition to obvious differences at the tail (associated with

[†] These features are also evident from the contour lines of equal velocity and of equal inclination to the direction of the undisturbed stream which were drawn by Bryant and Williams (*loc. cit.*).

[‡] *Phil. Trans.* A, **225** (1925), pp. 238–245.

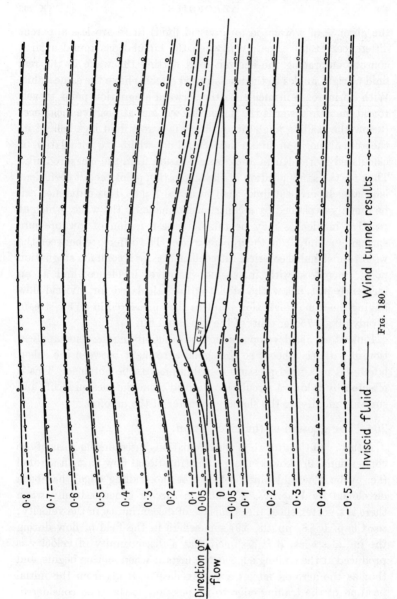

0·8
0·7
0·6
0·5
0·4
0·3
0·2
0·1
0·05
0
−0·05
−0·1
−0·2
−0·3
−0·4
−0·5

α = 7°

Direction of
flow

Inviscid fluid ———    Wind tunnel results --○--○--

FIG. 180.

the absence of a wake in an inviscid fluid) there are less apparent differences at the nose, even when the circulations round a large contour embracing the aerofoil (and cutting the wake in the real fluid at right angles to the undisturbed stream) have the same value. With an increase in incidence the forward stagnation point travels round the nose towards the under surface, and at positive incidences (below the stalling angle) its position in a real fluid lags behind its theoretical position in an inviscid fluid when the circulation is calculated by the Kutta-Joukowski relation from the measured lift.[†] Thus for the same position of the stagnation point the lift coefficient is smaller in inviscid fluid than in a real fluid. This difference can be roughly explained by a contribution made by the wake, which at positive incidences may be taken to be a region of low pressure springing mainly from the upper surface. The difference between the wind tunnel and theoretical lift coefficients, for a common stagnation point, increases with increasing (positive) incidence; and at an incidence just below the stall the lift calculated for inviscid flow having the same stagnation point as that measured in an air-stream is only about 75 per cent. of the measured lift.

Pinkerton[‡] has developed a method of modifying the usual application of potential theory so that good agreement between the calculated and measured pressure distributions can be obtained. This is effected by taking a circulation derived from the measured lift and suitably modifying the shape parameter of the aerofoil.

## 204. The growth of the circulation.

A theory of the growth of the circulation, based on the ideas of the conventional Joukowski theory for irrotational flow past an aerofoil (i.e. cyclic flow and finite velocity at the trailing edge), has been developed by Wagner.[||] Circulation can be generated only when there is a vortex-sheet (i.e. a surface of discontinuity in the velocity: see Chap. I, §8, pp. 28, 29) somewhere in the field of flow during the initial stages: it is assumed that a discontinuity of velocity is produced at the trailing edge at the instant when motion begins, and that as the aerofoil moves a vortex-sheet extends from the initial position of the trailing edge to its position at the time considered.

---

† Fage, *Phil. Trans.* A, **227** (1927), 1–19; *A.R.C. Reports and Memoranda*, No. 1097 (1927).

‡ *N.A.C.A. Report* No. 563 (1936).

|| *Zeitschr. f. angew. Math. u. Mech.* **5** (1925), 17–35.

The strength of the vortex-sheet at any point is determined from the fact that the circulation round a contour enclosing both the aerofoil and the vortex-sheet is always zero, together with the condition that the velocity at the sharp trailing edge should always be finite. The total vorticity in the sheet is equal and opposite to the circulation around the periphery of the aerofoil (the bound vorticity), and the rate at which vorticity is discharged into the fluid is governed by the condition that the velocity at the trailing edge due to the bound and free vorticity and to the cyclic flow is finite. The mathematical analysis is simplified by taking the aerofoil to be a flat plate (and later supposing the leading edge rounded so that the velocity there remains finite).

Wagner's theory relates essentially to the motion of an ideal inviscid fluid. The growth of circulation in a slightly viscous fluid has been discussed qualitatively by Howarth:[†] when viscosity is taken into account, a finite time-interval elapses before circulation begins to develop, namely the time-interval before separation of the forward flow begins in the boundary layer (see Chap. IV, §§ 65, 66). Separation begins first on the pressure side, and very shortly afterwards on the suction side. The rate at which vorticity is discharged into the main body of the fluid is the value of $\frac{1}{2}U_1^2$ at the position of separation, where $U_1$ is the velocity just outside the boundary layer (Chap. II, § 22, p. 66). The vorticity shed from the two sides is of opposite sign, and as soon as the position of separation on the suction side has moved forward into a position of sufficiently high velocity, the net rate of discharge of vorticity is sufficiently checked to produce a point of inflexion in the circulation-time curve. This inflexion occurs at a small value of the circulation.

The rate of growth of the circulation in the early stages of the motion of an aerofoil has been measured by Walker.[‡] The aerofoil had a symmetrical section (R.A.F. 30, see Fig. 179), and it was moved at an incidence of $7\frac{1}{2}°$ through water in a long tank having parallel sides 6 inches apart. Oil drops suspended in the water were illuminated by a bright beam of light, and their movements due to the passage of the aerofoil were photographically recorded. Photographs obtained in a short exposure of known duration revealed the instantaneous stream-lines relative to the undisturbed water, and the

† *Proc. Camb. Phil. Soc.* **31** (1935), 582–584.

‡ *A.R.C. Reports and Memoranda*, No. 1402 (1932). See also *Proc. Camb. Phil. Soc.* **30** (1934), 365–375.

velocity field was determined from the lengths of the short traces
made by the drops. The aerofoil was started suddenly from a known
position in the tank, and photographs of the flow were taken after
it had travelled about 1, 1·5, 2, 2·5, 3, 4·5, and 6 chords. The photo-

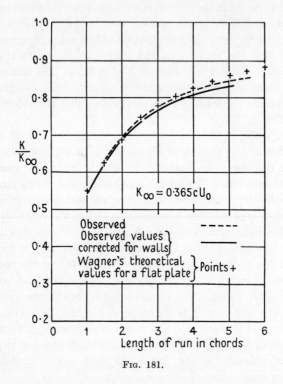

FIG. 181.

graph of Pl. 30a was obtained after the aerofoil had travelled about
one chord: a circulation round the aerofoil, and one in the opposite
sense in the wake, are clearly shown. The wake has a wavy form
and individual vortices are present at its end: these features are
not unexpected, because the vortex-sheet assumed in theory is unstable.

The cast-off vortices left behind in the water at first move down-
ward under the influence of the circulation round the aerofoil, and
ultimately become stationary; the rate of production of cast-off
vorticity becomes progressively smaller as the length of the run
increases. Pl. 30b is a photograph of the flow obtained after the

PLATE 30

*a*

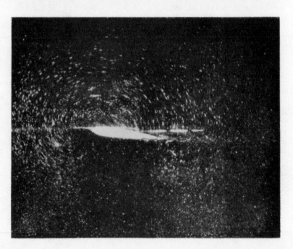

*b*

aerofoil has traversed six chords. The circulation round the aerofoil is clearly shown.

From photographs of the type of Pl. 30, Walker was able to compute the circulation ($K$) round the aerofoil at various stages in its run. These values of $K$, divided by the value ($K_\infty$) for steady motion, are plotted against length of run in Fig. 181. Included in this figure are theoretical values predicted by Wagner's theory for a flat plate: the experimental curve (corrected for the interference of the tank walls) is seen to be in fair agreement with the theoretical values, and this agreement leads to the conclusion that Wagner's theory accounts for the growth of circulation to the same extent as the Kutta-Joukowski theory accounts for steady circulation.

## 205. The intensity of skin-friction at the surface of an aerofoil.

Distributions of frictional intensity on the surface of a symmetrical aerofoil† are shown in Fig. 182. These distributions were obtained from observations of velocity taken by Fage and Falkner (at $U_0 c/\nu = 1\cdot25 \times 10^6$) with small Stanton tubes in the manner described in Chap. IX, § 179. On the upper surface the frictional intensity has a maximum value very near the nose, and a second and greater maximum value some distance beyond; the transition from laminar to turbulent flow in the boundary layer takes place in the region between these maxima. This transitional region moves towards the nose as the incidence increases ($-0\cdot18°$ to $5\cdot82°$). Distributions of normal pressure are included in Fig. 182, and these curves, taken in conjunction with those of frictional intensity, show that the transitional region is situated just beyond the point of lowest pressure. The distribution of frictional intensity on the lower surface resembles that on the upper surface, but the maximum values are smaller. On each surface the frictional intensity falls steadily towards the tail, but not to zero; at these small incidences the boundary layer has not separated from either the upper or the lower surface.

The contributions of the two surfaces to the frictional drag of the

---

† The distributions were actually found on the upper surface at incidences α = 5·82°, 2·82°, −0·18°, −3·18°, −6·18°. Since the shape is symmetrical, the results give the distributions on the lower surface at incidences α = −5·82°, −2·82°, 0·18°, 3·18°, and 6·18°.

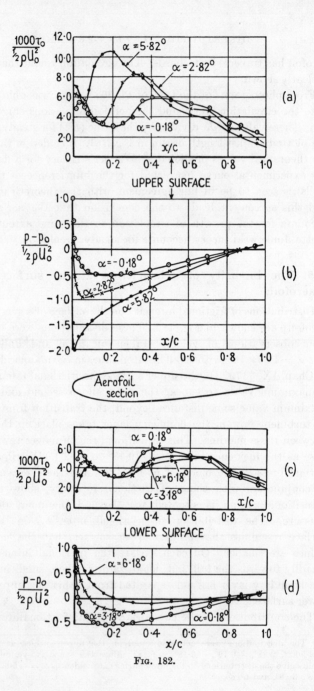

Fig. 182.

aerofoil are given in Table 16. The contribution of the upper surface increases with incidence, that of the lower surface decreases, and the total frictional drag coefficient increases over the range taken.

TABLE 16

| α | Contribution to the frictional drag coefficient | | Frictional drag coefficient |
| --- | --- | --- | --- |
| | Upper surface | Lower surface | |
| $-0\cdot18°$ | 0·0042 | 0·0043 | 0·0085 |
| $2\cdot82°$ | 0·0049 | 0·0041 | 0·0090 |
| $5\cdot82°$ | 0·0056 | 0·0039 | 0·0095 |

## 206. Effect of shape on the pressure distribution at the surface of an aerofoil.

It is not the purpose of the present work to discuss in detail the effect of shape on aerofoil characteristics, but there are some general features to which reference should be made. The behaviour of an aerofoil at positive incidences depends largely on the pressure distribution on the upper surface, and a comparatively small change in the contour of the upper surface at the nose is sufficient to affect the pressure distribution appreciably.[†] Calculations by Garrick[‡] of the theoretical pressure distributions for 20 aerofoils of different shapes and thicknesses show that the value of the negative pressure coefficient on the upper surface near the nose increases with increasing fineness of the nose, and that very high values are reached on thin sections. In Fig. 183[||] a comparison is shown of the theoretical pressure distributions for a thick section (Göttingen 387) and for a thin section (N.A.C.A. 2409) at the same calculated lift coefficient, and also at nearly the same angle of incidence. At $C_L = 1\cdot50$ the pressure on the thin aerofoil N.A.C.A. 2409 falls to about $p_0 - 2\cdot75\rho U_0^2$, whereas the fall for the thick aerofoil Göttingen 387 is to $p_0 - 1\cdot12\rho U_0^2$ approximately. The pressure gradient at the nose is very steep for a thin section, and usually much less steep for a thick section. The position of the maximum negative pressure depends both on the shape and on the incidence, and moves towards the leading edge as the incidence increases; it is nearer the nose on a thin section than on a thick section.

† Betz, *Luftfahrtforschung*, **11** (1934), 158–164.
‡ *N.A.C.A. Report* No. 465 (1933).
|| Lyon, *Aircraft Engineering*, **7** (1935), 35. The data used were taken from Garrick, *loc. cit.*

For all aerofoils the position of the maximum positive pressure is on the under surface close to the leading edge at positive incidences.

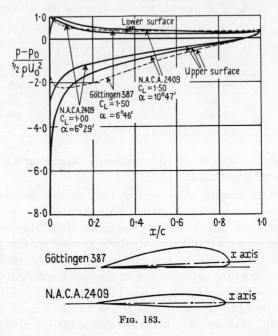

Göttingen 387

N.A.C.A. 2409

FIG. 183.

## 207. Influence of shape on the scale effect on the lift of an aerofoil.

With an increase in the Reynolds number a thin section shows a pronounced increase in the maximum lift coefficient, whereas a thick section shows at first a slight decrease (Fig. 167); and whereas the initial slope of the $(C_L, \alpha)$ curve is unaltered by increase of the Reynolds number for thin sections, the slope of the curve for a thicker section decreases with increase of the Reynolds number. The increase in the maximum lift coefficient in the case of a thin section is probably due to separation of a laminar boundary layer near the leading edge at the lower Reynolds numbers, with an earlier transition to turbulent flow at higher Reynolds numbers. For a thick section the pressure gradient along the surface is more uniform, and the separation is probably a turbulent one except at very low Reynolds numbers. The reason for a decrease in maximum lift coefficient with increasing Reynolds numbers is not known, but a possible explanation is that the earlier transition is accompanied by an earlier separation (see Chap. II, § 25, p. 78): this explanation is, however, not complete, for Fig. 169 shows that the thick section Göttingen 387 behaves in the

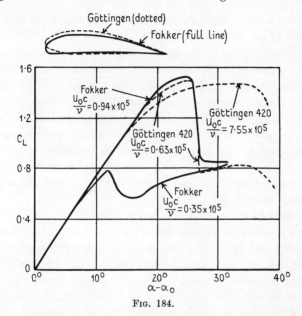

FIG. 184.

same way as thinner sections behind a turbulent screen in that there is an increase in maximum lift, and it is generally considered that turbulence in the free stream favours an earlier transition.

Lift curves for the thick sections Göttingen 420 and Fokker are shown in Fig. 184.[†] ($\alpha_0$ is the incidence at zero lift.) At $U_0 c/\nu = 0.35 \times 10^5$ a laminar separation near the leading edge of the Fokker section occurs at a low incidence, and the maximum $C_L$ reached is small: at $U_0 c/\nu = 0.94 \times 10^5$ a higher value of $C_L$ is reached before separation occurs, and the stall is sudden, probably because of abrupt separation from the forward part of the upper surface. At $U_0 c/\nu = 7.55 \times 10^5$ the stall for Göttingen 420 is gradual, and is probably associated with a gradual forward movement of the position of separation of a turbulent boundary layer.

It should be added that experiments[‡] made to find the effects on

† Lyon, *loc. cit.* The data used were taken from Bryant and Batson, *A.R.C. Reports and Memoranda*, No. 654 (1919); Fage and Cowley, *ibid.*, No. 935 (1925).

‡ Munk and Miller, 'Model Tests with a Systematic Series of 27 Wing Sections at

aerofoil behaviour of systematic changes of camber and thickness have shown that such effects depend (as would be expected) on the Reynolds number and on the turbulence in the free stream.

## 208. The velocity distribution in the boundary layer at the surface of an aerofoil. The transition to turbulence.

The boundary layer of an aerofoil resembles that of a flat plate in that at a sufficiently high Reynolds number there is a region of laminar flow at the nose, then a transitional region in which the distribution of mean velocity changes more or less erratically with time, and finally a region of fully developed turbulent motion. A further resemblance is that turbulence in the general stream, and irregularities on the surface near the nose, cause an earlier transition to turbulent flow. But separation eventually occurs on the upper surface of an aerofoil, no matter what the shape may be, when the angle of incidence is increased.

The relative extents of the laminar, transitional, and turbulent boundary layers, and the position of the separation (if any) depend on the shape and incidence of the aerofoil, on the Reynolds number, on surface roughness, and on the turbulence in the free stream.

In wind tunnels of ordinary type (compressed air and full-scale tunnels excluded) some difficulty may be experienced in obtaining details of the flow in the boundary layer of an aerofoil at large incidences, because when the chord is large compared with the width of the tunnel stream, the walls (for a closed jet) or free boundary (for an open jet) introduce a large interference with the flow. On the other hand, when the interference is lessened by making the chord relatively small, the boundary layer becomes too thin to permit reliable measurements. Further, flow phenomena in a wind tunnel have generally to be observed at a low Reynolds number; if in addition the flow in the tunnel is turbulent, the results may give misleading information on the behaviour of the same section in free flight at a higher Reynolds number—especially at the stall. For these reasons information on boundary layer flow has been sought not only on the model but also from flight experiments: in these experiments the flow is not two-dimensional, but results of great practical value are obtained.

The velocity distribution in the boundary layer of an aeroplane

Full Reynolds Number', *N.A.C.A. Report* No. 221 (1925); Jacobs and Pinkerton, 'Tests of N.A.C.A. Airfoils in the Variable-Density Wind Tunnel', *N.A.C.A. Technical Notes* Nos. 391 and 392 (1931).

wing has been measured by Cuno,† whose experiments were carried out on a Klemm L. 26 IIa aeroplane at a speed of 110 feet per second. The chord length of the section was approximately 6 feet and the incidence 1°. The thickness of the boundary layer on the upper surface increased continuously with distance from the nose to about 2·4 inches at the tail; on the lower surface the thickness was about 0·5 inch at the tail. Transition from laminar to turbulent flow in the boundary layer on the upper surface occurred just beyond the position of minimum pressure.

Stüper‡ has measured the distribution of velocity in the boundary layer of the wing of a monoplane (Klemm L. 26, Va) at a section having a maximum thickness $0·17c$ and situated about two-thirds of the semi-span from the wing tip, where the flow was undisturbed by the airscrew slipstream or the ailerons. These experiments were made at Reynolds numbers 2·82, 3·62, 4·26, and $4·88 \times 10^6$ (the corresponding lift coefficients being 0·91, 0·55, 0·40, and 0·31 respectively), with the aeroplane in horizontal flight and supporting a constant weight: the lift coefficient being then inversely proportional to the square of the flight speed (or of the Reynolds number), it follows that the conditions of the experiment were such that the combined effects of an increase of Reynolds number and of a decrease of lift coefficient were measured. These two influences affect the position of the transition from laminar to turbulent flow in opposite ways: increase of Reynolds number moves the transition region towards the nose, whereas decrease of lift coefficient is associated with a decrease of the pressure gradient along the chord, which tends to make the transition occur later. Stüper found the transition region on the upper surface at about $0·1c$ to $0·3c$ from the leading edge— a result in agreement with those obtained on models by van der Hegge Zijnen,‖ by Fage and Falkner,†† and by Gruschwitz‡‡: this result is typical of thick sections. Transition occurs just beyond the position of minimum pressure. (For a thin section this position is just behind the leading edge, so transition may be appreciably closer to the leading edge on a thin section than on Stüper's thick section.)

† *Zeitschr. f. Flugtechn. u. Motorluftschiffahrt*, **23** (1932), 189–192.
‡ *Luftfahrtforschung*, **11** (1934), 26–32.
‖ *Report of the IVth International Air Navigation Congress, Rome*, **4** (1928), 358–
381.                                    †† *Proc. Roy. Soc. A*, **129** (1930), 378–410.
‡‡ *Ingenieur-Archiv*, **2** (1931), 321–346.

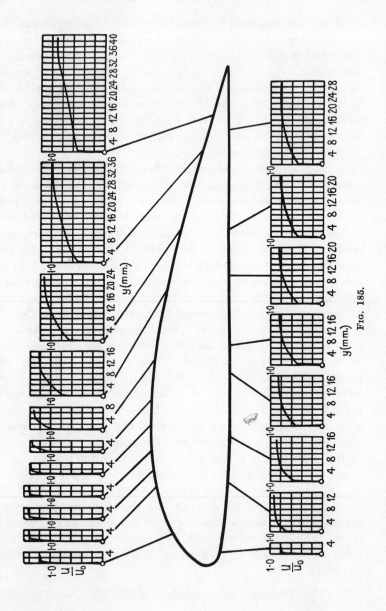

Fig. 185.

Stüper's observations also indicated that on the under surface the transition occurred earlier than on the upper surface, but the position could not be determined with precision.

The velocity graphs obtained by Stüper in the second series of measurements (Reynolds number $3 \cdot 62 \times 10^6$; lift coefficient $0 \cdot 55$) are reproduced in Fig. 185, where graphs of $U/U_0$ against $y$ are shown, $U$ being the velocity in the boundary layer at a distance $y$ from the surface and $U_0$ the velocity of the wing relative to the undisturbed air.

If $\vartheta$ is the momentum thickness defined as $\int_0^\delta (1 - U/U_1)(U/U_1)\, dy$, where $\delta$ is the thickness of the boundary layer and $U_1$ the velocity just outside the layer, then Stüper found that $U_1 \vartheta / \nu$ lies between 250 and 650 at the transition to turbulence.

## 209. The flow at the stall. Forward and rear separation. The 'bubble' of turbulence.†

On all aerofoils the first sign of an approaching stall is the formation of a thick region of low total head over the upper surface near the trailing edge (see the free-hand sketch in Fig. 186 $a$), but without definite separation of the stream from the surface. The formation of this region is associated with a progressive fall in the rate of increase of $C_L$ with incidence (before this region forms $C_L$ increases linearly with incidence), and with a slight rise in the rate of increase of drag; during or before this stage (which precedes the true stall) the airstream may separate from the upper surface near the leading edge, and then quickly rejoin the surface to form a kind of shallow 'bubble' in which the flow is turbulent. This 'bubble' of turbulence is probably due to a failure of the potential flow to force its way against a sharply rising pressure gradient, so that the stream-line which otherwise would coincide with the surface is separated from it for a short distance.

Some aerofoils show large discontinuities in the lift and drag curves at the stall: such discontinuities are due to a sudden and complete separation of the flow from the upper surface near the leading edge, the boundary of the stream passing well above the

† Comprehensive researches on the stalling of aerofoils, both in flight and on the model scale, have been made at the Aeronautics Laboratory, Cambridge. See B. M. Jones, *Journ. Roy. Aero. Soc.* **38** (1934), 753–770; Aeronautics Laboratory, Cambridge, *A.R.C. Reports and Memoranda*, No. 1588 (1934).

surface, with a region of permanent uniform low pressure occurring over practically the whole of it. There may be a small incidence range within which either the stalled state or the unstalled state of flow may occur and be maintained for a considerable time, with the corresponding parts of the lift curve (and of the drag curve) overlapping. The changes of flow may be sudden, and are then associated

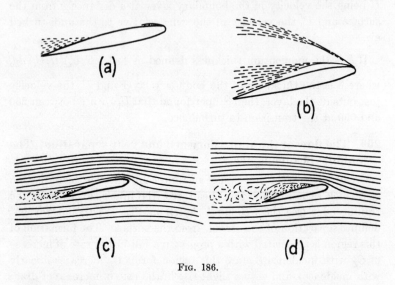

Fig. 186.

with large differences in the values of the lift and drag coefficients. When the differences are not large the boundary of the stream alternates between two fairly distinct forms, roughly outlined in Fig. 186 $b$: one of these forms represents the completely stalled state, whereas in the other the separated layer, clearly defined at first, remains above the surface over about half the chord, and finally becomes diffused and hard to define. Alternation between these two forms of flow is responsible for the violent fluctuations often observed on a model aerofoil in a wind tunnel. Apart from the turbulence bubble which may form at low incidences on the upper surface behind the leading edge, the first signs of separation from the leading edge are either discontinuities in the lift and drag curves or the onset of violent fluctuations: these phenomena do not necessarily happen at the incidence of maximum lift.

With some aerofoils, particularly when they are thick and the maximum thickness is well back from the leading edge, the first definite separation occurs near the trailing edge, and a dead-air region appears of the form illustrated in Fig. 186 c. With increasing incidence this dead-air region gradually extends until the form shown in Fig. 186 d is obtained, and during this growth the lift gradually falls without any serious discontinuities or fluctuations. Eventually, even in these circumstances, the flow separates suddenly from the leading edge and characteristic fluctuations occur, but at an incidence well beyond that of maximum lift.

It appears that as the incidence of an aerofoil of moderate thickness increases there are two positions in the boundary layer where the conditions may give rise to separation from the surface; one of these is near the leading edge and the other near the trailing edge. If separation occurs first at the front, violent fluctuations mark the attainment of maximum lift, whereas if rear separation occurs first, violent fluctuations are postponed to incidences well above that of maximum lift.

## 210. The hysteresis loop in the lift-incidence curve for a stalled aerofoil.

A balance has been designed by W. S. Farren for the Aeronautics Laboratory, Cambridge to record automatically the aerodynamic force acting on an aerofoil whose lift is rapidly changing, and researches† with this balance have shown that, when an aerofoil is started suddenly at an incidence well above that at which the stall occurs in steady motion, the flow remains unstalled during the first few chords of travel. This phenomenon suggests that the lift of an aerofoil whose angle of incidence is rising rapidly, from a value below the normal stalling angle to one well above it, may appreciably exceed that measured at a fixed angle of incidence above the stall; and this was found to be the case.

This latter phenomenon, and those associated with fluctuations of flow at the stall, were investigated on aerofoils of various shapes and thicknesses, under conditions of two-dimensional flow. It was found that a large hysteresis effect exists above the stall, the shape of the force-incidence curve depending both on the rate of change of incidence and on the sense of the change.‡ The general nature of this phenomenon

---

† *A.R.C. Reports and Memoranda*, No. 1648 (1935).

‡ A hysteresis loop in the lift-incidence curve at the stall has often been observed

is illustrated in Fig. 187. The lift curve for a very slow change of incidence is shown by a broken line, and that for the fluctuating region by a band. When the angle of incidence is changing 1° in the time the wind takes to travel 2·5 chords (the greatest rate of change which was investigated), the lift follows the full lines: no consistent difference is found in the lower part of the curve, but the whole of the stalled part of the curve lies above the mean curve for slow

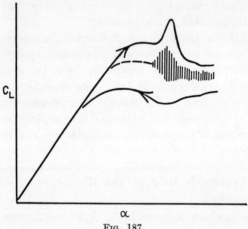

FIG. 187.

changes when the incidence is increasing, and below it when the incidence is decreasing. Near the fluctuating region there is, with increasing incidence, a notable 'peak' of high lift over a small incidence range: the position of this peak depends on the shape of the aerofoil section. In Farren's experiments the recorded lift exceeded the mean lift at fixed angles by 15 to 20 per cent. in a region covering this peak, and for half of the relevant period of time the excess was 30 per cent.; the maximum excess was 40 to 45 per cent.

Records of force fluctuations were taken by Farren at fixed angles of incidence near the stall: the fluctuations were nearly harmonic, the period for the slower oscillations being the time taken by the wind to travel a distance of about 13 chords; the fluctuations in lift were about ±15 per cent. of the mean value.

in wind tunnel and full-scale experiments. See, for example, Jacobs, *N.A.C.A. Report* No. 391 (1931); Kramer, *Zeitschr. f. Flugtechn. u. Motorluftschiffahrt,* **23** (1932), 185–189; White and Hood, *N.A.C.A. Report* No. 482 (1934); L. Sackman, *Comptes Rendus,* **202** (1936), 1019–1021; A. Lafay, *Publications scientifiques et techniques du ministère de l'air,* **79** (1935).

## 211. The effect of surface roughness on the drag and the lift of aerofoils.

A surface is aerodynamically rough when the irregularities affect the drag to a measurable extent: the effect depends not only on the size and the arrangement of the irregularities, but also on the conditions of flow in the boundary layer. A given surface is aerodynamically smooth for flow below a certain value of the Reynolds number and rough above this value; all surfaces become rough when the Reynolds number is sufficiently high. Surface roughness can in general be specified by three parameters: (a) a relative roughness, which can be taken as the ratio of the mean height of the excrescences to a linear dimension of the body, (b) a roughness density, and (c) a shape parameter; but when the excrescences are closely spaced, the first parameter (a) is usually sufficient to specify the roughness. It appears that excrescences affect the resistance only when they project beyond the thin laminar sub-layer: when their height is increased beyond the thickness of this sub-layer, vortices are formed behind them which cause an increase in resistance.[†] At high Reynolds numbers nearly all the loss of energy is due to vortex formation behind the projections, and the resistance is independent of viscosity and proportional to the square of the velocity.

The ratio of the mean height of the excrescences on the surface of an aerofoil to the thickness of the laminar sub-layer is not constant along the surface. If the boundary layer is turbulent over the whole surface, the ratio is relatively large near the leading edge, so the surface may be aerodynamically rough and the intensity of friction independent of viscosity. This region may be followed by one where both roughness and viscosity must be taken into account, and later by an aerodynamically smooth surface. The conditions are still more complicated when turbulent flow is preceded by a region of laminar flow near the leading edge.

Valuable information concerning the effect of surface roughness on aerofoils has been obtained by Schrenk[‡] from full-scale experiments on a monoplane (Junkers A. 20). The profile drag of the wing was deduced by the method due to Betz[||] from measurements of

[†] Cf. Chap. VIII, §§ 167, 168. The maximum permissible height there found for excrescences in order that the resistance may not be affected is approximately the same as the thickness of the laminar sub-layer.

[‡] *Luftfahrtforschung*, **2** (1928), 1–32.

[||] Chap. VI, § 115.

total head and pressure in the wake. The wing was tapered in plan form, and its maximum thickness varied from $0.176c$ at the body to $0.117c$ at the tip. Traverses of pressure tubes were made at a section about midway between the root and the tip. Various coverings (dural sheet, plywood varnished and polished, linen, plywood oiled, dural sheet with rivet heads, corrugated sheet, doped linen, coarse fabric, plywood roughened) were fitted at this section to give surfaces of different roughnesses. The curves in Fig. 188 show how the profile drag depends on the nature of the surface roughness. The smoothest surfaces have a profile drag which is close to that for turbulent flow along a smooth flat plate, but the roughest surfaces have a much higher drag. Whereas each of the smoother surfaces has an approximately constant profile drag up to large lift coefficients, the drag of each of the rougher surfaces rises at the higher lift coefficients; the rougher the surface, the lower is the lift coefficient at which the profile drag increases and the steeper is the rise. The roughest surfaces appear therefore to introduce an appreciable form drag apart from the increase of frictional drag.

The profile drag of the lower wing (R.A.F. 28 section) of a Hart biplane has been measured at Cambridge† by the pitot-traverse method. The technique developed allowed drag differences due to minor alterations of wing shape or surface roughness to be determined within 2 per cent. of the profile drag, and the absolute value of the local profile drag could probably be determined with similar accuracy. The profile drag coefficient of a very smooth wing, having 10·5 per cent. thickness/chord ratio and 2 per cent. camber, was found to be $0.0069$ at $R = 4.8 \times 10^6$, as compared with a mean value of $0.0090$ for a fabric-covered wing made in accordance with current practice, of substantially the same section. An appreciable part of this increase is due to the surface irregularities caused by the stitching of the fabric to the wings.

Experiments by Hooker‡ in the Variable Density Tunnel (U.S.A.), at a Reynolds number $U_0 c/\nu = 3.1 \times 10^6$, indicate that the forces on an aerofoil are very sensitive to isolated projections near the leading edge. Irregularities and scratches $0.0002c$ in depth and not more than $0.016c$ from the leading edge caused a notable drop in the

† The Cambridge University Aeronautics Laboratory, *A.R.C. Reports and Memoranda*, No. 1688 (1936).

‡ 'The Aerodynamic Characteristics of Airfoils as affected by Surface Roughness', *N.A.C.A. Technical Note* No. 457 (1933).

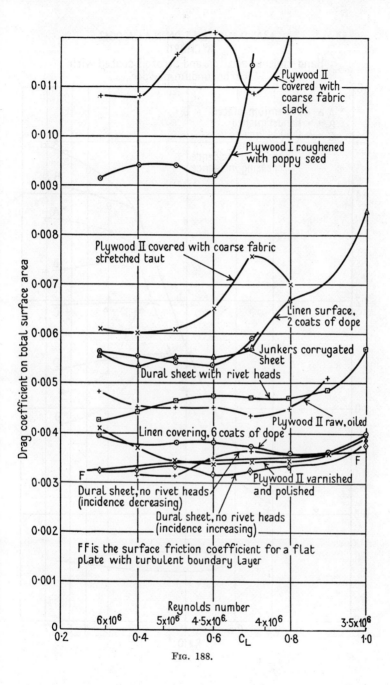

FIG. 188.

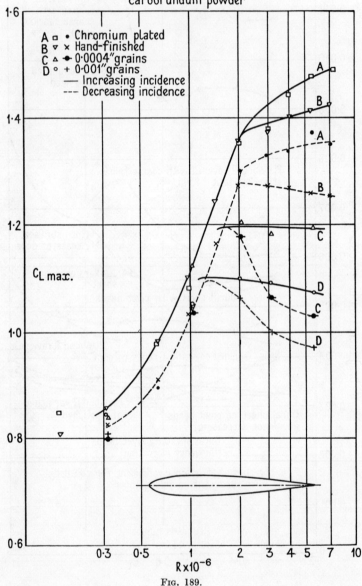

MAXIMUM LIFT ON N.A.C.A. 0012
8″ CHORD
Hand-finished aerofoil and aerofoil coated with
carborundum powder

A □ • Chromium plated
B ▽ × Hand-finished
C △ ● 0·0004″grains
D ○ + 0·001″grains
—— Increasing incidence
- - - Decreasing incidence

$C_{L}$ max.

R × 10⁻⁶

FIG. 189.

maximum lift coefficient without appreciably affecting the minimum drag. The effects of isolated protuberances on lift and drag have also been investigated by Jacobs,[†] by Schrenk,[‡] and by Dearborn.[||] Their experiments show that the effects of isolated protuberances depend appreciably on their position and size: when the protuberances

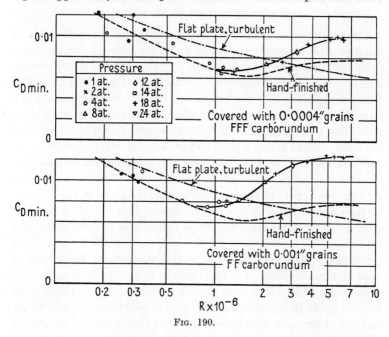

Fig. 190.

are numerous and relatively large (with a height greater than about $0.001c$), a large increase in drag is probable if they are situated near the leading edge, especially if they are on the upper surface; the effect on maximum lift depends on the aerofoil section, and a large effect is again probable when the protuberances are on the upper surface close to the leading edge.

Fig. 189 shows the effect of roughness on $C_{L\,\mathrm{max}}$ for the aerofoil N.A.C.A. 0012: the tests were made on a model ($c = 8$ inches, aspect ratio 6) in the C.A.T.[††] Roughening the surface with FFF carborundum ($0.0004$-inch grains) lowered $C_{L\,\mathrm{max}}$ to $1.20$ from its value $1.42$

† *N.A.C.A. Reports* No. 446 (1933), No. 449 (1933).
‡ *Loc. cit.* in footnote ‡ on p. 473.
|| *N.A.C.A. Technical Note* No. 461 (1933).
†† Jones and Williams, *A.R.C. Reports and Memoranda*, No. 1708 (1936).

for the hand-finished aerofoil at $R = 6 \times 10^6$; the coarser roughening
with FF carborundum (0·001-inch grains) reduced $C_{L\,\text{max}}$ to 1·08. Fig.
190 shows the effects of these roughnesses on $C_{D\,\text{min}}$ (aspect ratio 6).

## 212. Lift calculations by boundary layer theory for laminar boundary layers. Application to an elliptic cylinder.

The rate at which vorticity is discharged into the main body of
the fluid from a boundary layer in which the forward flow separates
from the surface is the value of $\frac{1}{2}U_1^2$ at the position of separation,
where $U_1$ is the velocity just outside the layer (Chap. II, § 22, p. 66).
When a steady state is reached in flow past an obstacle, the average
flux of vorticity across any fixed circuit embracing the obstacle is
zero (since the average flux of vorticity across a large circuit cutting
the wake at right angles is zero, and the average total strength of the
vorticity enclosed between this and any other circuit is constant).
Hence the average rate of discharge of vorticity must be the same
for both the upper and the lower surfaces of the obstacle, and $U_1$ must
be the same at the positions of separation. This condition provides
a criterion for calculating the value of the steady circulation that
will arise. The distribution of $U_1$ along the surface for any given
circulation is, however, required for this calculation. For a body of
good stream-line shape it is profitable to carry out the calculations
with values of $U_1$ found from the theory of the potential flow of an
inviscid fluid. Comparisons of measured pressure distributions at the
surfaces of aerofoils with those calculated from potential theory for
a circulation equal to the measured lift per unit length divided by
$\rho U_0$ (where $U_0$ is the undisturbed stream velocity)† are shown in
Chap. I, Fig. 12, and in Fig. 179. Comparisons, due to Fage,‡ for an
elliptic cylinder with axes in the ratio 5·4:1 (which is not such a good
stream-line shape as the aerofoils) are shown in Figs. 191 and 192
for incidences of 0° and 8·84°. It appears that (except in a region
at the rear which is affected by the wake) the calculated pressure
distributions are fair approximations to the measured ones, the
approach of theory and experiment being the closer the better the
stream-line shape of the body.

---

† This is also the measured value of the circulation in a circuit at some distance
from the body and cutting the wake at right angles to the undisturbed stream. See
p. 456 above.

‡ *A.R.C. Reports and Memoranda*, No. 1097 (1927).

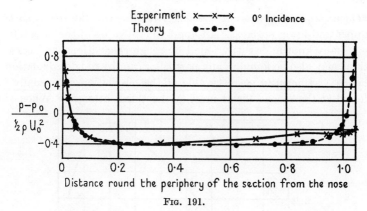

FIG. 191.

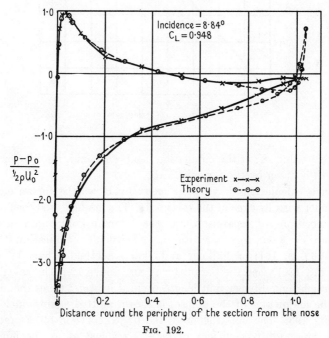

FIG. 192.

Calculations may be carried out by first finding the theoretical pressure distribution for any chosen value of the circulation; then using the methods of Chap. IV, §§ 60, 62, and 63 to calculate the position of separation of forward flow in the boundary layer with this pressure distribution; and finally adjusting the circulation until the velocity $U_1$ is the same at the positions of separation on both the upper and lower surfaces. An approximate value of the circulation is usually known, or easily found by rough calculations, and corrections may also be found in any particular case by rough calculations; hence a sufficiently correct value of the circulation can usually be found at the second (or at most at the third) attempt. When the

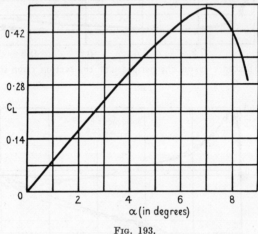

FIG. 193.

correct value, $K$, of the circulation has been found, the lift per unit length is given by the formula $K\rho U_0$.

Such calculations have been made by Howarth† for an elliptic cylinder with its axes in the ratio 6:1. (The method used for finding the positions of separation was the Kármán-Pohlhausen method (Chap. IV, § 60), together with either Dryden's modification (Chap. IV, § 60, p. 161) or Falkner and Skan's method (Chap. IV, § 64, pp. 178–180) when the parameter $\Lambda$ rose above 12.) The calculated lift coefficient (lift per unit length divided by $\frac{1}{2}\rho U_0^2$ times the length of the major axis of the ellipse) is shown plotted against incidence in Fig. 193. The initial slope of the $(C_L, \alpha)$ curve is 4·9 per radian; a maximum lift coefficient of 0·48 is reached at an incidence of about 7°.

At and near the stall the calculated wake will no longer be narrow, and the theoretical pressure distribution will cease to be a good approximation. Since, however, we begin with the pressure distribu-

† *Proc. Roy. Soc.* A, **149** (1935), 558–586.

tion and calculate the position of the points of separation, and since
the pressure distribution would be a good approximation with a
narrow wake, it follows that if a wide wake is obtained by calculation, the results are quantitatively wrong but
qualitatively right, in that a wide wake is
really to be expected. The results obtained
by Howarth for the positions of separation
$(S_1, S_2)$ are shown in Fig. 194, and indicate
that stalling probably occurs at about 7°, and
that the calculated numerical values for $C_L$
are probably reasonably accurate for $\alpha$ less
than 6°.

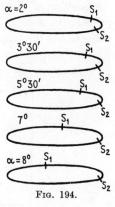

Fig. 194.

The pressure is the same at both points
of separation, and may roughly be taken to
stay constant between them over the rear of
the body. Both the skin-friction drag and
the form drag could then be calculated; but no such calculations have
yet been made.

## 213. Semi-empirical lift calculations for turbulent boundary layers. Application to an elliptic cylinder.

If the boundary layer is considered to be turbulent right from the
forward stagnation point, and if we are content to make the assumptions of Chap. IX, § 194, and Chap. VIII, § 166,—namely, that
$\delta_1/\vartheta$ and $\zeta$ are the functions of $\Gamma$ shown in Figs. 117 and 118 of
Chapter VIII, where $\delta_1$ and $\vartheta$ are the displacement and momentum
thicknesses, $\Gamma$ the shape parameter for the velocity graphs, and $\zeta$
the skin-friction parameter defined in equation (1) of Chap. IX,
§ 194,—then similar calculations to those in the preceding section
are easily carried out for turbulent boundary layers. It is simply a
matter of finding the position of separation for each assumed circulation by solving equation (9) of Chap. IX, § 194, separation being
supposed to occur at some fixed value of $\Gamma$. The results obtained by
Howarth (*loc. cit.*) for the elliptic cylinder of the preceding section,
with $\Gamma = -0.06$ at separation, are shown in Fig. 195, where the
lift coefficient is plotted against incidence. For $0° < \alpha < 20°$ the
graph is almost exactly a straight line, whose slope is 5·4 per
radian. The maximum lift coefficient is 1·95, and is reached at an
incidence of $22\frac{1}{2}°$.

When allowance is made for a laminar portion of the boundary layer near the forward stagnation point, the matter is more complicated. Assumptions must be made both for the conditions at, and the position of, the transition to turbulence. We may assume either (a) that the state of affairs in the boundary layer downstream of the transition is the same as if the flow had been turbulent right from the forward stagnation point, or (b) that the momentum thickness $\vartheta$ is continuous at transition. (See Chap. VII, § 151, p. 329.) As

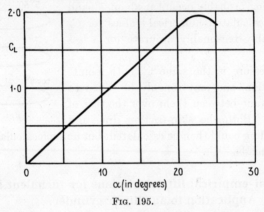

$\alpha$ (in degrees)

FIG. 195.

regards the position of transition to turbulence, if the calculations in the laminar portion of the boundary layer are carried out by the Kármán-Pohlhausen method (Chap. IV, § 60), then a definite value is obtained for the thickness $\delta$ of the boundary layer, and the simplest assumption is that $U_1 \delta / \nu = C$ at transition, where $U_1$ is the velocity just outside the boundary layer, and $C$ a number depending on the amount of turbulence in the main stream. $C$ appears to lie between 1,300 and 6,000 for wind tunnel conditions.† Alternative, and probably preferable, assumptions would be that $U_1 \vartheta / \nu = C_1$ or $U_1 \delta_1 / \nu = C_2$, where $\vartheta$ and $\delta_1$ are the momentum and displacement thicknesses, and $C_1$ and $C_2$ are again numbers depending on the amount of turbulence in the main stream. The values obtained by Gruschwitz‡ for $U_1 \vartheta / \nu$ in experiments in the 2·24-m. wind tunnel of the Göttingen aerodynamical laboratory, both on a model of the aerofoil Göttingen 387 and on a flat plate at zero incidence with a considerable variety

† See Fage, *A.R.C. Reports and Memoranda*, No. 1765 (1937).

‡ *Ingenieur-Archiv*, **2** (1931), 321–346; *Zeitschr. f. Flugtechn. u. Motorluftschiffahrt*, **23** (1932), 308–312.

of pressure distributions along the plate (obtained by varying the shape of the wall of the wind tunnel opposite the plate), lay between 360 and 680; for flow along a cylindrical surface Schmidbauer[†] found values between 500 and 790. The values obtained by Stüper on an aerofoil in free flight (see p. 469) lay between 250 and 650; and in experiments on a symmetrical aerofoil (maximum thickness $\frac{1}{6}$th of the chord at $\frac{1}{3}$rd of the chord from the leading edge) at the Massachusetts Institute of Technology, Peters[‡] found that transition took place consistently for values of $U_1\vartheta/\nu$ between 600 and 650. The values of $U_1\delta_1/\nu$ at transition have been examined by Fage[||] for a flat plate from Dryden's results (Chap. VII, § 151, p. 326) obtained at the National Bureau of Standards, and for an aerofoil (Joukowski section No. 3 of Chap. IX, Fig. 136), a circular cylinder and a sphere in various N.P.L. tunnels. The values lie between about 500 and 1,000 for values of $u/U_0$ between 0·01 and 0·05 (where $u$ is the root-mean-square of the longitudinal component of turbulent velocity in the stream).

With any of the above assumptions concerning the position of the transition to turbulence, there will, for each value of $C$ (or $C_1$ or $C_2$), be a critical value, $R_c$, of the Reynolds number $R$ of the flow past the body[††] for which, with any given circulation round the boundary, the position of the transition to turbulence and the position of separation of a laminar boundary layer coincide. This value can be calculated at once for all values of $C$ if the values of $U_1$ and $\delta$ (or $\delta_1$ or $\vartheta$) in terms of $R$ have been found at the position of separation of the laminar layer. For $R < R_c$, there will be no transition to turbulence before laminar separation; for $R > R_c$, there will be a transition to turbulence and no laminar separation. Further, with assumption (a) as to the conditions at transition (the state of affairs in the turbulent portion of a boundary layer the same as that at the same position in an everywhere turbulent layer), turbulent separation for $R > R_c$ will always take place at the same position as for the completely turbulent layer, and will not depend on the position of the transition to turbulence; hence the assumption $U_1\delta/\nu = C$ (or $U_1\vartheta/\nu = C_1$ or $U_1\delta_1/\nu = C_2$) at transition must be applied, in order to find $R_c$, only at the position of separation of the laminar layer.

† *Luftfahrtforschung*, **13** (1936), 160–162.

‡ *Journ. Aero. Sciences*, **3** (1935), 7–12.

|| *A.R.C. Reports and Memoranda*, No. 1765 (1937).

†† $R = (U_0/\nu)$ multiplied by a characteristic length of the solid body.

For any incidence we may calculate the values $R_{cL}$ and $R_{cT}$ of $R_c$ corresponding to the correct circulations for a laminar layer and for a fully turbulent layer, respectively: for $R < R_{cL}$ the circulation and the lift coefficient calculated for a laminar boundary layer will be correct; for $R > R_{cT}$ the lift coefficient calculated for a fully turbulent layer will be correct on assumption (a). On assumption (b) further calculation of the turbulent portion of the layer is necessary, and it appears that the circulation and the lift coefficient rise suddenly at a certain value of $R$ to values very nearly but not quite equal to those for a completely turbulent layer, which, as $R$ further increases, are then approached asymptotically.

From such calculations the maximum lift coefficient, $C_{L\max}$, is known for each Reynolds number, and curves of $C_{L\max}$ against $R$ may be drawn. There will be one such curve for each value of $C$ (or $C_1$ or $C_2$) in the assumed equation for the position of transition. The value of $C_{L\max}$ will be that for a completely laminar boundary layer until $R$ reaches the value at which transition and laminar separation coincide at the incidence at which the turbulent lift is equal to the maximum laminar lift; then $C_{L\max}$ will rise: on assumption (a) it will reach the value for a completely turbulent layer as soon as $R$ reaches the value at which transition and laminar separation occur simultaneously at the incidence of maximum (completely turbulent) lift: on assumption (b) it will approach the completely turbulent value asymptotically as $R$ increases.

Calculations carried out in this way by Howarth (loc. cit.) for the elliptic cylinder of the previous section (the calculations being performed by the Kármán-Pohlhausen method for the laminar portion of the boundary layer, the condition $U_1 \delta/\nu = C$ being applied for the position of the transition to turbulence, and the condition $\Gamma = -0.06$ for turbulent separation) resulted in the curves of $C_{L\max}$ against $R$ reproduced in Fig. 196. ($C_L$ is defined as the lift per unit length divided by the product of $\frac{1}{2}\rho U_0^2$ and the length of the major axis, and $R$ is $U_0/\nu$ times half the distance between the foci, which is $0.986$ times half the major axis.) It was found that there was very little difference in the results on assumption (a) and on assumption (b) except in the neighbourhood of the value (1·95) of $C_{L\max}$ for a completely turbulent layer. Assumption (a) was therefore used in obtaining Fig. 196 except in that neighbourhood, where assumption (b) was used in order to ensure that the value 1·95 should be approached asymptotically, and not reached for a finite value of $R$.

On assumption (a) the slope of the $(C_L, \alpha)$ curve at the origin with a boundary layer which is partly laminar and partly turbulent has the value (5·4) for a completely turbulent layer; on assumption (b) it probably rises suddenly at a certain Reynolds number to a value

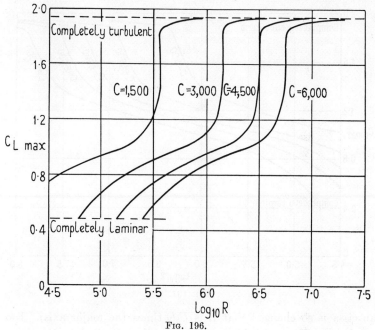

FIG. 196.

between 4·9 and 5·4 and then rises asymptotically to the value 5·4 as $R$ increases.

At any given incidence, as $R$ increases and the transition to turbulence moves forward, it is found that the positions of turbulent separation also move forward. The displacement of the position of separation is greater on the top than on the bottom, but for an elliptic cylinder slight changes also occur on the lower surface; they take place in a region of considerable pressure gradient, and are sufficient to make the maximum lift coefficient rise with an increase of the Reynolds number. (Contrast § 207, pp. 464, 465.)

The very rapid increase in $C_{L\max}$ near a particular Reynolds number in Fig. 196 is not very plausible: it arises from the fact that, with $U_1 \delta/\nu = C$ at transition, the calculated values of $R_c$ are nearly constant (for each value of $C$) for incidences between 15° and 22°; and it disappears if the condition $U_1 \delta/\nu = C$ is replaced by $U_1 \delta/\nu = (C^2 + DR)^{\frac{1}{2}}$, with a suitable choice of $D$. Resulting curves of $C_{L\max}$ when $D = 6.25$ are shown in Fig. 197; the curve with $C = 1,535$ is shown dashed in Fig. 167 (where the scale of the

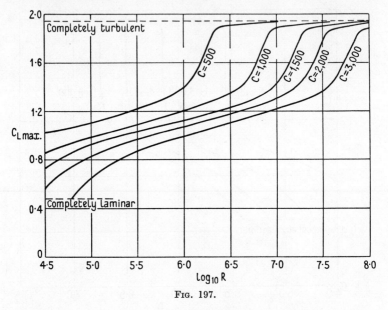

FIG. 197.

abscissa is so changed that $R$ is $U_0/\nu$ times the major axis). Too much importance should, however, not be attached to the comparison in Fig. 167, because of the difference in the shapes of the aerofoils and the elliptic cylinder, and because the calculations, though promising, are still very crude.†

† Similar, but less developed, calculations were made by Kármán and Millikan, *Journ. Applied Mechanics*, **2** (1935), 21–27. (See also Millikan and Klein, *Aircraft Engineering*, **5** (1933), 169–174; Kármán, *Chambre Syndicale des Industries Aéronautiques, Journées Techniques Internationales de l'Aéronautique* (Paris, 1932), pp. 1–26.) They applied the condition $U_1 \delta/\nu = C$ at transition, but assumed simply that the turbulent boundary layer would not separate from the surface. For any chosen value of $C_L$, if the laminar layer separates before the transition to turbulence, the aerofoil is considered stalled in their method, and the value of $C_L$ is not attained; otherwise that value of $C_L$ can be reached. The assumption that the turbulent boundary layer does not separate from the surface is unsatisfactory in that it implies that stalling would never occur at sufficiently high Reynolds numbers, and infinitely large values of $C_{L\max}$ could theoretically be reached.

A different method of considering turbulent boundary layers has been suggested by Gruschwitz.† He plots $U/U_1$ against $y/\vartheta$, where $U$ is the velocity in the boundary layer at a distance $y$ from the wall; and suggests that this gives a one-parameter family of curves. The value of $U/U_1$ at a particular value of $y/\vartheta$ then fixes each curve. Actually the parameter used is

$$\eta = 1 - \left[\frac{U}{U_1}\right]^2_{y=\vartheta}.$$  (1)

Then, since

$$\int_0^{\delta/\vartheta} \left(1 - \frac{U}{U_1}\right) d\left(\frac{y}{\vartheta}\right) = \frac{\delta_1}{\vartheta},$$  (2)

$\delta_1/\vartheta$ should be a function of $\eta$. Moreover, according to Gruschwitz,

$$-\frac{\vartheta}{\frac{1}{2}\rho U_1^2} \frac{d}{dx}[\frac{1}{2}\rho U_1^2 \eta] = 0 \cdot 00894\eta - 0 \cdot 00461$$  (3)

very nearly for $\eta < 0 \cdot 8$, according to experimental results. This equation, and the equation of momentum

$$\frac{\tau_0}{\rho U_1^2} = \frac{d\vartheta}{dx} + \frac{\vartheta}{U_1}\frac{dU_1}{dx}\left[2 + \frac{\delta_1}{\vartheta}\right]$$  (4)

(equation (38) of Chap. IV), are then a pair of simultaneous equations for $\vartheta$ and $\eta$ if $\tau_0/(\rho U_1^2)$ is known. For $\tau_0/(\rho U_1^2)$ Gruschwitz is content to take a constant value, or a value obtained from the approximate (power-law) formula for a flat plate (Chap. VIII, § 163, equation (62)); and he suggests that separation takes place at or soon after $\eta = 0 \cdot 8$. At the transition to turbulence, the conditions are taken to be that $\vartheta$ is continuous and $\eta = 0 \cdot 1$. (It is stated that the form of the integral curves for $\eta$ is such that the value at transition is of minor importance.)

It appears that the relations obtained by Gruschwitz, and others calculated therefrom, hold with fair accuracy so long as no separation takes place or the position of separation is not too closely approached. Thus Stüper‡ obtained fair agreement with the results of his observations in free flight, in which no separation of the forward flow occurred; Peters‖ found that the relations and calculated results held with fair accuracy for incidences at which no separation occurred, but for other incidences departed more and more from the measured values as the position of separation was approached. The measured value of $\eta$ at separation was about $0 \cdot 8$, but the calculated values did not rise above $0 \cdot 7$. A linear relation between $-[\vartheta/(\frac{1}{2}\rho U_1^2)] d[\frac{1}{2}\rho U_1^2 \eta]/dx$ and $\eta$ was found to be a rough approximation when no separation occurred, but even this rough approximation was not valid when a position of separation was approached.

The curves for $\delta_1/\vartheta$ and $-[\vartheta/(\frac{1}{2}\rho U_1^2)] d[\frac{1}{2}\rho U_1^2 \eta]/dx$ obtained by Gruschwitz are shown in Figs. 198 and 199. The plotted points were obtained from observations in boundary layers at a flat surface, though with a considerable variety of pressure distributions: they do not take into account either curvature of the surface or widely different degrees of turbulence in the main stream.

---

† *Loc. cit.* in footnote ‡ on p. 482.      ‡ *Loc. cit.* in footnote ‡ on p. 467.
‖ *Loc. cit.* in footnote ‡ on p. 483.

For flow along the convex side of a curved surface, Schmidbauer,† as a result of experiments on flow along surfaces of constant curvature, suggested that (3) should be generalized to

$$-\frac{\vartheta}{\frac{1}{2}\rho U_1^2}\frac{d}{dx}(\tfrac{1}{2}\rho U_1^2\eta) = (\eta+0\cdot1)(0\cdot00894-0\cdot315\kappa\vartheta)-0\cdot0055, \qquad (5)$$

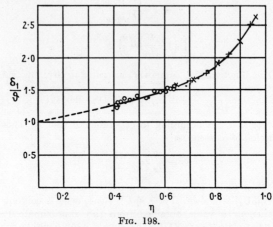

Fig. 198.

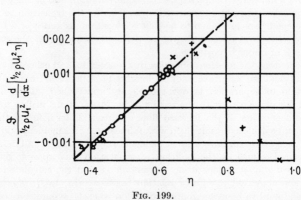

Fig. 199.

where $\kappa$ is the curvature (the reciprocal of the radius of curvature). The curve of $\delta_1/\vartheta$ against $\eta$ was unchanged; and according to Schmidbauer $\vartheta$ is still to be taken as continuous and $\eta = 0\cdot1$ at transition; but with a moderate pressure gradient in the last part of the pressure rise before separation, separation is not to be anticipated until $\eta = 0\cdot95$.

Gruschwitz's method has been applied by Stuper‡ to calculate the effect of the boundary layer and wake in lessening the lift of a Joukowski aerofoil

† *Loc. cit.* in footnote † on p. 483.
‡ *Zeitschr. f. Flugtechn. u. Motorluftschiffahrt*, **24** (1933), 439–441.

from the value calculated by the method of Kutta-Joukowski. The pressure
distribution having been first found in the usual way, the circulation being
chosen to prevent infinite velocity at the trailing edge, Gruschwitz's method,
with $U_1 \vartheta/\nu = 650$ at the transition to turbulence, was applied to calculate the
boundary layer characteristics, the Kármán-Pohlhausen method being used
for the laminar portion. A new calculation was then made for a contour which
was that of the original aerofoil displaced outwards through a distance $\delta_1$, and so
on. Various approximate methods were used in order to shorten the calculation.
The fifth approximation was calculated for a Reynolds number $7 \times 10^6$. The
only experimental results related to Reynolds numbers of $2 \times 10^5$ and $4 \times 10^5$,
so that a proper comparison was not possible. The calculated values actually
agreed well with the experimental values at the much lower Reynolds numbers.

## ADDITIONAL REFERENCES

*Treatises and collected accounts.*

F. W. LANCHESTER, *Aerodynamics* (London, 1907).

L. PRANDTL, Applications of Modern Hydrodynamics to Aeronautics,
*N.A.C.A. Report* No. 116 (1921).

N. E. JOUKOWSKI, Lectures on Hydrodynamics, *Trans. Centr. Aero-
Hydrodyn. Inst., Moscow,* No. 40 (1929).

H. GLAUERT, *The Elements of Aerofoil and Airscrew Theory* (Cambridge,
1926).

L. PRANDTL and A. BETZ, *Vier Abhandlungen zur Hydro- und Aerodynamik* (Göttingen, 1927).

A. BETZ, *Handbuch der Physik,* **7** (Berlin, 1927), 215–288.

E. PISTOLESI, *Aerodynamica* (Turin, 1932).

R. SEIFERTH and A. BETZ, *Handbuch der Experimentalphysik,* **4**, part 2
(Leipzig, 1932), 107–206.

H. GLAUERT, Wind Tunnel Interference on Wings, Bodies and Airscrews,
*A.R.C. Reports and Memoranda,* No. 1566 (1933).

R. FUCHS, L. HOPF, and FR. SEEWALD, *Aerodynamik* (Berlin, 1934).

TH. VON KÁRMÁN and J. M. BURGERS, *Aerodynamic Theory* (edited by W. F.
Durand), **2** (Berlin, 1935).

A. BETZ, *Aerodynamic Theory* (edited by W. F. Durand), **4** (Berlin, 1935),
1–129.

*Other works.*

PRANDTL, Tragflächenauftrieb und Widerstand in der Theorie, *Jahr. Wiss.
Gesellsch. Luftfahrt,* **5** (1920), 37–65.

MUNK, General Theory of Thin Wing Sections, *N.A.C.A. Report* No. 142
(1922).

BETZ, Eine Verallgemeinerung der Joukowskyschen Flügelabbildung,
*Zeitschr. f. Flugtechn. u. Motorluftschiffahrt,* **15** (1924), 100.

GLAUERT, A Generalized Type of Joukowski Aerofoil, *A.R.C. Reports and
Memoranda,* No. 911 (1924).

LANCHESTER, Sustentation in Flight, *Journ. Roy. Aero. Soc.* **30** (1926),
587–606.

The Göttingen Laboratory. Messungen von Joukowsky-Profilen, *Ergebnisse der Aerodynamischen Versuchsanstalt zu Göttingen*, **3** (1927), 59–77.

PERRING, The Theoretical Pressure Distribution around Joukowsky Aerofoils, *A.R.C. Reports and Memoranda*, No. 1106 (1927).

ROY, A propos des théories aérodynamiques et de l'hydrodynamique rationnelle, *L'Aérophile*, **35** (1927), 185–187 and 215–217.

SCHRENK, Systematische Untersuchungen an Joukowsky-Profilen, *Zeitschr. f. Flugtechn. u. Motorluftschiffahrt*, **18** (1927), 225–232.

FAGE and JOHANSEN, The Connection between Lift and Circulation for an Inclined Flat Plate, *A.R.C. Reports and Memoranda*, No. 1139 (1928).

PIERCY and RICHARDSON, On the Flow of Air adjacent to the Surface of an Aerofoil, *A.R.C. Reports and Memoranda*, No. 1224 (1929).

BICKLEY, The Effect of Rotation upon the Lift and Moment of a Joukowsky Aerofoil, *Proc. Roy. Soc.* A, **127** (1930), 186–196.

PISTOLESI, Sull'origine della portanza, *5th Internat. Congress for Aerial Navigation, The Hague*, **1**, Section B (1930), 488–496.

WILCKEN, Turbulente Grenzschichten an gewölbten Flächen, *Ingenieur-Archiv*, **1** (1930), 357–376.

BETZ, Theoretische Berechnung von Tragflügelprofilen, *Zeitschr. f. Flugtechn. u. Motorluftschiffahrt*, **24** (1933), 437–439.

RANDISI, L'Effetto della turbolenza sulle caratteristiche aerodinamiche delle ali, *L'Aerotecnica*, **13** (1933), 867–889.

THEODORSEN and GARRICK, General Potential Theory of Arbitrary Wing Sections, *N.A.C.A. Report* No. 452 (1933).

MILLIKAN, Further Experiments on the Variation of the Maximum Lift Coefficient with Turbulence and Reynolds Number, *Trans. Amer. Soc. Mechanical Engineers*, 56 (1934), 815–825.

BETZ and KEUNE, Verallgemeinerte Kármán-Trefftz-Profile, *Luftfahrtforschung*, **13** (1936), 336–345.

PLATT, Turbulence Factors of N.A.C.A. Wind Tunnels as Determined by Sphere Tests, *N.A.C.A. Report* No. 558 (1936), pp. 13–18.

FEDIAEVSKY, Turbulent Boundary Layer of an Aerofoil, *Journ. Aero. Sciences*, **4** (1937), 491–498.

DOETSCH, Profilwiderstandsmessungen im grossen Windkanal der DVL, *Luftfahrtforschung*, **14** (1937), 173–178; 367.

JONES, Profile Drag, *Journ. Roy. Aero. Soc.* **41** (1937), 339–368.

JONES, Flight Experiments on the Boundary Layer, *Journ. Aero. Sciences*, **5** (1938), 81–101,

WILLIAMS and BROWN, Experiments on an Elliptic Cylinder in the Compressed Air Tunnel, *A.R.C Reports and Memoranda*, No. 1817 (1938).

Other references will be found in *Hydrodynamics* (Bulletin No. 84 of the National Research Council, Washington, 1932), by Dryden, Murnaghan, and Bateman, pp. 83–88.

# FLOW PAST SOLID BODIES OF REVOLUTION

## 214. Introduction.

In this chapter we consider flow past a sphere and past a solid body of symmetrical stream-line shape, such as an airship hull. A large amount of work has been done on bodies of both these types, from which it is possible to draw a number of general conclusions. Investigations of three-dimensional flow past other solid bodies, such as disks or rectangular plates, have not been so extensive. Some of these investigations, in so far as they were concerned with flow in the wake, are reported in Chap. XIII, § 250; in the present chapter references are confined to the list on pp. 526–528. This list also contains references to some of the few published works on flow past ellipsoids, symmetrical stream-line bodies with their axes inclined to the direction of flow, and rotating bodies of revolution.

SECTION 1

SPHERES

## 215. Drag at low Reynolds numbers.

Investigations on the drag of spheres have been made over a very wide range of Reynolds numbers. At low Reynolds numbers the drag is determined from tests on small spheres—often of microscopic size—falling through viscous fluids, and at higher Reynolds numbers on much larger spheres, dropped or towed through air or water, or mounted in the artificial stream of a wind tunnel.

The drag at very low Reynolds number was predicted by Stokes[†] on the assumption that the inertia terms in the equations of motion of a viscous fluid can be neglected in comparison with the terms involving the viscosity $\mu$. The value of the drag obtained is $3\pi\mu du_0$, where $d$ is the diameter of the sphere and $u_0$ the relative velocity of the sphere and the fluid at a great distance. The corresponding value of the drag coefficient $C_D$ (obtained by dividing the drag by the product of $\frac{1}{2}\rho u_0^2$ and the cross-sectional area $\frac{1}{4}\pi d^2$) is $24R^{-1}$, where $R$ is the Reynolds number $u_0 d/\nu$.

Attempts have been made to deduce a more general relation for the drag of a sphere at small Reynolds numbers than that obtained

† Lamb, *Hydrodynamics* (Cambridge, 1932), pp. 597–604.

by Stokes, by taking some account of the inertia terms in the equations of motion. The earlier attempts are described in a paper by Weyssenhoff.† Oseen‡ pointed out that at a great distance from the sphere the inertia terms become more important than the viscous terms, and suggested a new approximation. The solution of Oseen's equations obtained by Goldstein‖ leads to the series formula

$$C_D = 24R^{-1}\left\{1 + \frac{3}{16}R - \frac{19}{1280}R^2 + \frac{71}{20480}R^3\right.$$
$$\left. - \frac{30179}{34406400}R^4 + \frac{122519}{550502400}R^5 - ...\right\}.$$

The value of $C_D$ for values of $R$ up to and including 2 may be calculated from this series, but for larger values of $R$ the series does not converge sufficiently rapidly, and a numerical method of evaluation is necessary. If the neglected terms in the equations of motion could be retained, only the first two terms in the expansion of $C_D$ would be unaltered; Goldstein's solution represents therefore only a first step towards obtaining such an expansion.

The first term in the series $24R^{-1}$ is the value obtained by Stokes, and the second approximation $24R^{-1}+4\cdot5$ was obtained by Oseen.

The experimental determination of the rate of fall of a sphere affords a convenient method of finding the value of the viscosity of a fluid, and to establish this method on a satisfactory basis many experiments with small spheres of different sizes have been made. Castleman,†† of the Bureau of Standards, U.S.A., has analysed the results of some of these experiments to determine the range of Reynolds numbers over which Stokes's relation holds with good accuracy. The results considered were those of Arnold‡‡ (obtained with metal spheres in colza oil), of Liebster and Schiller‖‖ (with steel spheres in glycerine, etc.), and of Allen††† (with air and paraffin spheres in aniline, and air bubbles in water‡‡‡). Castleman's mean curve is reproduced in Fig. 200, and meets the line representing Stokes's relation at $\log_{10} R = -0\cdot3$—that is, at $R = 0\cdot5$ approximately.

---

† *Ann. d. Phys.* **62** (1920), 1–45.

‡ *Ark. f. mat., astr. och fys.* **6**, No. 29 (1910).

‖ *Proc. Roy. Soc.* A, **123** (1929), 225–235. [For the correction of a numerical error see D. Shanks, *Journ. Math. and Physics,* **34** (1955), 36.]

†† *N.A.C.A. Technical Note* No. 231 (1925).

‡‡ *Phil. Mag.* (6), **22** (1911), 755–775.    ‖‖ *Physik. Zeitschr.* **25** (1924), 670–672.

††† *Phil. Mag.* (5), **50** (1900), 323–338 and 519–534.

‡‡‡ When the sphere is itself fluid Stokes's formula must be modified. See Lamb, *op. cit.*, pp. 600, 601.

(The importance of the establishment of Stokes's relation arises from its use to determine the diameters of small spheres from their rates of fall in fluids of known viscosity, to determine the diameters of droplets of water in mists or fogs, and to determine the dimensions of microscopic particles in sedimentation analysis.†)

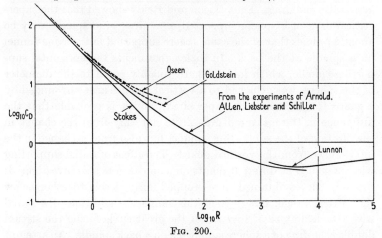

Fig. 200.

Curves plotted from the solutions of Oseen and of Goldstein, also shown in Fig. 200, lie above the experimental curve.‡

## 216. Drag. Interference. The critical range of Reynolds numbers.

Measurements of the drag of spheres mounted in the artificial stream of a wind tunnel have been made over a wide range of Reynolds numbers, and in many countries. The earlier results obtained in and beyond the critical range of $R$—i.e. the range over which the drag coefficient experiences a large fall (see p. 495)—were in considerable disagreement, and this caused some concern, for the differences were often much greater than could be attributed to experimental error. Subsequent researches threw light on the nature of these disagreements, and it is now known that the drag of a sphere depends not only on the Reynolds number, but also on the method of support, on surface roughness, on flow disturbances due to mechanical and acoustical vibrations, and on the turbulence in the stream. At one

† For a critical analysis of experimental results on the drag of small spheres and a description of the falling sphere method of measuring viscosity see Barr, *Monograph of Viscosity* (Oxford, 1931).   ‡ For the curve marked Lunnon in Fig. 200 see p. 496.

time it was the common practice to support a sphere in a wind tunnel either on a spindle normal to the air-stream (cross-wind spindle) or on a system of wires of which one at least was attached to the sphere at or upstream of the great circle facing the oncoming stream. Such methods of support may greatly affect the flow past the sphere and hence also the drag. Thus Bacon and Reid[†] showed that the super-critical drag of a sphere supported on a cross-wind spindle may be about 2·5 times that of the same sphere supported in the same tunnel on a spindle at the back. In their experiments the horizontal supporting spindle, which had a length roughly equal to the diameter of the sphere, was attached to a vertical balance arm surrounded by a guard (see Fig. 5 of the paper cited), and it is probable that the difference in the measured drag arose not only from the change in the position of the supporting spindle, but also from a change in the interference from the vertical guard. The effect of radial supporting wires attached at different points on a sphere was also investigated: the drag curves obtained were grouped fairly closely together below the critical range, but beyond this range the addition of a radial wire attached at 22·5° forward of the great circle facing the stream doubled the drag of a sphere supported on a back spindle. An attempt to investigate interference phenomena by measuring drag when the end of a wire approached the surface had to be abandoned, because violent oscillations of the sphere, due to instability of flow, occurred when the end was brought nearer to the surface than one-twelfth of a diameter, so drag measurements could not be made. The evidence available from such experiments clearly indicates that supporting wires and spindles must not disturb the continuity of the boundary layer, and tests are now made with the sphere supported on a spindle completely immersed in the wake at the back.[‡]

Representative drag curves obtained from tests in two modern wind tunnels—the Variable Density Tunnel, U.S.A.,[‖] and the 10-foot

[†] N.A.C.A. Report No. 185 (1924). See also Flachsbart, Physik. Zeitschr. 28 (1927), 461–468; and Hoerner, Luftfahrtforschung, 12 (1935), 42–54.

[‡] References to papers describing earlier experiments on fairly large spheres by Eiffel, Loukianof, Riabouchinsky, Maurain, Costanzi, Shakespear, Prandtl, Wieselsberger, Toussaint and Hayer, Hesselberg and Birkeland, with critical discussions of the results obtained, are to be found in papers by Pannell (Advisory Committee for Aeronautics, Reports and Memoranda, No. 190 (1916)), by Bacon and Reid (loc. cit.), and by Hoerner (loc. cit.).

[‖] Jacobs and Abbott, N.A.C.A. Report No. 416 (1932); Stack, N.A.C.A. Technical Note No. 364 (1930).

tunnel at the Guggenheim Aeronautical Laboratory, California Institute of Technology† (hereinafter referred to as the Galcit tunnel)—are shown in Fig. 201, curves 1 and 2, respectively. In each case there is a critical range of Reynolds numbers over which the drag coefficient experiences a large fall, but the critical ranges are different in the two tunnels. The reason for the fall is in each case the same as that given earlier for the large drop in the drag coefficient

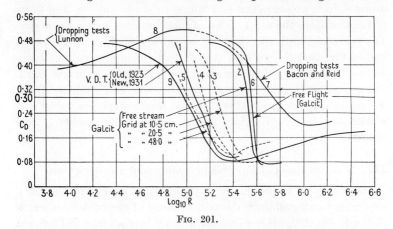

Fig. 201.

of a circular cylinder: at low Reynolds numbers the flow in the boundary layer is laminar up to the position of separation, but at higher Reynolds numbers a transition to turbulent flow occurs, with the result that separation is delayed, the wake is smaller, and the form drag is reduced. The drag due to skin-friction is increased, but its magnitude is too small compared with the form drag to affect the total drag appreciably. The difference in the critical ranges in the two tunnels is due to different amounts of turbulence in the wind tunnel streams.‡

Measurements of the drag of a sphere moving through relatively calm air in the atmosphere have been made by Klein‖ with special apparatus designed for use on an aeroplane. The mean curve drawn

† Millikan and Klein, *Aircraft Engineering*, **5** (1933), 169–174.

‡ See Chap. II, § 24, pp. 72, 73. The effect of turbulence in the main stream on the critical range of $R$ is further discussed in § 219, where reference is also made to curves 3, 4, and 5 in Fig. 201.

‖ *Aircraft Engineering*, **5** (1933), 167–174. Measurements in flight have also been made at the R.A.E. (Serby and Morgan, *A.R.C. Reports and Memoranda*, No. 1725 (1936)), and in Germany (Hoerner, *Luftfahrtforschung*, **12** (1935), 42–54).

through these observations taken in free flight is reproduced as curve 6 in Fig. 201; the shape of this curve resembles that of the curves measured in a wind tunnel.

Included in Fig. 201 are also curves obtained from dropping tests on large spheres by Bacon and Reid† (curve 7) and by Lunnon‡ (curve 8). The results of Bacon and Reid were obtained from the rates of fall of wax, rubber, and wooden spheres of known weight dropped from an aeroplane and falling through calm air, and those of Lunnon from the rates of fall of metal spheres in air and water. The curve of Bacon and Reid can be extrapolated (as in the dotted line in Fig. 201) to join smoothly on to Lunnon's curve, but has a different shape from those obtained in wind tunnels or in free flight. The reason for this difference in shape has not been established, but it may be connected with the fact that whereas spheres tested in a wind tunnel are held rigidly in the stream, falling spheres have freedom to rotate.

### 217. Distributions of normal pressure and skin-friction.

The change in the character of the pressure distribution round a sphere on passing through the critical range of Reynolds numbers is shown in Fig. 202, where $\theta$ denotes angular distance from the forward stagnation point. These curves were obtained by Fage‖ from experiments on a 6-inch sphere mounted in an open jet tunnel (N.P.L.).

The critical range of $R$ was from $1\cdot4\times10^5$ to $3\cdot3\times10^5$, so the distributions for $R = 1\cdot57\times10^5$ and $R = 2\cdot98\times10^5$ can be regarded as representative of the beginning and the end of the critical range, respectively. The difference in the shapes of these two curves is very striking: at the lower value of $R$ there is an extensive region of low pressure at the back of the sphere, whereas at the higher value the pressure increases steadily towards the back and eventually reaches a positive value. There is also a fall in the minimum pressure on the front of the sphere and a backward traverse of the point where this pressure occurs as $R$ increases over the critical range.

Measurements of skin-friction showed that the well-defined inflexions (marked $T_2$, $T_3$, $T_4$) in the pressure curves within and beyond the critical range occurred at positions where the intensity of laminar

† *N.A.C.A. Report* No. 185 (1924).
‡ *Proc. Roy. Soc.* A, **118** (1928), 680–694; *ibid.* **110** (1926), 304.
‖ *A.R.C. Reports and Memoranda*, No. 1766 (1937).

friction fell to a minimum and the transition to turbulent flow in the boundary layer began. At the lower end of the critical range ($R = 1\cdot57 \times 10^5$ in these experiments) the intensity of friction rose to a maximum near $\theta = 55°$, and then fell rapidly to zero at $\theta = 83°$

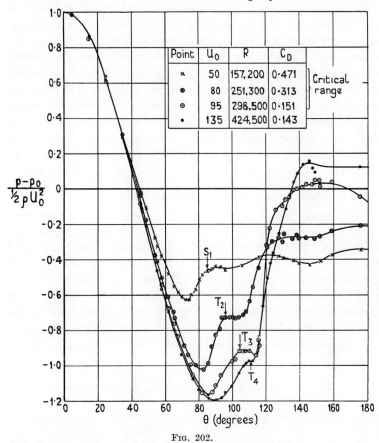

| Point | $U_0$ | R | $C_D$ | |
|---|---|---|---|---|
| × | 50 | 157,200 | 0·471 | Critical |
| ◉ | 80 | 251,300 | 0·313 | range |
| ◎ | 95 | 298,500 | 0·151 | |
| • | 135 | 424,500 | 0·143 | |

$$\frac{p - p_0}{\tfrac{1}{2}\rho U_0^2}$$

Fig. 202.

(i.e. at the point marked $S_1$ in Fig. 202), where the laminar layer separated from the surface before a transition to turbulence had occurred. At the upper end of the range the curve of frictional intensity has two maxima: the first, with laminar flow in the boundary layer, is on the front of the sphere, and the second, in the region of turbulent flow, is at the back. The transition to turbulence moves in the direction of increasing $\theta$ as the Reynolds number is increased over the critical range. At the middle of the critical range, defined as the Reynolds number at which $C_D = 0.3$, transition begins at $\theta = 95°$ approximately.

The experiments included explorations of pressure round the 6-inch sphere mounted in the stream of a closed jet tunnel, and also in a very turbulent stream close behind a high-resistance honey-comb. An analysis of the results showed that the non-dimensional distribution of normal pressure for any selected value of $C_D$ within the critical range (the values of $R$ in the critical range depending on the turbulence in the free stream) is independent, within the accuracy of measurement, both of the wind speed and of the degree of free-stream turbulence. This conclusion is in agreement with the fact that the greater part of the drag is due to the normal pressure on the surface, and only a small contribution is made by the skin-friction.

## 218. Laminar boundary layer calculations.

The laminar boundary layer on the surface of a sphere has been studied theoretically by Tomotika,[†] who used the momentum equation for a body of revolution, and, assuming a quartic form for the velocity distribution across the layer, obtained a solution of the differential equation for the thickness ($\delta$) of the layer. For a sphere of radius $a$ the distance ($x$) along the surface from the forward stagnation point is $a\theta$, where $\theta$ is the angle subtended at the centre of the sphere, and the radius $r_0$ of a cross-section is $a \sin \theta$. On substitution in equation (149) of Chapter IV (p. 163) the momentum equation becomes

$$\frac{dZ_*}{d\theta} = \frac{u_0}{u_1} g(\Lambda) + Z_*^2 \frac{d^2 u_1/d\theta^2}{u_0} h(\Lambda) - \frac{u_0}{du_1/d\theta} h^*(\Lambda) \cot \theta,$$

where
$$Z_* = \frac{u_0}{a} \frac{\delta^2}{\nu}, \qquad \Lambda = \frac{\delta^2}{a\nu} \frac{du_1}{d\theta},$$

$u_0$ is the undisturbed velocity of the stream, and $u_1$ is the velocity just outside the boundary layer. The values of $g(\Lambda)$, $h(\Lambda)$, and $h^*(\Lambda)$ are tabulated in Chap. IV, pp. 160, 164. Solutions of the above equation were obtained for the theoretical velocity distribution $u_1 = \frac{3}{2} u_0 \sin \theta$, and for the velocity distribution obtained from a

[†] A.R.C. Reports and Memoranda, No. 1678 (1936). The growth of the boundary layer, for motion started impulsively from rest, has been studied by Boltze (see Chap. IV, § 65, p. 186).

pressure distribution measured at a Reynolds number just below the critical range. In the latter case the value of $u_1 \delta/\nu$ at the position of minimum pressure ($\theta = 75°$ approximately) was $3 \cdot 8(u_0 d/\nu)^{\frac{1}{2}}$, and at the position of separation ($\theta = 81°$) it was $6 \cdot 8(u_0 d/\nu)^{\frac{1}{2}}$.

## 219. The effect of turbulence on the critical range of Reynolds numbers. The sphere as a turbulence indicator.

The position and extent of the critical range of Reynolds numbers, in which the large drop of drag coefficient occurs, depends on the turbulence in the free stream. Turbulence in the stream of a wind tunnel can be artificially created by a grid normal to the stream, and the dotted curves 3, 4, and 5 of Fig. 201 (p. 495) were obtained at the California Institute of Technology[†] from tests on a sphere mounted at three distances behind such a grid. They show that the critical range of $R$ occurs earlier as the distance behind the grid decreases—that is, as the turbulence increases. Hence we conclude from the relative positions of the curves 1 and 2, for example, that the turbulence in the air-stream of the Variable Density Tunnel is very considerably greater than in the 10-foot tunnel at the California Institute of Technology: on the basis of results such as those in Fig. 201 it has in fact been suggested[‡] that the sphere should be used for the quantitative measurement of turbulence, the turbulence number of the stream being defined as the Reynolds number at which the drag coefficient of a sphere is $0 \cdot 3$.

The critical Reynolds numbers for modern large wind tunnels, when determined in this way, lie between the extreme values 119,000 for the Variable Density Tunnel and 360,000 for the 1·2 metres D.V.L. Tunnel (1935 form);[‖] and the sequence of the critical Reynolds numbers can be used to determine an order of steadiness of the flow in the various tunnels.[††] A useful conclusion from such a sequence is that tunnels which have a large area contraction at the entry, with a honeycomb at the larger end, are less turbulent than those which have a small contraction and the honeycomb at the smaller end. In other words, disturbances flowing into a honeycomb, and

† Millikan and Klein, *Aircraft Engineering*, **5** (1933), 169–174.

‡ Dryden, *N.A.C.A. Report* No. 342 (1929).

‖ Silverstein, *ibid.*, No. 502 (1934).

†† See Hoerner, *loc. cit.* in footnote ‖ on p. 495; Harris and Graham, *A.R.C. Reports and Memoranda*, No. 1662 (1935); Fage, *ibid.*, No. 1370 (1931); Dryden, *N.A.C.A. Report* No. 392 (1931).

those created by the honeycomb itself, can be largely damped out if the stream undergoes a large contraction before the working section is reached. (See Chap. V, § 78.) The decay of turbulence in a wind tunnel with increasing distance from the honeycomb (see Chap. V, § 94) has also been demonstrated—at any rate qualitatively—by sphere drag measurements in the manner here contemplated.

The value of $R_{\text{crit}}$ in the atmosphere obtained by Klein in flight experiments (*loc. cit.* in footnote ‖ on p. 495) was $3 \cdot 65 \times 10^5$: the closest approximations to this value so far measured in wind tunnels are the values $3 \cdot 35 \times 10^5$ in the Galcit 10-foot tunnel, $3 \cdot 4 \times 10^5$ in the N.A.C.A. Full-Scale Tunnel, and $3 \cdot 6 \times 10^5$ in the 1·2 metres D.V.L. Tunnel (1935 form); but turbulence in the atmosphere, as indicated by the effect on the drag of a sphere, is in general appreciably smaller than in an artificial wind tunnel stream.

It has been suggested, because of the ease of measurement, that a constant value of a pressure coefficient should be taken as a criterion for the determination of the turbulence number, instead of a constant drag coefficient. If a sphere is mounted on a hollow back spindle, and the pressure coefficient is taken as the pressure difference between an impact hole at the front stagnation point and small holes in the spindle just behind the sphere divided by $\frac{1}{2}\rho U_0^2$, then the turbulence number is practically the same for a pressure coefficient 1·20 as for $C_D = 0 \cdot 30$.†

Dryden and Kuethe‡ have suggested that the ratio of the root-mean-square value $u$ of the longitudinal turbulent velocity component to the mean velocity $U_0$ should be taken as a measure of turbulence. The measurements of $u$ can be made with a hot wire anemometer in conjunction with an amplifier and apparatus to compensate for the lag of the hot wire.

Measurements of the temperature distribution close behind a heated wire across the stream lead to a determination of $v/U_0$, where $v$ is the root-mean-square turbulent velocity component at right angles to the stream and to the wire (see Chap. V, § 88, p. 219). The temperature measurements are comparatively simple, and it has been suggested,‖ therefore, that $v/U_0$ would provide a

---

† Hoerner, *loc. cit.*; Fage, *loc. cit.* in footnote ‖ on p. 496.

‡ *N.A.C.A. Report* No. 342 (1929).

‖ Dryden, *Journ. Washington Acad. Sci.* **25** (1935), 101–122.

satisfactory measure of turbulence. This method requires rather
simpler apparatus than measurements of $u/U_0$.[†]

Values of $u/U_0$ at several distances behind the honeycombs of wind
tunnels (from measurements by Dryden and Kuethe at the Bureau
of Standards, and by Wattendorf and Kuethe[‡] in the 10-foot wind
tunnel at the California Institute of Technology), and also a value in
the open air (free flight), are plotted in Fig. 203 against the corresponding values of $R$ at which $C_D = 0.3$ for a sphere. Both series
of observations show that $R_{crit}$ can be correlated with $u/U_0$, $R_{crit}$

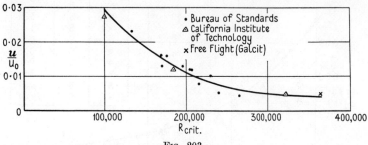

FIG. 203.

increasing as $u/U_0$ decreases. More extensive observations, however,
show that the correlation between the two parameters is not satisfactory: this arises from the fact that $R_{crit}$ depends also on the
ratio of a characteristic length of the turbulence-producing mechanism to the diameter of the sphere—in other words, on the relative
scale of the turbulence. Thus $R_{crit}$ is a function both of $u/U_0$ and of
$d/M$, where $d$ is the sphere diameter and $M$ a characteristic length
of the turbulence-producing mechanism—for example, the mesh of
a grid or the cell diameter of a honeycomb. Some insight into the
relations involved has been obtained by Taylor[‖] on the theory
explained in Chap. VII, § 151 (p. 328), where it was applied to
transition in the boundary layer along a flat plate.

If $U_1$ is the velocity just outside the boundary layer at a sphere,
$\partial p/\partial x$ the mean pressure gradient along the surface, and $\partial p/\partial x$ the
root-mean-square of the fluctuating pressure gradient along the

[†] A complete specification of the turbulence in a wind tunnel would require the
determination of the magnitude and 'frequency spectrum' of the velocity fluctuations
(see Chap. VI, § 121), and their spatial changes.

[‡] *Physics*, **5** (1934), 153–164.

[‖] *Proc. Roy. Soc. A*, **156** (1936), 307–317.

surface, then it is assumed that the velocity distribution in the boundary layer depends on $\Lambda + \Lambda'$, where

$$\Lambda = \frac{\delta^2}{\rho \nu U_1} \frac{\partial p}{\partial x}, \qquad \Lambda' = -\frac{\delta^2}{\rho \nu U_1} \partial p / \partial x,$$

and, as in equation (57) of Chapter VII (p. 328),

$$\partial p / \partial x = \text{constant}\, \rho u^{\frac{5}{3}} M^{-\frac{1}{2}} \nu^{-\frac{1}{2}}.$$

The non-dimensional pressure distribution depends on the value of

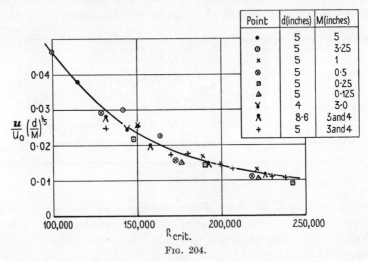

| Point | d(inches) | M(inches) |
|-------|-----------|-----------|
| ● | 5 | 5 |
| ⊙ | 5 | 3·25 |
| × | 5 | 1 |
| ⊗ | 5 | 0·5 |
| ▣ | 5 | 0·25 |
| △ | 5 | 0·125 |
| ⅄ | 4 | 3·0 |
| ⋏ | 8·6 | 3 and 4 |
| + | 5 | 3 and 4 |

FIG. 204.

$C_D$ and not on the value of $R$ (see § 217), and transition at $R_{\text{crit}}$—that is, at $C_D = 0 \cdot 3$—occurs at a fixed value of the angular distance $\theta$ from the forward stagnation point. Hence, on the approximate Kármán-Pohlhausen theory of laminar boundary layers, $\Lambda$ is constant at the transition point. Now the formula for $\Lambda'$ can be written, on substitution for $\partial p / \partial x$, in the form

$$\Lambda' = -\text{constant}\left(\frac{u}{U_0}\right)^{\frac{5}{2}}\left(\frac{\nu}{U_0 d}\right)^{\frac{1}{2}}\left(\frac{d}{M}\right)^{\frac{1}{2}}\left(\frac{U_0}{U_1}\right)^3\left(\frac{U_1 \delta}{\nu}\right)^2,$$

and at transition, since $\theta$ is fixed, $(U_0/U_1)$ is constant and $(U_1 \delta/\nu)$ is a function of $R\ (= U_0 d/\nu)$. Hence, on the assumption that $\Lambda'$ at transition is a function of $U_1 \delta/\nu$, it follows that $(u/U_0)(d/M)^{\frac{1}{2}}$ is a function of $U_0 d/\nu$, and therefore that $R_{\text{crit}}$ is a function of $(u/U_0)(d/M)^{\frac{1}{2}}$. Values of $(u/U_0)(d/M)^{\frac{1}{2}}$ calculated from Dryden's observations[†] on

† Dryden, *Journ. Washington Acad. Sci.* **25** (1935), 101–122; *N.A.C.A. Report* No. 392 (1931); Dryden and Kuethe, *ibid.*, No. 342 (1929). For $d = 5$ inches and

4-, 5-, and 8·6-inch spheres in streams covering a wide range of values of $M$ are plotted against $R_{crit}$ in Fig. 204, and a single curve can be drawn to pass closely through them. It should be specially noted that $d/M$ varies from 40 to 1 in Fig. 204; and that, if $u/U_0$ had been plotted against $R_{crit}$, the results would have shown a huge scatter.† (The variation of $d/M$ for the wind tunnel results in Fig. 203 was less than 5 to 1.)

## 220. Surface roughness.

The effect of surface roughness on the drag of a sphere has been investigated by Hoerner.‡ Tests on 7-, 14-, and 28-cm. spheres in free air (aeroplane flight tests) gave values of $R_{crit}$ equal to 3·48, 3·72, and $3·92 \times 10^5$, respectively, whence it was inferred (since the surfaces were the same and the relative roughness decreases with an increase in diameter) that roughness lowers the critical Reynolds number. Wind tunnel tests were made on the 28-cm. sphere coated with shellac, the surface being made systematically rougher with sand grains. The effect on drag was similar to that which occurs when the surface of a circular cylinder is roughened (Chap. IX, § 192, p. 432): with an increase in roughness the critical drop in $C_D$ begins at a lower Reynolds number, the fall in $C_D$ becomes smaller, and the supercritical $C_D$ becomes greater. When the roughness is very coarse the critical drop in $C_D$ tends to disappear, and the value of $C_D$ to become independent of Reynolds number.

It was found that waviness of a smooth surface also produced a decrease in the values of the Reynolds numbers for the critical range, and in some cases the change of $C_D$ with Reynolds number within the range became erratic.

## 221. Lift and drag coefficients of a rotating sphere.

Measurements have been made by Maccoll‖ of the forces acting on a sphere of diameter 6 inches rotating about a cross-wind axis in a wind tunnel. The results obtained at three values of $R$ are shown

---

all values of $M$ except 3 and 4 inches, $R_{crit}$ in Fig. 204 is the value at which the pressure difference between the forward stagnation point and a ring of pressure holes at $\theta = 157·5°$ is $1·22 \times \frac{1}{2}\rho U_0^2$, the criterion in the other cases being $C_D = 0·3$.

† Further observations over a wide range of values of $d/M$ have lately been published by Dryden, Schubauer, Mock, and Skramstad (*N.A.C.A. Report* No. 581 (1937), pp. 14–19), and good agreement found with Taylor's relation.

‡ *Luftfahrtforschung*, **12** (1935), 42–54.

‖ *Journ. Roy. Aero. Soc.* **32** (1928), 777.

in Fig. 205, where values of $C_D$ and $C_L$ (obtained by dividing the drag and the lift, respectively, by the product of $\frac{1}{2}\rho U_0^2$ and the cross-sectional area $\frac{1}{4}\pi d^2$) are plotted against the ratio of the equatorial speed to the wind speed. With an increase in this ratio the drag coefficient at first falls and then slowly rises. Special features are the

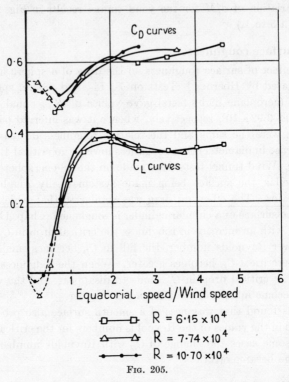

FIG. 205.

negative lift coefficient at values of the speed ratio less than 0·5, and the steep rise of the lift coefficient beyond.† Maccoll's experiments were made at values of $R$ of the order of $10^5$, and although no evidence is available, it is probable that some of the measurements were made over the critical range: if this is so, these measurements will not be representative of those which would be obtained over a different range of $R$.

Luthander and Rydberg‡ have made wind tunnel experiments on

† For a description of the phenomena for flow past a rotating cylinder see Chap. XII, § 239.

‡ *Physik. Zeitschr.* **36** (1935), 552–558.

a sphere rotating about an axis parallel to the wind direction. The diameter of the sphere (20 cm.) was selected so that, with the sphere at rest, the critical range of Reynolds numbers occurred well within the range of tunnel speed (3·5 to 47·5 m./s.). The range of rotational speed covered was 0 to 3,800 r.p.m. The flow past the sphere, particularly in the separation region and the boundary of the dead-air region, was made visible by titanium tetrachloride conveyed to the surface by a capillary tube; and the changes in drag arising from the rotation of the sphere were explained with the aid of these flow patterns by consideration of the effects of centrifugal force and of the transition to turbulence in the boundary layer. The effects of rotation on drag were complicated, the drag in some cases increasing with rotational speed and in others decreasing. The effects on the position of separation were also markedly different at different Reynolds numbers.

SECTION 2

AIRSHIP SHAPES

## 222. The transition to turbulence in the boundary layer.

The discussion in this section relates to stream-line bodies which are solids of revolution whose generators consist of smooth continuous curves. Their essential characteristic is that separation of the forward flow in the boundary layer occurs, if at all, only towards the extreme tail. The drag therefore arises mainly from skin-friction, and hence is determined very largely by the type of flow in the boundary layer, being greater when the layer is turbulent than when it is laminar.

In many respects the boundary layer on a stream-line body resembles that on a flat plate along the direction of flow. Thus a laminar portion near the nose is succeeded by a transition to the turbulent state farther back; and the position of transition varies with the turbulence in the main stream and with the Reynolds number.[†] It varies also, presumably, with the roughness of the surface of the solid body. No direct experimental evidence on this last point has been obtained from tests on stream-line bodies, but certain drag measurements (referred to in § 229) leave little room for doubt. Indeed, it would be surprising, in view of the general resemblance of

---

† The linear dimension in the Reynolds number is the axial length unless otherwise stated.

the flow to that past a flat plate, if surface roughness had no effect. Usually a stream-line model tested in a wind tunnel has a surface either of polished metal or of smooth varnished wood; it is to this type of surface that the whole of this section, except § 229, relates.

An increase of turbulence or of Reynolds number causes the transition region to move towards the nose, and *vice versa*. For the degree of turbulence present in the air-stream of the average wind tunnel the transition occurs at a value of $R_T$ which lies between about $5 \times 10^5$ and $1 \cdot 5 \times 10^6$, where $R_T$ is the Reynolds number formed from the axial distance of the transition point from the nose.†

Experience shows that comparatively small changes in the turbulence of the general stream—changes which may occur from day to day in any particular wind tunnel without apparent cause—are enough to produce appreciable changes in $R_T$ within the range mentioned. For most practical purposes it is therefore of little use to experiment with a stream-line body unless the Reynolds number based on the total length of the body is considerably above $1 \cdot 5 \times 10^6$: otherwise the uncertainty regarding the state of the boundary layer holds for an appreciable part of the whole surface of the body; and since the intensity of turbulent skin-friction is many times greater than that of laminar skin-friction, the corresponding variations in the measured drag may be large. To obtain results of general practical utility the Reynolds number of the experiment should be sufficiently high to ensure that substantially the whole of the boundary layer may be regarded as turbulent, with laminar flow occurring over only a small fraction of the surface near the nose, so that any movement in the transition region due to a change in general turbulence has an insignificant effect.

These considerations were first clearly stated by B. M. Jones in 1929.‡ Before that date much experimental work had been carried out on stream-line bodies, but (for lack of facilities as well as of knowledge) the Reynolds numbers were as a rule so low that the results are of little practical value. It is for this reason that

---

† See Simmons, 'Experiments relating to the Flow in the Boundary Layer of an Airship Model', *A.R.C. Reports and Memoranda*, No. 1268 (1930); Ower and Hutton, 'Investigation of the Boundary Layers and the Drags of two Streamline Bodies', *ibid.*, No. 1271 (1930); Freeman, 'Measurements of Flow in the Boundary Layer of a 1/40-scale Model of the U.S. Airship "Akron"', *N.A.C.A. Report* No. 430 (1932); Simmons and Brown, 'An Experimental Investigation of Boundary Layer Flow', *A.R.C. Reports and Memoranda*, No. 1547 (1935).

‡ *Ibid.*, No. 1199 (1929).

in the subsequent discussion we ignore most of the early experimental results obtained with stream-line shapes.

### 223. Form drag.

We restrict ourselves to consideration of bodies for which separation does not occur. As already stated, the drag then arises mainly from skin-friction, but there is usually an additional form drag resulting from the distribution of normal pressure over the surface. Usually this form drag is very small, but there are some stream-line bodies for which it is appreciable despite the absence of separation. The evidence on this point is scanty, and the accuracy of measurement of form drag is poor, but it is probable that the form drag tends to increase as the fineness ratio—that is, the ratio of the axial length to the maximum diameter—is decreased beyond a certain limit. Thus, whereas the form drag of a particular stream-line body of 5·45 to 1 fineness ratio was almost negligible, a body of 3 to 1 fineness ratio had a form drag about 27 per cent. of its total drag.† It is probable that relatively high form drag such as this indicates an approach to the condition for separation, caused by the rapid convergence of the after portion of a body of low fineness ratio.

When allowance is made for the form drag of a stream-line body, the residual coefficient (expressed on surface area) of the drag due to skin-friction approximates to the skin-friction drag coefficient at the same Reynolds number for a flat plate lying along the stream. This matter is discussed in more detail in §§ 225, 226. For the present we merely note it as an experimental result which is of some help in assessing the value of any particular experiment with a stream-line body: thus, if the curve of drag coefficient against Reynolds number falls close to the curve for the flat plate in turbulent flow, we can be reasonably sure that the boundary layer is turbulent over practically the whole of the surface of the body, and the results are then unlikely to depend critically on the degree of turbulence in the tunnel stream.

### 224. The effect of fineness ratio on drag.

It seems probable that there is an optimum fineness ratio for stream-line bodies. For evidently, if the fineness ratio is too small, the drag will be high on account of separation (as for a sphere: fineness ratio 1): on the other hand, once the length is enough to

† Lock and Johansen, *ibid.*, No. 1452 (1933), pp. 4–6.

ensure a sufficiently gradual taper of the tail to prevent separation or high form drag, any additional increase in length will merely increase the total drag because of the greater surface. In most of the fairly numerous experiments made to determine the effect of the variation of fineness ratio the Reynolds number has been of the order of $10^5$ or $10^6$, so the boundary layer has in all probability been partly laminar and partly turbulent. Moreover, these experiments

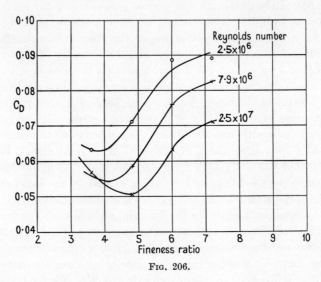

Fig. 206.

have usually consisted of drag measurements on bodies of constant maximum diameter but of different lengths, and the fineness ratio for minimum drag under these conditions is generally lower than would be found from tests at considerably lower or considerably higher Reynolds numbers, at which the boundary layer could, for practical purposes, be regarded as entirely laminar or entirely turbulent. Other experiments have been unsatisfactory because the fineness ratio has been altered by the insertion of different lengths of cylindrical body between the nose and the tail. It is not clear what purpose such tests are expected to serve, since, given a well-shaped nose and tail and junction between them, it seems that the intro-duction of a cylindrical portion must increase the drag.

The scanty experimental data not open to any of these objections point to a value about 5 to 1 as the fineness ratio which, at high

Reynolds numbers, corresponds to the minimum drag coefficient $C_D$ ($C_D = \text{drag}/\frac{1}{2}\rho U_0^2 A$, where $A$ is the maximum cross-sectional area of the body). It must be emphasized that the evidence is by no means sufficient to enable this conclusion to be accepted as final: it is, in fact, based mainly on the results of the single set of American experiments[†] plotted in Fig. 206. In support, we may refer (see § 223) to the high form drag measured on a large body of considerably smaller fineness ratio (3 to 1), and the negligible form drag of the longer body (fineness ratio 5·45 to 1).

## 225. Skin-friction drag.

As stated in § 223, the skin-friction drag coefficient of a stream-line body approximates to the skin-friction drag coefficient of a flat plate at the same Reynolds number, the latter being formed for the stream-line body from its axial length. This experimental fact is illustrated by the results of a number of measurements of the total drag of stream-line bodies, collected in Figs. 207, 208, and 209. The curves $AA$ and $BB$ in Figs. 207 and 209 represent the accepted values of the drag coefficient of a flat plate along the stream when the boundary layer is entirely turbulent and entirely laminar, respectively: the equations of the two curves are[‡]

$$C_f = 0.455(\log_{10} R)^{-2.58}$$

and

$$C_f = 1.328 R^{-\frac{1}{2}},$$

respectively, where $C_f$ is the drag coefficient based on the total surface area.

It will be seen that at Reynolds numbers of the order of $5 \times 10^6$ and upwards the results all approach the turbulent curve for the flat plate ($AA$). At lower Reynolds numbers the results sometimes fall near the curve $AA$ and sometimes below it. Experimental evidence shows that the turbulence in the tunnel is responsible for these variations: the more turbulent the tunnel the lower the value of the Reynolds number at which the drag coefficient of a stream-line body approaches the turbulent drag coefficient of a flat plate. The balance of evidence also suggests that once the turbulence of the stream is enough to make practically the whole of the boundary layer turbulent,

---

† Abbott, *N.A.C.A. Report* No. 394 (1931).

‡ See Chap. VIII, § 163, p. 365, and Chap. IV, equation (58), p. 138. ($C_f$ has half the value of $C_D$ as given in this latter equation, since the total surface area is used in forming $C_f$.)

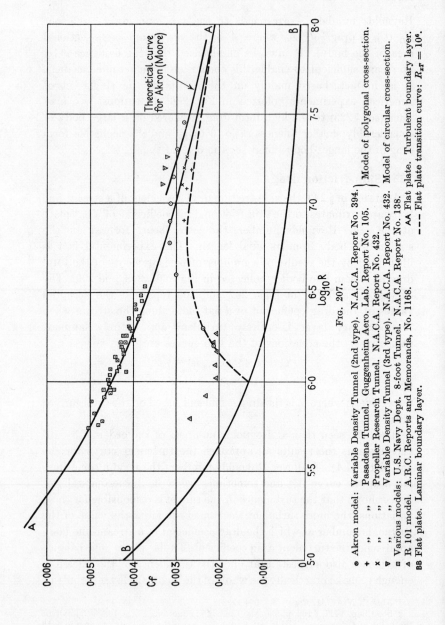

FIG. 207.

Akron model: Variable Density Tunnel (2nd type). N.A.C.A. Report No. 394.  
+    ,,    ,,    Pasadena Tunnel. Guggenheim Aero. Lab. Report No. 105.  ⎫ Model of polygonal cross-section.  
×    ,,    ,,    Propeller Research Tunnel. N.A.C.A. Report No. 432.  ⎭  
▽    ,,    ,,    Variable Density Tunnel (3rd type). N.A.C.A. Report No. 432. Model of circular cross-section.  
⊡ Various models: U.S. Navy Dept. 8-foot Tunnel. N.A.C.A. Report No. 138.  
▲ R 101 model. A.R.C. Reports and Memoranda, No. 1168.  AA Flat plate. Turbulent boundary layer.  
BB Flat plate. Laminar boundary layer.  --- Flat plate transition curve: $R_{xT} = 10^6$.

a further increase of turbulence does not appreciably increase the drag. More experimental information is required before this last statement can be regarded as established, but it certainly appears to hold good for the range of turbulence ordinarily found in wind tunnels.

Within what we may term the critical range of Reynolds numbers (i.e. about $5 \times 10^5$ to $2 \times 10^6$), where the boundary layer is partly laminar and partly turbulent, we should expect the curves of drag coefficient against Reynolds number to lie between the curves $AA$ and $BB$. Figs. 207, 208, and 209, which are typical of all stream-line bodies, show that this expectation is fulfilled, except in so far as for fully developed turbulence the results lie above the curve $AA$. This latter behaviour can also be predicted theoretically (see § 226).

It has been shown in Chap. VIII, § 163 (p. 366) that the drag coefficient (based on total surface area) of a flat plate within the critical range can be represented by the equation

$$C_f = C_{fT} - A/R,$$

where $C_{fT}$ is the drag coefficient of the plate with wholly turbulent boundary layer. The value of $A$ depends on $R_{xT}$, the critical value of $R$ at which transition takes place. The dotted curve in Fig. 207 is the 'transition' curve obtained by assuming that $R_{xT} = 10^6$, and is similar in shape to the curve through the experimental points for an airship model (Akron) tested in the American Variable Density Tunnel.

Fig. 208 illustrates the extent to which the characteristics of the wind tunnel can influence the drag of a stream-line body. All the experimental curves in this diagram relate to a particular airship hull (that of the British rigid airship known as R. 33), a single model of which, made in metal and very carefully treated to avoid damage, was tested in several British and American wind tunnels.† With the exception of those in the U.S.A. Variable Density Tunnel all the experiments were carried out in atmospheric tunnels within the critical range of Reynolds numbers, and we see that the results at the same Reynolds number in different tunnels differ very considerably—in some cases by over 100 per cent. It is worthy of note that the highest experimental values in an atmospheric tunnel were

† The detailed results of these tests have not been published; a diagram of the results has lately been published by Jacobs, *Aerodynamic Theory* (edited by Durand), **3** (Berlin, 1935), 328.

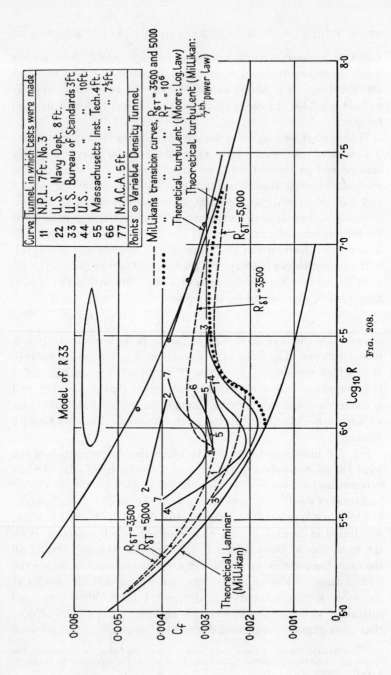

FIG. 208.

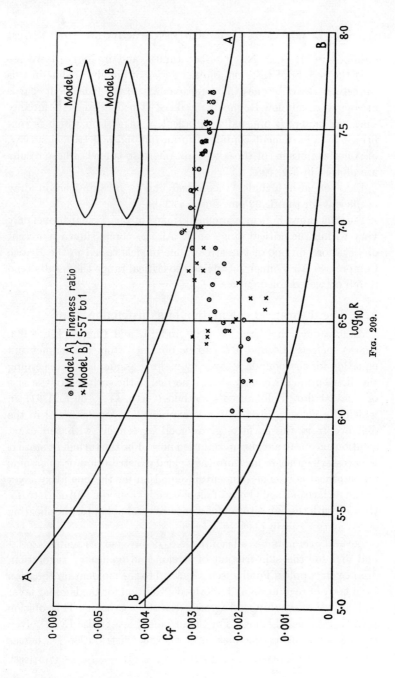

Fig. 209.

obtained in the U.S. Navy 8-foot tunnel, and in Fig. 207 we see that the results of tests on another series of airship models in this tunnel are the only ones that approximate to the 'turbulent' curve at comparatively low Reynolds numbers. This fact suggests forcibly that this tunnel is unusually turbulent, a suggestion which is confirmed by a statement contained in an unpublished critical survey, of American origin, of those tests in American tunnels whose results are plotted in Fig. 208.

Fig. 208 also includes theoretical curves of skin-friction drag coefficient for the R. 33 model (see § 226).

Fig. 209 shows how measurements made in a particular tunnel may vary within the critical range of $R$. All the points shown represent observations on one of two stream-line models tested in the British Compressed Air Tunnel.† Outside the critical range the results tend to fall on well-defined curves.

## 226. The theoretical calculation of skin-friction drag.

As explained in Chap. IV, §§ 60 and 61, and Chap. VIII, § 163, curves such as $AA$ and $BB$ can be obtained from the momentum equation for the boundary layer by making assumptions concerning the distribution of velocity along a normal to the surface. In the case of laminar flow Pohlhausen's method (Chap. IV, §§ 60 and 61), in which the velocity distribution is expressed as a polynomial in the distance from the surface, gives good agreement with the exact solution except in a region of retarded flow. For the turbulent state it is necessary either to measure the velocity distribution or to assume an empirical law. The momentum equation for the boundary layer at the surface of a solid body of revolution is then eqn. (40) of Chap. IV (p. 134) with the substitution of a general symbol $(\tau_0)$ for the shearing stress at the wall in place of $\mu(\partial u/\partial y)_{y=0}$.

Several attempts have been made to deduce curves similar to $AA$ and $BB$ for the skin-friction of stream-line bodies of revolution. Miss Lyon‡ applied Pohlhausen's method to the momentum equation for a body of revolution and obtained a solution for the laminar layer which gives reasonably good agreement with experiment. This solution is further discussed below. For the turbulent layer Miss Lyon solved the equation using the measured velocity distribution,‡ whereas

† R. Jones, Williams, and Brown, *A.R.C. Reports and Memoranda*, No. 1710 (1936).
‡ *Ibid.*, No. 1622 (1935).

Dryden and Kuethe,[†] Millikan,[‡] Fediaevsky,[||] and Moore[††] assumed that the velocity distribution is the same as at a flat plate. This assumption can at best be only approximately correct: the available experimental data on the velocity distributions in the boundary layers of stream-line bodies are insufficient to show how far it departs from the facts, but it has been found to lead (after somewhat laborious calculations) to theoretical estimates of the skin-friction drag which do not differ seriously from measured values in the two cases for which the results have been worked out (see p. 516 *et. seq.*). The theory shows that the skin-friction drag of a stream-line body is somewhat higher than that of a flat plate of the same area at the same Reynolds number, a result which is in agreement with experimental results outside the critical range.

The practical value of the calculations, as distinct from their theoretical interest, is doubtful. Miss Lyon, using measured velocity distributions (whose determination involves lengthy experiments), was able to obtain only an approximate solution for the turbulent boundary layer, and commented on the need for more experimental data. On the other hand, the assumption that the velocity distribution is the same as at a flat plate can, as we have already remarked, be only an approximation. Moreover, whichever method is used, the calculations are heavy and the accuracy is not likely to be very high, partly because differentiation of the velocity distribution outside the boundary layer will be necessary. Also the calculations have to be carried through afresh for each different shape. Hence, and in view of the fact that for an average stream-line body the skin-friction drag appears to be only some 10 or 15 per cent. above that for the flat plate, the labour of the calculations seems scarcely justified from a practical point of view. For most practical purposes it would seem to be sufficiently accurate to use the skin-friction drag curves for the flat plate to predict the skin-friction drag of a stream-line body.

We proceed to discuss briefly the various theoretical treatments of the problem. The method used by Miss Lyon has already been indicated. Of the remaining investigations, that of Dryden and

† *N.A.C.A. Report* No. 342 (1930), pp. 12–14.

‡ *Trans. Amer. Soc. Mechanical Engineers*, Applied Mechanics Section, **54** (1932), 29–39.

|| *Trans. Centr. Aero-Hydrodyn. Inst., Moscow*, No. 179 (1934); *Journ. Aero. Sciences*, **3** (1935), 17–20.

†† *Guggenheim Airship Institute (Akron, Ohio)*, Publication No. 2 (1935), pp. 21–31; *Journ. Aero. Sciences*, **2** (1935), 32–34.

Kuethe is mainly of interest as representing the first published attempt to obtain a quantitative comparison between theory and the measured drag of stream-line bodies. They calculated the drag in two parts. For the forward portion of the body with a laminar boundary layer they assumed that the skin-friction was the same as in two-dimensional flow past a flat plate with a longitudinal pressure gradient equal to that measured on the model airship, and to calculate this skin-friction they assumed simply a linear velocity distribution across the laminar boundary layer. For the turbulent portion of the boundary layer they neglected the pressure gradient and assumed a velocity distribution of the form $U/U_0 = (y/\delta)^{\frac{1}{7}}$. A further assumption was that Prandtl's condition† held at the transition from laminar to turbulent flow.

Millikan's analysis‡ went considerably farther. He included the terms allowing for the radius of curvature of the generating curve of the stream-line body—i.e. he used equation (40) of Chapter IV, whereas Dryden and Kuethe had used equation (33). He assumed a parabolic velocity distribution in the laminar part of the boundary layer and the same law as Dryden and Kuethe in the turbulent portion, but took into account the axial velocity gradient which they had neglected.

The $\frac{1}{7}$th power law of turbulent velocity distribution was the law accepted for flat plates at the time when the work of Dryden and Kuethe was published. Subsequently this law was shown not to agree with experiment at sufficiently high Reynolds numbers. Fediaevsky‡ obtained a solution similar to that of Millikan, but substituted a general index $1/n$ for the index $\frac{1}{7}$ in the power law, and used an empirical relation, deduced from Nikuradse's tests‖ in pipes, to determine the value of $n$ for any particular value of $R$. This relation is given by Fediaevsky in the form of a curve from which the following table has been compiled.

TABLE 17

| $\log R$ | $n$ |
|:---:|:---:|
| 6·5 | 6·6 |
| 7·2 | 7·0 |
| 7·9 | 8·0 |
| 8·4 | 9·0 |
| 8·83 | 10·0 |

† See Chap. VII, § 151, p. 329.                                        ‡ *Loc. cit.*
‖ *Ver. deutsch. Ing., Forschungsheft* 356 (1932).

Moore's analysis[†] relates only to high values of the Reynolds number, for which in tunnel tests the boundary layer may be assumed to be completely turbulent. His solution again is similar to that of Millikan, but the logarithmic law of velocity distribution within the boundary layer replaces the power law. (Cf. Chap. VIII, § 163.)

The results of calculations made by Millikan and by Moore for the model of R. 33 are compared in Fig. 210 with the flat plate curves

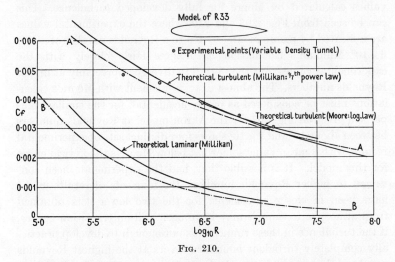

FIG. 210.

$AA$ and $BB$, and in Fig. 208 with experimental results for the same model. Included in Fig. 208 are Millikan's theoretical 'laminar' curve for the model, and three 'transition' curves, which are discussed below. It will be seen that Millikan's laminar curve and both his and Moore's turbulent curves lie above the corresponding curves for the flat plate.[‡] This can be explained, without resort to the mathematical derivation of the frictional drag of bodies of revolution, by the physical consideration that the average velocity over the surface of such a body is higher than that along a flat plate in a stream having the same velocity at infinity. The same argument indicates that different curves will be obtained for bodies of different shape, and, more particularly, that the greater the fineness ratio the smaller

[†] *Loc. cit.*
[‡] Millikan's turbulent curve, for which he gives the equation $C_f = 0.077R^{-\frac{1}{5}}$, should be compared not with the curve $AA$ but with the curve derived for a flat plate from the $\frac{1}{7}$th power law of velocity distribution, viz. $C_f = 0.074R^{-\frac{1}{5}}$ (Chap. VIII, § 163, p. 362).

will be the amount by which the curve for the stream-line body lies above that for the flat plate. This point is illustrated by the fact that Moore's theoretical curve for R. 33 (fineness ratio 8·2) lies closer to the turbulent curve for the flat plate (Fig. 210) than does that for the Akron model (fineness ratio 5·9) in Fig. 207.

The drag figures obtained for the R. 33 model in the Variable Density Tunnel of the N.A.C.A. agree remarkably well with the values calculated by Moore for fully developed turbulence. This can be seen from Figs. 208 and 210, where the experimental values are indicated by points in circles. Up to a Reynolds number of about $6 \times 10^6$ Millikan's calculations also agree fairly closely with the experimental values, but they begin to diverge appreciably at higher Reynolds numbers. The almost exact agreement with Moore's calculations must be considered as partly accidental, for the similar comparison given in Fig. 207 for the Akron model at Reynolds numbers between about $10^7$ and $2 \times 10^7$ shows four distinct sets of experimental results, with a mean value somewhat below Moore's theoretical curve for this model. It is possible that, had the experiments been continued to higher Reynolds numbers, the four sets of results might have been in closer agreement; for the two lower sets obtained from atmospheric tunnels may have been still on transition curves, if the turbulence in these tunnels was not enough to develop practically completely turbulent boundary layers at the highest Reynolds numbers that could be reached. It is known that the Pasadena tunnel, in which the lowest of the four curves was obtained, is very nearly as turbulence-free (according to sphere drag tests) as the atmosphere, and it is probable that the Propeller Research Tunnel is a good deal less turbulent than either of the two types of Variable Density Tunnel in which the two higher sets of results were obtained.

Millikan's theory enables transition curves to be calculated for the change of drag coefficient with Reynolds number when the boundary layer is partly laminar and partly turbulent. Millikan adopted the same assumption regarding the nature of the transition that Prandtl made in calculating the transition curve for the flat plate.† He assumed also that the transition took place at a constant value of the Reynolds number ($R_{\delta_T}$) formed from the thickness of the laminar boundary layer. Prandtl, on the other hand, assumed that the criterion was the value of the Reynolds number formed from the

† See Chap. VII, § 151, p. 329.

distance of the transition point from the front of the plate. For a flat plate, with no longitudinal velocity gradient outside the boundary layer, either assumption implies the other, but this is not so for a stream-line body, for which one would perhaps expect Millikan's assumption to be nearer the truth. In Fig. 208 three transition curves are plotted for the model of R. 33: two (shown by dashes) are based on Millikan's assumption, and relate to values of $R_{\delta T}$ of 3,500 and 5,000; the third (shown dotted) is calculated from the data given in Millikan's paper previously cited, but on the assumption that the transition occurs at a value of $R_T = 10^6$, where $R_T$ is the Reynolds number formed from the axial distance of the transition point from the nose of the model. On the whole the latter type of transition curve resembles the experimental curves more closely than the former type, but one must recognize that the theory contains assumptions that at best can be only approximations to fact. In particular, the change from laminar flow to fully developed turbulence in the boundary layer does not occur at a point: the transition takes place over an appreciable length.

## 227. Explorations of the flow in the boundary layer.

The flow in the boundary layer of a stream-line body has been studied experimentally by several workers,† and their results have been compared with the assumptions and calculated results of theories previously mentioned. The velocity distribution normal to the surface resembles in general form that for a flat plate. As shown in Fig. 211, which is based on Freeman's measurements,‡ two distinct types of non-dimensional velocity distribution are found, characterized essentially by different slopes at the origin. The steeper slope, which occurs at higher values of the Reynolds number, is associated with a turbulent boundary layer and the correspondingly higher local intensity of skin-friction. The other type of distribution corresponds to laminar flow in the boundary layer.

The velocity distribution curve changes shape when the transition from laminar to turbulent flow takes place, and Simmons and Brown‡ have shown that the transition region determined from this change of shape agrees in position with that obtained by other means,

---

† See the works cited in footnote † on p. 506; also Klemperer, *Abhandl. Aero. Inst. Aachen*, **12** (1932), 47–51; Miss Lyon, *A.R.C. Reports and Memoranda*, No. 1622 (1935).

‡ *Loc. cit.* on p. 506.

including the use of hot wires to detect velocity fluctuations in the turbulent layer. The resistance changes of the hot wires may be made audible in telephone receivers or loud speakers by suitable electrical apparatus, or they may be recorded by means of an oscillograph such as an Einthoven galvanometer. Records obtained in this way show marked differences according to whether the hot

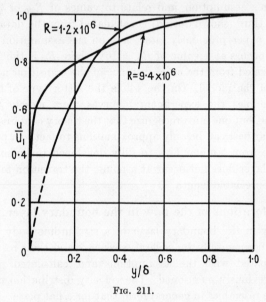

FIG. 211.

wire is in a laminar or a turbulent region of the boundary layer (Chap. VI, §119). In another method† of detecting the transition region the total head was measured at a constant small distance from the surface of the model at various positions along a generating line. It was found that the total head decreased along the length of the model to a minimum value whose position coincided with the beginning of the change from laminar to turbulent flow. The total head then increased, over an appreciable length of the model, to a maximum value which presumably occurred when the change to turbulent flow was complete. When this method is used the total-head tube must be close enough to the surface to be within the boundary layer in the laminar portion, so that, since the boundary layer decreases in thickness with an increase in the Reynolds number,

† Simmons and Brown, *loc. cit.* in the footnote on p. 506.

the higher the Reynolds number the nearer to the surface must the total-head exploration be made. In the experiments of Simmons and Brown the total head was measured 0·03 inch from the surface, the Reynolds number of the flow at the beginning of the transition region being about $1·3 \times 10^6$, at which Reynolds number a laminar boundary layer in two-dimensional flow in the absence of a pressure gradient has a thickness of about 0·06 inch.

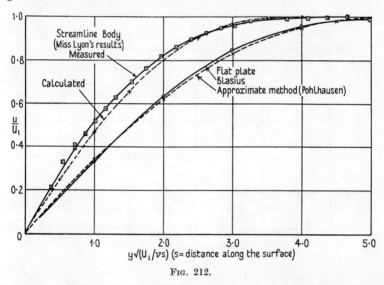

Fig. 212.

As mentioned on p. 514, Miss Lyon[†] has adapted Pohlhausen's method to obtain an approximate solution of the laminar boundary layer equations for a solid of revolution. Assuming a quartic velocity distribution in the boundary layer, she obtained the velocity distribution shown in Fig. 212, where it is compared with a measured distribution: the agreement is reasonably good, though not as good as that between the laminar distributions calculated by Blasius and by Pohlhausen for the flat plate. (These latter distributions, for the same Reynolds number as those for the stream-line body, are also shown in Fig. 212).

Freeman's explorations[‡] of velocity in the boundary layer at various sections of the model of the Akron are the most complete that exist for a stream-line body. From them Freeman deduced that

† *Loc. cit.* in footnote ‡ on p. 514.        ‡ *Loc. cit.* in the footnote on p. 506.

the thickness δ of the boundary layer ranged from less than 0·1 inch at a section 1·3 feet from the nose to about 10 inches at the extreme tail, 19·6 feet from the nose. The transition to turbulence was observed to occur at a Reynolds number of about $10^6$.

Freeman also compared his measured boundary layer thicknesses with the values calculated from Millikan's theory, and obtained

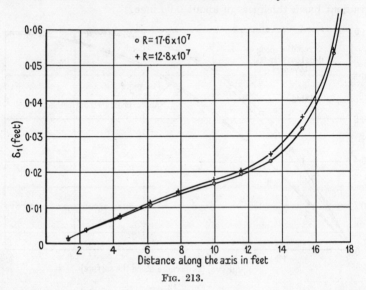

FIG. 213.

quite good agreement for the forward 85 per cent. or so of the length of the model. It should be observed, however, that the boundary layer thickness cannot be accurately deduced from total-head or velocity explorations. The thickness is taken as the distance from the boundary at which the velocity or the total head ceases to rise, but as this condition is reached asymptotically, the precise thickness is not well defined. A quantity that can be determined much more accurately from the observations is the displacement thickness, denoted by $\delta_1$ and defined as

$$\delta_1 = \int\limits_0^\delta \left(1 - \frac{U}{U_1}\right) dy.$$

(Cf. Chap. IV. eqn. (15), p. 123.) $\delta_1$ has been calculated from Freeman's measurements for the Akron model, and the results are plotted against distance from the nose of the model in Fig. 213. The values of $\delta_1$ are, for about the forward 60 per cent. of the body, approximately the same as those calculated for a flat plate on the assumption of the $\frac{1}{7}$th power law of velocity distribution. Farther aft the boundary layer thickens much faster than on a plate. This thickening is a characteristic feature of the boundary layer of a stream-line body, and is attributable to the geometrical contraction of the tail aft of the section of maximum diameter.[†]

A notion of the order of thickness of the boundary layer may be obtained by multiplying the values of $\delta_1$ in Fig. 213 by 8, since $\delta = 8\delta_1$ is the relation for the $\frac{1}{7}$th power law of velocity distribution (Chap. VIII, eq. (63)).

Moore[‡] also has calculated the boundary layer thickness for the Akron, and has obtained approximate agreement with Freeman's measurements.

### 228. The distribution of normal pressure, and its effect on transition.

It is found experimentally[||] that the measured distribution of normal pressure on the surface of a stream-line body agrees closely over most of the surface with that calculated from ideal fluid theory. The two begin to diverge as the boundary layer thickens towards the tail of the body, the actual pressure at the extreme tail being less than the theoretical value $\frac{1}{2}\rho U_0^2$. Fig. 214 is reproduced from Jones's results for a prolate spheroid whose major axis was four times as long as the minor axis: the theoretical pressures are shown by the full curve and the observations by the points. Remarkably close agreement is found over about 95 per cent. of the length of the body. The Reynolds number of this experiment was about $5 \times 10^5$.

Miss Lyon[††] has shown that the shape of the nose of the stream-line body may cause an early transition at low Reynolds numbers if it produces a pressure gradient along the surface tending to promote instability of the laminar boundary layer. Fig. 215 shows the pressure distributions measured by Miss Lyon on two stream-line bodies of the same length $l$ and maximum diameter $D$. The bluff

---

† Ower and Hutton, *loc. cit.* in the footnote on p. 506.

‡ *Loc. cit.* in footnote †† on p. 515.

|| Fuhrmann, *Zeitschr. f. Flugtechn. u. Motorluftschiffahrt*, **11** (1911), 165, 166; *Jahrbuch der Motorluftschiff-Studiengesellschaft*, **5** (1921), 65–123; R. Jones, *A.R.C. Reports and Memoranda*, No. 1061 (1927); *Phil. Trans.* A, **226** (1926), 231–266.

†† *A.R.C. Reports and Memoranda*, No. 1622 (1935), pp. 1, 2 and 22.

nose of the body shown as model $B$ in Fig. 215 ($b$) produced a rising
pressure gradient well forward, which seems to have caused the
transition to occur earlier than on model $A$, as shown in the following
table.  ($R$ is the Reynolds number for the full length (5·83 feet) of

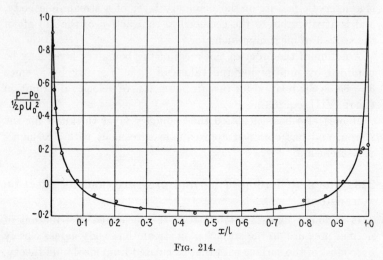

FIG. 214.

the model, and $x$ is the axial distance from the nose at which the
transition took place.)

TABLE 18

| Model A | | Model B | |
|---|---|---|---|
| $R \times 10^{-6}$ | $x$ (feet) | $R \times 10^{-6}$ | $x$ (feet) |
| 2·04 | 3·5 | 2·08 | 1·95 |
| 3·06 | 3·1 | 3·11 | 1·7 |

## 229. The effect of the state of the surface on the skin-friction drag.

The effect of surface roughness on turbulent flow in pipes and along
flat plates has been discussed in Chap. VIII. Abbott† has shown that
a similar effect exists in the case of a stream-line body. Fig. 216
shows curves plotted through points read off the rather small-scale
diagrams given by Abbott. The drag of the model with the 'normal'
surface—rubbed varnish finish—is only slightly higher than that of
the same model with a highly polished surface, but a large increase of

† *N.A.C.A. Report* No. 394 (1931).

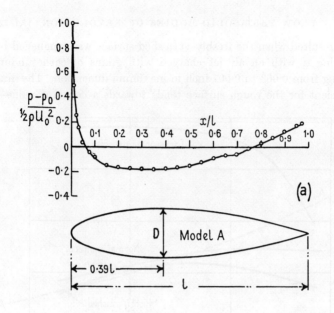

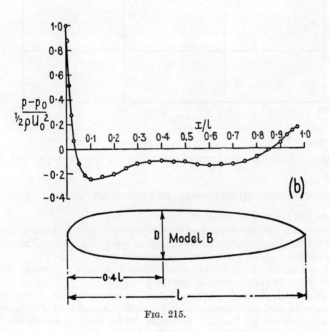

Fig. 215.

drag resulted when the freshly varnished surface was roughened by spraying it with an air jet charged with grains of carborundum ranging from 0·003 to 0·007 inch in maximum dimension. The drag coefficient for the rough surface tends towards a constant value—

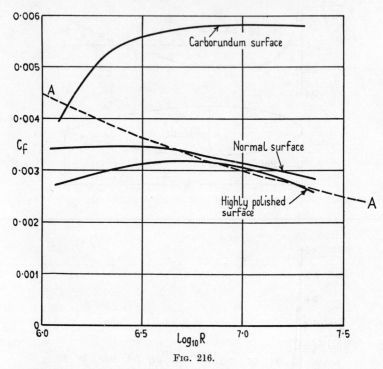

Fig. 216.

a feature which has also been observed in the case of pipes and flat plates.

## ADDITIONAL REFERENCES

*Treatises and collected accounts.*

H. Muttray, *Handbuch der Experimentalphysik*, **4**, part 2 (Leipzig, 1932), 233–336. (Drag determinations on bodies experiencing no lift. The article deals with two- as well as three-dimensional flow.)

L. Schiller, *Handbuch der Experimentalphysik*, **4**, part 2 (Leipzig, 1932), 337–387. (Dropping tests on spheres, disks, etc.)

For discussions of airship performance and of the application of ideal fluid theory to airships reference may be made to articles in *Aerodynamic Theory* (edited by Durand), **6** (Berlin, 1936), by K. Arnstein and W. Klemperer on pp. 49–133 and by M. Munk on pp. 32–48.

*Disks and rectangular plates perpendicular to the direction of flow.*

EIFFEL, *La résistance de l'air et l'aviation* (Paris, 1910), p. 73 (disks); pp. 39–43 (square and rectangular plates).

EIFFEL, *Nouvelles recherches sur la résistance de l'air et l'aviation* (Paris, 1914), p. 93 (disks).

PRANDTL, *Ergebnisse der Aerodynamischen Versuchsanstalt zu Göttingen*, **2** (1923), 28–34 (disks and rectangular plates).

SCHMIEDEL, *Physik. Zeitschr.* **29** (1928), 593–596; 604–610 (disks).

SCHILLER and SCHMIEDEL, *Zeitschr. f. Flugtechn. u. Motorluftschiffahrt*, **19** (1928), 497–501 (disks).

SIMMONS and DEWEY, *A.R.C. Reports and Memoranda*, No. 1334 (1931) (disks).

STANTON and MARSHALL, *A.R.C. Reports and Memoranda*, No. 1358 (1932); *Proc. Roy. Soc.* A, **130** (1931), 295–301 (disks).

FLACHSBART, *Ergebnisse der Aerodynamischen Versuchsanstalt zu Göttingen*, **4** (1932), 96–100 (disks and rectangular plates).

SCHUBAUER and DRYDEN, *N.A.C.A. Report* No. 546 (1935) (disks and rectangular plates).

*Spheres.*

EIFFEL, *Nouvelles recherches sur la résistance de l'air et l'aviation* (Paris, 1914), pp. 87–93.

PRANDTL, *Ergebnisse der Aerodynamischen Versuchsanstalt zu Göttingen*, **2** (1923), 28–31.

ERMISCH, *Abhandl. Aero. Inst. Aachen*, No. 6 (1927), pp. 48–50.

SCHMIEDEL, *Physik. Zeitschr.* **29** (1928), 593–604.

SCHILLER and SCHMIEDEL, *Zeitschr. f. Flugtechn. u. Motorluftschiffahrt*, **19** (1928), 497–501.

*The sphere as a turbulence indicator.*

PLATT, *N.A.C.A. Report* No. 558 (1936).

SERBY and MORGAN, *A.R.C. Reports and Memoranda*, No. 1725 (1936).

ROBINSON, *Journ. Aero. Sciences*, **4** (1937), 199–201.

*Ellipsoids.*

PRANDTL, *Ergebnisse der Aerodynamischen Versuchsanstalt zu Göttingen*, **2** (1923), 28–31.

R. JONES, *Phil. Trans.* A, **226** (1927), 231–266; *A.R.C. Reports and Memoranda* No. 1061 (1927).

*Stream-line bodies; axial flow.*

GURJIENKO, *Trans. Centr. Aero-Hydrodyn. Inst., Moscow*, No. 257 (1936) (theoretical).

*Stream-line bodies; non-axial flow.*

R. JONES and WILLIAMS, *A.R.C. Reports and Memoranda*, No. 600 (1919) (partly theoretical).

R. JONES, WILLIAMS, and BELL, *A.R.C. Reports and Memoranda*, No. 802 (1923), pp. 1–6, 16, 21–26.

MUNK, *N.A.C.A. Report* No. 184 (1924) (theoretical).

VON KÁRMÁN, *Abhandl. Aero. Inst. Aachen*, No. 6 (1927), pp. 1–18 (theoretical).

R. JONES, *Phil. Trans.* A, **226** (1927), 231–266; *A.R.C. Reports and Memoranda*, No. 1061 (1927) (partly theoretical).

R. JONES and BELL, *A.R.C. Reports and Memoranda*, No. 1168 (1929), pp. 1–4, 13–16.

R. JONES and BELL, *A.R.C. Reports and Memoranda*, No. 1169 (1929), pp. 1–3, 11–16.

UPSON and KLIKOFF, *N.A.C.A. Report* No. 405 (1931), pp. 1–12 (theoretical).

FREEMAN, *N.A.C.A. Report* No. 432 (1932).

FREEMAN, *N.A.C.A. Report* No. 443 (1932).

R. JONES and BELL, *A.R.C. Reports and Memoranda*, No. 1400 (1932), pp. 5, 22.

HARRINGTON, *Guggenheim Airship Institute (Akron, Ohio)*, Publication No. 2 (1935), pp. 32–52.

*Rotating bodies of revolution.*

R. JONES, *Phil. Trans.* A, **226** (1927), 231–266; *A.R.C. Reports and Memoranda*, No. 1061 (1927).

WIESELSBERGER, *Physik. Zeitschr.* **28** (1927), 84–88.

## BOUNDARY LAYER CONTROL

### 230. Introduction. Suction.

RECOGNITION of the large increase in drag which is produced by separation of the boundary layer from the surface of a solid body has led to many attempts to prevent such separation. An account has been given in Chap. II, §18 of the way in which separation arises, and it has been shown in §24 that when the boundary layer is turbulent separation occurs farther back than when the layer is laminar. Indeed, if the tail of the body tapers slowly, so that the rising pressure gradient is sufficiently gradual for the rate at which energy can be absorbed from outside to balance the rate of loss in overcoming the pressure gradient and the frictional resistance, the layer adheres to the surface right up to the extreme rear. This occurs in stream-line bodies. In general, however, separation takes place somewhere behind the position of minimum pressure even when the boundary layer is turbulent, unless means are used to prevent it.

According to the physical concept of the boundary layer, it should be possible to delay or even to prevent separation by removing the accumulation of fluid, caused by the reversal of flow, in the region where separation normally develops. This possibility was recognized by Prandtl when he first put forward his theory, and his original paper† contains a brief description of an experiment which established the validity of his reasoning. He immersed a circular cylinder in a stream of water, with its axis perpendicular to the direction of flow. A slit cut in the surface of the cylinder, parallel to the axis, communicated directly with the hollow interior, to which suction was applied. Water was thus withdrawn from the boundary layer, and was discharged at one end of the cylinder. The position of the slit was well aft of the position at which the boundary layer would ordinarily have separated from the surface, but under the influence of the suction separation did not occur until the flow had passed the slit.

This first essay in boundary layer control was undertaken by

---

† *Verhandlungen des dritten internationalen Mathematiker-Kongresses, Heidelberg,* 1904 (Leipzig, 1905), pp. 484–491; reprinted in *Vier Abhandlungen zur Hydrodynamik und Aerodynamik*, L. Prandtl and A. Betz (Göttingen, 1927). See also *Journ. Roy. Aero. Soc.* **31** (1927), 729.

Prandtl—amongst other experiments—with the sole purpose of pro-
viding support for his theoretical arguments. He contented himself
with observing the phenomenon by making the flow visible: no
quantitative measurements of the drag of the cylinder were made,
nor was there any attempt to find the best position of the slit, or to
see whether the turbulent wake, which was appreciably reduced in
size by the single slit, could be entirely suppressed by the use of
several slits suitably spaced. There would seem, however, to be no
fundamental reason why this should not be feasible: Ackeret† cites
an instance in which an air-stream was deflected through 180° round
a cylinder by such means.

## 231. Introduction. The supply of additional energy to the boundary layer.

The prevention by suction of the thickening of the boundary layer
is not the only possible method of reducing the size of the eddying
wake. An alternative is to supply additional energy to the boundary layer, to enable it to proceed farther against an adverse
pressure gradient. An obvious method of doing this is the reverse
of the method of suction just discussed—namely, to introduce into
the boundary layer additional fluid carrying with it energy of motion
from a separate source. The moving fluid so introduced should be
discharged parallel to the surface in order that the best use may be
made of its kinetic energy.

The idea of discharging fluid into the boundary layer was not
investigated by Prandtl, but other workers have subsequently found
it effective (see § 236). The fluid can either be supplied by means of
a separate pumping system, or it can be conducted to the proper
points of discharge by connecting them, by means of internal ducts,
to regions of higher pressure on the body itself. An outstanding
instance of an effective application of the latter alternative, which
has obvious advantages from the operational standpoint, is furnished
by the slotted wing. Here the slot forms a short, well-shaped duct
from the high-pressure side of an aeroplane wing to a region of
high suction on the upper surface somewhat forward of the maximum camber, and the flow of air through the slot delays the stall
of the wing (which is, of course, a manifestation of separation)
to a considerably higher angle of incidence than that at which it

† *Zeitschr. des Vereines deutscher Ingenieure*, **70** (1926), 1155.

would otherwise occur. It should, however, be emphasized that although the wing slot undoubtedly delays the separation of the flow from the upper surface of a wing, it does so mainly by virtue of the powerful downwash proceeding from the forward slat. Its action in supplying additional energy to the boundary layer is probably very small by comparison. This important point of difference between the slotted wing and boundary layer control by discharge is discussed more fully in § 237.

Apart from the slotted wing, there is little experimental evidence as to the efficacy for the prevention of separation of the discharge of air taken in at regions of high pressure; but from the unpublished results of two different sets of tentative experiments made at the N.P.L. it appears unlikely that much can be done on these lines. The chief difficulty is to get enough air to the region of discharge, owing partly to the loss of head due to turbulence and to friction in the internal passages if these are not straight and short, and partly to the restricted area of the regions of high pressure available as sources for the air supply. The slotted wing is structurally much more favourable to this method of boundary layer control without auxiliary power than any other arrangement that has yet been tried.

### 232. Introduction. Motion of the solid wall.

If the solid boundary where the boundary layer normally forms is set in motion in the same direction as the fluid and at the maximum speed of the fluid, it is clear that no retardation of the flow by surface friction can occur. This was realized by Prandtl, who, about 1907,† showed that by rotating a cylinder immersed in a stream of water he could prevent separation on the side advancing with the fluid. (Actually, in order to avoid the accentuated separation on the side moving against the main stream, Prandtl used two cylinders rotating in opposite directions, with parallel axes and adjacent sides almost touching.)

### 233. The flow produced with boundary layer control.

Before proceeding to discuss practical results, we shall give some attention to the type of flow produced by boundary layer control. Since its effect is to prevent separation there is a tendency to infer

---

† *Die Naturwissenschaften*, **13** (1925), 100. See also *Journ. Roy. Aero. Soc.* **31** (1927), 729, 730.

that it produces the flow pattern of ideal fluid flow. Little thought
is, however, needed to show that this view is not correct: in the case
of ideal fluid flow the walls of the body past which the fluid is streaming are surfaces containing stream-lines, whereas with boundary layer
control, either by suction or by pressure, fluid flows through the
walls. Nevertheless, certain of the experimental results available
indicate that the flow does approach the potential type. Thus with
suction Schrenk,[†] Bamber,[‡] and Perring and Douglas[||] found that
the lift of an aerofoil section approached the theoretical value,
although in no case, even with the highest suction, did the measured
and theoretical lifts actually coincide: the former was always less
than the latter. The explanation of the approach to perfect flow,
in spite of the dissimilar conditions that must prevail, probably
lies in the fact that it is necessary to withdraw or introduce only
a small quantity of fluid to prevent separation of the boundary
layer.

Prandtl[††] and Schrenk[‡‡] have shown that when the suction apertures are arranged, as is customary, so that the indrawn air flows
normally through the boundary, there is a 'sink effect' superim-
posed on the effect due to the removal of the retarded air in the
boundary layer. Each suction aperture being treated as a sink, it
is possible to calculate the pressure distribution of the potential flow
with suction, and it is found to be of the form shown by the full lines
in the lower part of Fig. 217, the dashed curve representing the pres-
sure distribution for the undisturbed potential flow.[||||] It is assumed
in Fig. 217 that the suction is powerful enough to produce a stagnation point $P$ on the downstream side of the suction slit, and to draw
with it parts of the potential flow hitherto unaffected by friction.
Comparing the pressure distributions with and without suction we
see that in the former case the adverse pressure gradient has been
relieved on both the upstream and the downstream sides of the slit.
It is because this pressure effect is felt upstream of the slit that, as
shown experimentally by Schrenk, separation can be prevented by
suction at an aperture well downstream of the region where it would

† *Zeitschr. f. Flugtechn. u. Motorluftschiffahrt,* **22** (1931), 259–264.
‡ *N.A.C.A. Report* No. 385 (1931).
|| *A.R.C. Reports and Memoranda,* No. 1100 (1927).
†† *Aerodynamic Theory* (edited by Durand), **3** (Berlin, 1935), 116, 117.
‡‡ *Luftfahrtforschung,* **12** (1935), 10–27.
|||| Prandtl, *Aerodynamic Theory* (edited by Durand), **3** (Berlin, 1935), 117.

normally occur. On the other hand, if the suction aperture is situated downstream in this way, more air has to be removed to obtain the same effect.

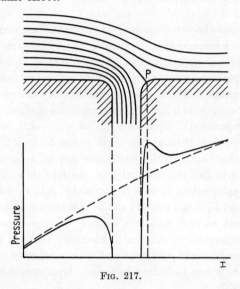

FIG. 217.

## 234. Pressure gradients and air quantities.

Using Pohlhausen's method (Chap. IV, § 60), Prandtl† has made an interesting investigation of the order of the pressure gradient against which a laminar boundary layer can flow without separation. He chooses as his initial conditions a boundary layer defined by $\Lambda = -10$ in Pohlhausen's equations (that is to say, a boundary layer that has developed nearly up to separation), and then considers what the maximum permissible pressure gradient can be if the value of $\Lambda$ is to remain unchanged downstream. He finds that the pressure gradient is given by

$$\frac{\delta(\partial p/\partial x)}{\frac{1}{2}\rho u_1^2} = \frac{20}{R_\delta},$$

where $u_1$ is the velocity of the potential flow just outside the boundary layer, $\delta$ is the boundary layer thickness, and $R_\delta$ is $u_1 \delta/\nu$. Since values of $R_\delta$ of the order of $10^3$ can easily occur in fluids of small viscosity, this expression shows that the allowable pressure gradient for laminar flow is very small (except at low Reynolds numbers).

† *Aerodynamic Theory* (edited by Durand), **3** (Berlin, 1935), 112–114.

Higher pressure gradients than the limiting value indicated above occur without separation when the layer has become turbulent, so that there is an interchange of energy between it and the free stream.

Prandtl[†] uses the same method to estimate the effects produced by suction on a laminar boundary layer. As an example, he assumes that the surface is porous and that the suction is adjusted over the region of increasing pressure in such a way that the boundary layer thickness remains constant. Taking the velocity distribution everywhere as the quartic in $y$ (the distance from the surface) which corresponds to the position of separation (i.e. to $\Lambda = -12$), he finds that $c$, the velocity of flow through the surface, has to be $2 \cdot 2(-\nu \, du_1/dx)^{\frac{1}{2}}$.

Applying this result to the particular case of a circular cylinder of diameter $d$ in two-dimensional flow, we find that the value of $c$ at the rear stagnation point is given by $c/U_0 = 4 \cdot 4R^{-\frac{1}{2}}$, where $R = U_0 d/\nu$ and $U_0$ is the velocity of the free stream at infinity. It is thus clear that we may expect a suitably arranged suction system to prevent separation of a laminar boundary layer in the presence of quite large pressure gradients and with a relatively small withdrawal of air. For a turbulent boundary layer conditions are even more favourable.

The final velocity distribution attained with suction on a surface of infinite extent can be found as follows from the general equations of motion. The analysis is due to Griffith and Meredith. The condition that the final profile has been reached is $\partial u/\partial x = 0$, and hence the equation of continuity reduces to $\partial v/\partial y = 0$. Thus $v$, the normal velocity due to suction, is constant. The equation of motion becomes

$$v \frac{\partial u}{\partial y} = \nu \frac{\partial^2 u}{\partial y^2},$$

which, when integrated with the proper boundary conditions, gives

$$u/U_0 = 1 - e^{vy/\nu}.$$

In this equation $v$ is negative, since the transverse velocity is directed towards the surface.

## 235. Control with auxiliary power. Diffusers.

The only application of boundary layer suction which we shall discuss before passing to a consideration of the results obtained with aerofoils, is the improvement of the efficiency of diffusers.† It is well known that too rapid a rate of expansion in such devices leads to serious loss of total pressure because they do not 'run full'. Defining the efficiency of a diffuser as the ratio of the difference in static pressure at throat and outlet to the difference between the kinetic pressures at these two sections, and taking into account the work done in suction, Ackeret found that the efficiency of a particular diffuser was increased from 50 per cent. without boundary layer removal to 81 per cent. by withdrawing, through one annular slit intermediate between throat and outlet, a quantity of water equal to 5 per cent. of the total quantity passing the throat. On the assumption that the efficiency of the pumping system was 75 per cent., the work required to remove this quantity of water was 3·4 per cent. of the kinetic energy at the throat of the diffuser. Ackeret suggested that this method of improving diffuser efficiency might have important applications in hydraulic or ventilating engineering. The results quoted were obtained with a small diffuser, only 14 mm. in diameter at the throat, and Ackeret pointed out that as the scale is increased the relative thickness of the boundary layer becomes less, so the results should be even more favourable on the larger scale. This relative decrease in thickness of the boundary layer as the scale becomes larger must be borne in mind generally in considering the application of model results to full-scale systems.

## 236. Control with auxiliary power. Suction or discharge under pressure at aerofoils.

As soon as the question of boundary layer control received any prominence, the improvement of wing efficiency and maximum lift appeared an attractive possibility, which was investigated by a number of workers. Many of the experiments, however, were of a tentative or exploratory nature and, being rather incomplete, need not receive detailed consideration here: references are given in the list on p. 549. We shall discuss only the two most comprehensive series of experiments of which an account has yet been published—those of Schrenk‡ and of Bamber.‖

† See Ackeret, *Zeitschr. des Vereines deutscher Ingenieure,* **70** (1926), 1155–1157. The deflexion of an air-stream through 180° was noted on p. 530, and reference is made in the list of additional references (p. 549) to some tests by Schrenk on spheres.

‡ *Loc. cit.* in footnotes † and ‡‡ on p. 532; the second paper includes some results for aerofoils with flaps.                    ‖ *Loc. cit.* in footnote ‡ on p. 532.

In what follows we shall frequently have to compare the two types of boundary layer control, one in which air is discharged under pressure through apertures in the upper wing surface, and the other in which it is withdrawn from the boundary layer by suction. For brevity we shall henceforth designate them as the 'pressure' and the 'suction' systems or types of control, respectively, basing this nomenclature on the sign of the pressure within the wing in relation to the static pressure in the free stream outside.

In order to judge the practical value of any improvements that may be effected in wing characteristics, we must take into account both the power required for the pumping system, and the reactions produced by the method of taking in the air in the pressure type of control and of discharging it in the suction type. For the method of intake or discharge we are entitled to select the most favourable scheme. Thus in the suction system the air sucked in is assumed to be discharged backwards in the direction of motion from some convenient place in the structure, and at the optimum velocity, which is shown by analysis[†] to be equal to $U_0$, the main air speed. Similarly for the pressure system the air is assumed to be taken in at a velocity $U_0$ by intakes pointing forward in the direction of motion. Further, in order to obtain results applicable to practical cases, the assumption is made that the wing, as forming part of an aeroplane, is to be driven through the air by means of an engine and airscrew, and the efficiency of the latter is taken as equal to that of the pumping system for the boundary layer control. In this way one obtains an 'effective drag coefficient' (denoted in the results to be quoted by $C'_D$) with which the drag coefficient $C_D$ without control has to be compared.

Turning now to practical results, we shall consider first those obtained by Schrenk.[†] In his work only the suction system was investigated. The model used was a very thick highly cambered aerofoil (see Fig. 218), which was one of a series of profiles developed by Kármán and Trefftz[‡] from theoretical considerations, so its potential flow pattern and theoretical lift were calculable. Its aerodynamic characteristics, however, were distinctly bad: in particular, the profile drag coefficient is stated to have varied between 0·05 and 0·10 (the latter, presumably, at the stall), so that its minimum value was about four times that of a good modern moderately thick

---

† Schrenk, *loc. cit.* in footnote † on p. 532.
‡ *Zeitschr. f. Flugtechn. u. Motorluftschiffahrt,* **9** (1918), 111–116.

wing section. The maximum lift coefficient of the section was 1·0. (It is probable that the experimental method in view had a decided influence on the choice of the section: it was thought desirable to

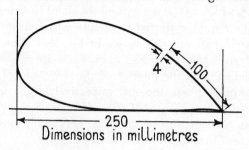

Dimensions in millimetres

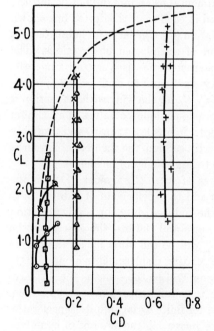

Key

| Points | V m./sec. | $C_Q$ | $C_p$ |
|---|---|---|---|
| + | 20 | 0·038 | 16·0 |
| × | 20 | 0·022 | 8·2 |
| △ | 30 | 0·024 | 8·0 |
| ⊡ | 30 | 0·014 | 3·9 |
| ⊠ | 20 | 0·011 | 2·0 |
| ⊙ | 30 | 0·005 | 1·0 |

Results corrected to infinite aspect ratio

Fig. 218.

accommodate the suction fans, of which there were two, and their driving motors inside the wing, and for this purpose a thick section was obviously necessary.) The aerofoil had an aspect ratio of 4 to 1 and was fitted with end plates. Its profile drag was measured by the pitot-traverse method (see Chap. VI, § 115).

Schrenk investigated various suction slits, including some with rounded edges, in several positions on the upper surface of his aerofoil. Best results were obtained with the sharp-edged slit shown in Fig. 218, which was 60 per cent. of the chord back from the leading edge.† The results, corrected to infinite aspect ratio (i.e. neglecting the induced drag) are also shown in Fig. 218. The quantities $C_p$ and $C_Q$ given in the key to Fig. 218 are non-dimensional coefficients of the suction $p$ in the interior of the wing and of the quantity of air flowing through the slit, respectively: they are defined by $C_p = p/(\frac{1}{2}\rho U_0^2)$ and $C_Q = Q/(U_0 S)$, where $U_0$ is the velocity of the air-stream, $S$ is the area of the plan form of the aerofoil, and $Q$ is the volume of the air flowing through the slit.

Schrenk states that the experimental procedure did not permit the optimum condition to be reached in each case; the dotted line shown in Fig. 218 is the estimated optimum $(C_L, C_D')$ curve for infinite aspect ratio obtained by extrapolation of the plotted observations. The improvement of the profile characteristics is very striking: the maximum value of $C_L/C_D'$ is between 40 and 50 at $C_L = 0.8$ approximately; the maximum value of $C_L/C_D$ for the wing without suction is not stated, but, judged by the previously quoted values of maximum lift and minimum drag coefficients, it was less than 20 and probably nearer 12. Fig. 219 shows how, with suction, the theoretical lift curve of the section is approached up to large angles of incidence.

In order to obtain the curves of total $C_D'$ against $C_L$ from the curves of Fig. 218, we have to add the coefficients of induced drag to the plotted values. Since the induced drag for a given section depends only upon the aspect ratio and the lift, the abscissa to be added at any given value of $C_L$ will be the same for all the curves shown in Fig. 218. Hence the relative displacements along the $C_D'$ axis of the curves for a finite wing will be the same as the displacements as plotted, which relate to infinite aspect ratio. It is, then, evident from these curves that, for finite as for infinite aspect ratio, there is nothing to be gained by increasing the amount of boundary layer absorption if a lower absorption suffices to prevent the stall:

† Later experiments by Schrenk (footnote ‡‡ on p. 532), which included some measurements with a flapped aerofoil, substantially confirmed the results here quoted, but showed that a rather lower equivalent drag coefficient could be obtained with a wider slit at the position on the aerofoil surface shown in Fig. 218. These later experiments also show that there is an optimum width of slit, which for the aerofoil of Fig. 218 with slit in the position shown was found to be about 3·8 per cent. of the aerofoil chord.

on the contrary, the effective drag coefficient is increased at a given lift. It is only when a given amount of absorption ceases to prevent the stall that there is any advantage to be gained by increasing the suction. The dotted line in Fig. 218 enables the most economical suction for any $C_L$ to be estimated.

The Reynolds numbers for the results shown in Figs. 218 and 219

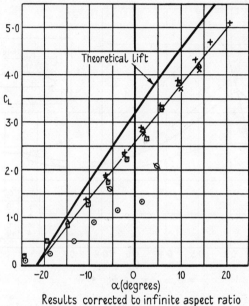

Results corrected to infinite aspect ratio

FIG. 219.

lay between $3\cdot4\times10^5$ and $5\cdot2\times10^5$. Neglect to work at a sufficiently high Reynolds number probably accounts for the comparatively poor results obtained by certain other experimenters, for it follows from general considerations that the results obtained from boundary layer control when the layer is turbulent should be better than those for a laminar layer. This was confirmed by some tests made by Schrenk at lower Reynolds numbers.

Bamber's experiments were made with a N.A.C.A. M. 84 aerofoil (whose shape is shown in Fig. 220), of 15 inches chord and 25·25 inches span; the Reynolds number was $4\cdot45\times10^5$. The aerofoil was tested between circular end plates, but this method of test does not by any means reproduce the characteristics of infinite aspect ratio when the model aerofoil is itself of low aspect ratio, and it is evident from an inspection of the results given by Bamber for the aerofoil without pressure or suction that induced drag was present. It may be recalled that Schrenk also tested his aerofoil between end plates, but measured the profile drag by the pitot-traverse method. There-fore, in comparing the results obtained by Bamber with those of

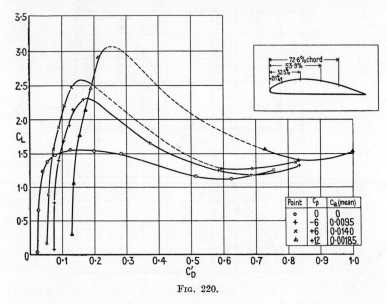

Fig. 220.

Schrenk we must remember that the former do not relate to infinite aspect ratio and are likely to be on this account, apart from any other possible causes, inferior to the latter.

Bamber investigated both the pressure and the suction systems, and found that on the whole the best results for either were obtained with a slit 53·9 per cent. of the chord back from the leading edge: this agrees roughly with the best position found by Schrenk, and it is probably not greatly different for any normal wing. The greatest slit width—of four tested by Bamber—was 0·67 per cent. of the chord, as compared with Schrenk's optimum width of about 3·8 per cent. It is probable, therefore, that Bamber might have obtained better results by using a wider slit: this view is supported, as he himself points out, by the type of curve obtained when his results are plotted on a base of width of slot.

The range of $C_p$ covered by Bamber's tests was from $-6$ to $+12$; the curves of $C_L$ against $C_D'$ for the best slit position and width (see above) are shown in Fig. 220 for $C_p = 12$, 6, and $-6$, and for the smooth section without a slit. At first sight it would seem from

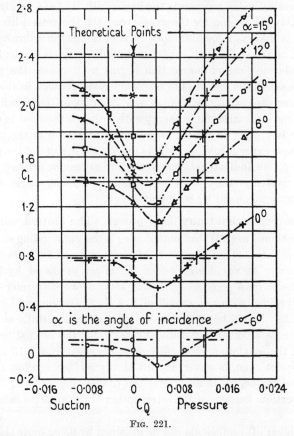

Fɪɢ. 221.

Fig. 220 that the pressure is more efficient than the suction system, but this conclusion is not justified from the evidence of these tests, since Bamber used a backward opening slit and did not alter its shape when he changed over from pressure to suction. It is evident from the drawings of the apparatus, and from the figures given for the quantities of air flowing under suction and pressure, that the slit offered a higher resistance to inward than to outward flow.

On the other hand, as is shown by Fig. 221 (reproduced with slight modification from the original report), it is possible by the use of the pressure system to reach a lift in excess of the theoretical value at a given incidence; whereas on the suction system Bamber's results confirm those of Schrenk and other investigators in suggesting that the theoretical value represents the highest lift that can be attained. Presumably the reason for the excess over the theoretical lift under pressure is that, as more and more air is blown out in a downstream direction on the upper surface of the aerofoil, a stage is reached when the circulation around the section begins to increase; the phenomenon ceases to be one of pure boundary layer control, in the sense here under discussion, as soon as the quantity of air flowing out through the slot carries more energy with it than is required to enable the boundary layer to adhere to the surface right up to the trailing edge. These experiments, and others mentioned in some of the papers listed on p. 549, indicate no limit to the lift that can be reached on the pressure system. They show also that the stalling angle can be considerably increased by both systems.

### 237. Control without auxiliary power. The slotted wing.

As mentioned in § 231, the slotted wing is the outstanding example of the application of automatic boundary layer control in a practically useful form. As an illustration of the effect produced by a well designed slot on a good modern wing section, we may refer to the results obtained with an aerofoil of R.A.F. 31 section† and plotted in Fig. 222. It will be seen that the slotted section stalls at about 23° incidence as compared with about 12° for the normal aerofoil, and that the maximum lift coefficient is increased from 1·15 to 1·82. With the slot open, however, the lift at any incidence below that at which stalling begins is lower and the drag is higher than for the normal section, because of the interruption of the surfaces caused by the slot.

Still higher lift coefficients can be obtained by using more than one slot. Handley Page‡ quotes a case in which a R.A.F. 19 aerofoil was divided into eight sections to give seven slots about equally spaced along the chord length. The maximum lift coefficient of the unslotted section (1·64 approximately) was thereby raised to about 3·4, and the stalling angle from 14° to 43°. With only six of the slots open the maximum lift coefficient was 3·92 at 45° stalling angle.

† Irving, Batson, and Williams, *A.R.C. Reports and Memoranda*, No. 1063 (1927).
‡ *Journ. Roy. Aero. Soc.* 25 (1921), 271.

As stated on p. 531 there is a distinct difference between the behaviour of a slotted wing and that of a wing whose boundary layer is
controlled either by pressure or by suction. With control by auxiliary
power the lift at an angle below the normal stall can be increased,

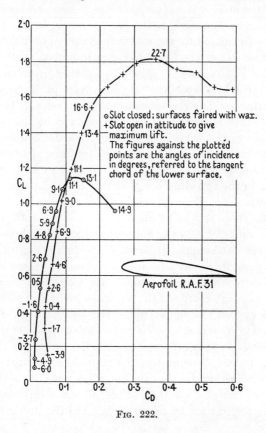

Fig. 222.

whereas the effect of the slot is mainly to continue the linear part of
the lift-incidence curve of the unslotted section to a higher incidence
with the same slope. At incidences below the stall the lift for a slotted
wing is about equal to, or slightly less than, that for the unslotted
section. This difference in behaviour may be partly related to a
difference in position between the slot and the pressure or suction
apertures hitherto tried: the slot is always at the extreme nose of the
main aerofoil, where the boundary layer is laminar and very thin,

whereas neither Bamber† nor Perring and Douglas† used a pressure slit nearer to the leading edge than 0·13 wing chords, where the boundary layer is appreciably thicker. Further, considerably more air flows through the slot than the quantity discharged in the tests of either of the investigators just mentioned. From measurements of velocity made by Townend‡ in the slot passage of R.A.F. 31, values of $C_Q$ can be calculated for the slot in the position to which Fig. 222 relates. These values are found to range from 0·05 at 6° incidence to 0·08 at 27°. The values of $C_Q$ used by Bamber (given on Fig. 221) do not exceed 0·019, while the maximum rate of discharge in the tests of Perring and Douglas corresponded to a value 0·014 of $C_Q$.

It is clear then that most of the air delivered by the slot to the upper surface of the aerofoil remains outside the boundary layer, whereas the air discharged through the pressure apertures in the above-mentioned experiments was mainly retained within the layer. The actions in the two cases are therefore not strictly comparable. Whereas the pressure discharge system is a case of pure boundary layer control in the sense here under discussion, the action of the slot seems to be rather different. By the admission of air through the slot the leading flap or slat becomes an aerofoil, the high lift on which is reflected in a powerful downwash which forces the air behind the slot to cling to the surface of the main aerofoil. The action is not so much one of supplying enough energy to the boundary layer to prevent it from leaving the surface, as one in which the boundary layer is, as it were, pressed into contact with the surface by the action of the stream outside, which has been deflected towards the surface by the action of the slat. Indeed, Townend in the paper already cited describes experiments which suggest that the boundary layer of a slotted wing separates well forward on the surface of the main wing, but is immediately forced back by the momentum of the stream outside.

## 238. Control without auxiliary power. The Townend ring.

Very similar in its action to the slat of a slotted wing is the Townend ring, which is extensively used to reduce the drag of aeroplane bodies or nacelles fitted with radial air-cooled engines. The ring is in effect an endless slat bent into circular form, and it is fitted near the nose of the body or nacelle, round the heads of the engine cylinders. Like the slat, the ring is of aerofoil section and is arranged

† *Loc. cit.* on p. 532.    ‡ *Journ. Roy. Aero. Soc.* **35** (1931), 713–718; 733, 734.

to be at a positive angle of incidence to the relative wind direction at the region where it is fitted. The ring therefore produces a down-wash, which reduces the spread of the disturbance caused by the cylinders projecting into the air-stream.

There are no reliable full-scale measurements of the reduction of drag produced by a Townend ring, but model experiments[†] have shown reductions up to 60 per cent. of the drag of the body and engine. A few tests made in the Compressed Air Tunnel at the N.P.L.[‡] show that the reduction observed in an atmospheric tunnel is maintained at high Reynolds numbers. It should be added that the effectiveness of the ring depends rather critically on its angular setting: in order to produce a large effect the incidence of the aerofoil section must be fairly large, but at the same time the section must not be stalled. Hence the useful angle range is somewhat restricted.

### 239. Moving surfaces. Rotating cylinders.

Apart from the visual experiments of Prandtl mentioned on p. 531, and some similar work carried out later by Prandtl and Tietjens,[‖] very few investigations seem to have been made on the effect on separation of moving surfaces. What information there is comes from work undertaken primarily with a quite different object, namely to investigate the cross-wind force or lift produced on a circular cylinder rotating about its axis in a stream of fluid. (The phenomenon was discussed theoretically in Chap. II, § 27.)

Measurements of this kind, as well as of drag, have been made on cylinders rotating in air by Reid[††] and by Betz.[‡‡] Their results are summarized in Figs. 223 and 224, in which the lift and drag coefficients are plotted against the ratio $V_0/U_0$, $V_0$ being the peripheral speed of the cylinder and $U_0$ the wind speed. Considering first the results for lift, we see that they agree in indicating that there is no lift for small rotational speeds when $V_0/U_0$ is less than about 0·5; lift then begins suddenly and rises as $V_0/U_0$ increases. The value of the lift coefficient that can be attained depends upon the end conditions. Betz remarks that two-dimensional flow is not achieved by

† Townend, *A.R.C. Reports and Memoranda*, No. 1267 (1930).
‡ Relf, *Journ. Roy. Aero. Soc.* **39** (1935), 6, 7.
‖ *Die Naturwissenschaften*, **13** (1925), 1049–1053.
†† *N.A.C.A. Technical Note* No. 209 (1924).
‡‡ *Zeitschr. des Vereines deutscher Ingenieure*, **69** (1925), 11–14. See also Prandtl, *Die Naturwissenschaften*, **13** (1925), 104–107.

having the cylinder spanning the tunnel completely, and accordingly he fitted to the cylinder end disks of diameter 1·7 times that of the cylinder (the aspect ratio of which was 4·7). These disks raised the maximum lift coefficient from just over 4 to about 9. Reid, with an

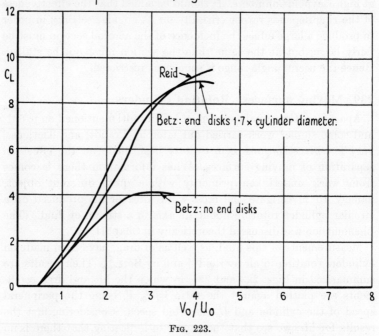

Lift on Rotating Cylinders

Reid: Aspect ratio 13·3; Reynolds number 3·9×10⁴ to 1·16×10⁵; no end disks.

Betz: Aspect ratio 4·7; Reynolds number 5·2×10⁴

FIG. 223.

aspect ratio of 13·3, and without end disks, obtained results as good as those of Betz with end disks on the shorter cylinder.

The results for drag (Fig. 224) suggest that Betz's conditions did not approach very closely to the two-dimensional, since the value of $C_D$ he obtained at zero rotational speed is considerably lower than that of a long cylinder at the Reynolds number of his experiments. Reid's results cannot be analysed in this way, as the Reynolds number at which his experiments were carried out at the lower

rotational speeds (viz. $1·16 \times 10^5$) is close to that at which a sharp
fall of drag occurs with a non-rotating cylinder.

As the rotational speed of the cylinder increases, there is at first

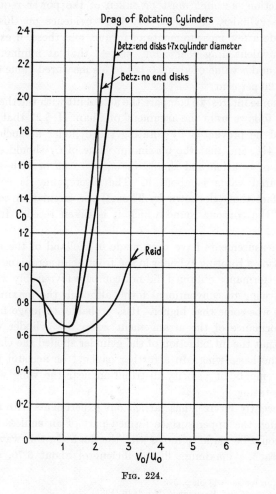

FIG. 224.

a fairly rapid drop in $C_D$, but after a minimum is reached the drag
rises steeply again. In both cases the minimum value of $C_D$ is between
one and a half times and twice that obtained with a non-rotating
cylinder at Reynolds numbers greater than that at which the sharp
fall in $C_D$ is complete (cf. p. 419). There is no experimental evidence

to show whether the drag coefficient of a cylinder could be further reduced by rotation at these high Reynolds numbers.

In order to estimate the true advantage of rotation as regards drag reduction, account must be taken of the power required to rotate the cylinder. There is very little evidence on this point: Reid[†] made a few measurements of torque, and these, if expressed as an equivalent drag coefficient, suggest that at minimum drag we should add a value of about $0\cdot124$ to his measured value of $0\cdot6$— i.e. about 20 per cent.

It is of some interest to compare the actual lift with the theoretical value. It follows from the argument of Chap. II, § 27 that separation would be prevented by rotating the cylinder at such a rate that $V_0 = 4U_0$, and that the maximum value of $C_L$ should occur at this speed of rotation and should be equal to $4\pi$ or $12\cdot6$, whereas the measured value is about 9. The discrepancy is connected with the fact that (as shown by Thom[‡]) the circulation continues to fall as the contour round which it is taken recedes from the cylinder.[||]

Some measurements have been made in Holland of the effect of incorporating a rotating cylinder in the front of an aeroplane wing,[††] but the performance obtained did not compare favourably with that of a good wing of conventional form, although the maximum lift coefficient was somewhat higher. It is probable, to judge from the poor performance of the arrangement when the cylinder was not rotating, that the introduction of the cylinder spoiled the characteristics of the basic wing. In a further paper[‡‡] an account is given of measurements of velocity distribution near the wing with the cylinder rotating.

More recently Favre[||||] has carried out experiments with an aerofoil in which the upper surface formed part of an endless band of silk and could be made to move with the fluid. In this way he was able to reach a maximum lift coefficient of about $3\cdot70$, and the

† *Loc. cit.* in footnote †† on p. 545.

‡ *A.R.C. Reports and Memoranda*, No. 1410 (1931).

|| For an attempt to simulate the lift-drag curves for a rotating cylinder by consideration of the motion of an inviscid fluid with a concentrated line-vortex in the fluid and a circulation round the cylinder, reference may be made to Bickley, *Proc. Roy. Soc.* A, **119** (1928), 151–156.

†† Wolff and Koning, *De Ingenieur*, **41** (1926), 181–190.

‡‡ van der Hegge Zijnen, *ibid.*, pp. 878–883.

|||| *Comptes Rendus*, **202** (1936), 634–636.

stalling angle was considerably delayed. The paper describing this work contains no information regarding the amount of power expended in moving the surface.

## ADDITIONAL REFERENCES

J. ACKERET, A. BETZ, and O. SCHRENK, Versuche an einem Flügel mit Grenzschichtabsaugung, *Vorläufige Mitteilungen der Aerodyn. Versuchs-anstalt zu Göttingen*, No. 4 (1925).

O. SCHRENK, Versuche an einer Kugel mit Grenzschichtabsaugung, *Zeitschr. f. Flugtechn. u. Motorluftschiffahrt*, **17** (1926), 366–372.

H. ZICKENDRAHT, Magnuseffekt, Flettner-Rotor und Düsenflügel, *Schweiz. Techn. Zeitschr.* **1** (1926), 837–843; 855–857.

H. ZICKENDRAHT and K. WIELAND, Propriétés aérodynamiques de surfaces portantes munies d'ajutages, *Arch. des Sciences Phys. et Nat. de Genève* (5), **8** (1926), 145–147.

K. WIELAND, Untersuchungen an einem neuartigen Düsenflügel, *Zeitschr. f. Flugtechn. u. Motorluftschiffahrt*, **18** (1927), 346–350.

F. SEEWALD, Die Erhöhung des Auftriebs durch Ausblasen von Druckluft an der Saugseite eines Tragflügels, *Zeitschr. f. Flugtechn. u. Motorluftschiffahrt*, **18** (1927), 350–353.

E. N. FALES and L. V. KERBER, Tests of Pneumatic Means for raising Airfoil Lift and Critical Angle, *Journ. Soc. Automotive Eng.* **20** (1) (1927), 575–581.

E. G. REID and M. J. BAMBER, Preliminary Investigation on Boundary Layer Control by means of Suction and Pressure with the U.S.A. 1927 Airfoil, *N.A.C.A. Technical Note* No. 286 (1928).

M. KNIGHT and M. J. BAMBER, Wind Tunnel Tests on Airfoil Boundary Layer Control using a Backward Opening Slot, *N.A.C.A Technical Note* No. 323 (1929).

F. HAUS, Portances élevées et profils hypersustentateurs, *L'Aéronautique*, **13** (1931), 125–131.

F. HAUS, L'Utilisation pratique des procédés d'hypersustentation, *L'Aéronautique*, **13** (1931), 205–213.

E. A. STALKER, Reduction of Drag by Boundary Layer Control, *Journ. Aero. Sciences*, **2** (1935), 66, 67.

R. B. MALOY, Engineering Applications of Boundary Layer Control, *Journ. Aero. Sciences*, **3** (1936), 407–409.

G. J. HIGGINS, An Airfoil fitted with a Slotted Flap, *Journ. Aero. Sciences*, **3** (1936), 431–433.

# XIII

# WAKES

## 240. Small Reynolds numbers.

FOR very small Reynolds numbers calculations may be made with the approximations of Stokes or Oseen.[†] The stream-lines obtained by the method of Stokes do not show a wake. The first stage in the development of a wake appears in the broadening out behind an obstacle (for example, behind a sphere[‡] or behind a circular cylinder[||]) of the stream-lines calculated by the method of Oseen. In this method it is assumed that, if the velocity components are written as $U_0+u$, $v$, $w$, where $U_0$ is the undisturbed velocity of the stream, then squares and products of $u$, $v$, $w$ and their derivatives may be neglected.[††] The effect of this is to destroy the symmetry; the motion is no longer reversible; the stream-lines behind the body broaden out, and the vorticity is appreciable only in a wake bounded by a vaguely defined surface of parabolic form.

Another method of calculation, due to Thom,[‡‡] which yields interesting results, is a numerical 'difference' method of solving the exact equations of motion of a viscous fluid, and is accurate up to and including third-order differences. The system is built up, rather laboriously, from the network of potential and stream-line curves appropriate to the boundaries considered, these curves being obtained from ideal fluid theory. The drag coefficients found by the method are in good agreement with experiment. Fig. 225, which is taken from Thom's paper, shows the stream-lines for flow past a circular cylinder at $R = 10$. ($R = U_0 d/\nu$, where $d$ is the cylinder diameter.)

[†] Lamb, *Hydrodynamics* (1932), §§ 335–343, pp. 594–616.

[‡] Oseen, *Ark. f. mat., astr. och fys.* **6** (1910), No. 29; *ibid.* **9** (1913), No. 16; *Vorträge aus dem Gebiete der Hydro- und Aerodynamik, Innsbruck,* 1922 (Berlin, 1924), pp. 127–129; Goldstein, *Proc. Roy. Soc.* A, **123** (1929), 225–235.

[||] Lamb, *Phil. Mag.* (6), **21** (1911), 120, 121; Bairstow, Cave, and Lang, *Phil. Trans.* A, **223** (1923), 383–432; Faxén, *Nova Acta Regiae Societatis Scientiarum Upsaliensis, Volumen extra ordinem* (1927), pp. 3–55. For flow past a flat plate along the stream, see Piercy and Winny, *Proc. Roy. Soc.* A, **140** (1933), 543–561; also Burgers, *Proc. Kon. Akad. v. Wetensch. Amsterdam,* **33** (1930), 1–9.

[††] The method may therefore be used to discover the nature of the flow at great distances (where $u$, $v$, $w$ would in any case be expected to be small) for a steady (non-turbulent) flow (see § 249). The restriction to very small Reynolds numbers arises when the method is employed up to the surface of the immersed solid body. The solution at a great distance in the wake is, however, of theoretical interest only, since in practice the motion is turbulent.

[‡‡] *Proc. Roy. Soc.* A, **141** (1933), 651–669.

PLATE 31

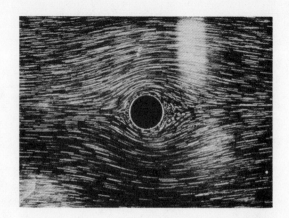

(a) $R = 3.9$

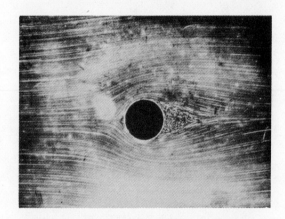

(b) $R = 18.6$

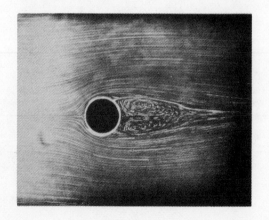

(c) $R = 33.5$

Thom's method can be applied over a fairly wide range of Reynolds numbers, but the amount of computation involved is very great.

The stream-lines for flow past a circular cylinder at $R = 20$, calculated by Thom, and a composite picture derived from actual flows at about the same Reynolds number, are shown in Figs. 26 and 27, Chapter II (p. 64). A vortex-pair has developed behind the cylinder. These figures should be compared with Pl. 29 (b), (c), (d), and

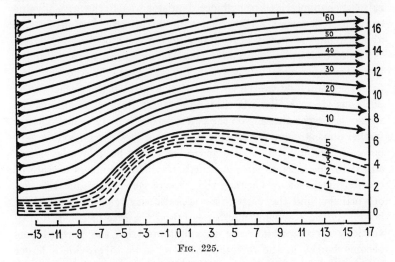

FIG. 225.

(e) of Chapter VI (facing p. 296), obtained by Fage[†] by ultramicro-scopic photography. The elongation of the vortices in the direction of flow with increasing Reynolds numbers is noticeable. Further photographs of flow past a circular cylinder are shown in Pl. 31,[‡] including one (Pl. 31a) at the very small Reynolds number of 3·9.

The state of affairs shown in the figures and plates referred to is characteristic of bluff bodies. As the Reynolds number is increased, the stream-lines widen out more and more. They form a closed region behind the obstacle, and within this region there is an inflow along the axis of the wake and a flow in the general direction of motion in the outer portions. This circulatory motion constitutes a vortex-pair. (The inflow along the axis of the wake, and the outflow, exist even at small values of $R$, but as $R$ increases the effect becomes larger and more noticeable.) The vortices become more and more elongated

---

[†] *Proc. Roy. Soc.* A, **144** (1934), 381–386.
[‡] F. Homann, *Forschung auf dem Gebiete des Ingenieurwesens,* **7** (1936), 1–10.

in the direction of flow. Above some critical Reynolds number (depending on the shape of the obstacle, the degree of turbulence in the main stream, and also on the proximity of the channel walls) the vortices become asymmetrical shortly after the beginning of the motion, leave the obstacle and move downstream. (See Chap. II, Pls. 7 and 8, and Pl. 32.†) Thom (*loc. cit.*), investigating the motion of a circular cylinder along the axis of a channel, found that when the ratio of channel width to cylinder diameter was 40 the critical value of $R$ was little more than 30, when it was 20 the critical value of $R$ was about 46, and when it was 10 eddies did not come off below $R = 62$.

The mechanism of the motion may be explained in terms of the boundary layer and vortex-layers. Separating the fluid in the wake from the main flow are the vortex-layers which left the surface at the positions of separation. At moderately small Reynolds numbers the vortex-layers, originating on the two sides of the body, come together farther downstream. Vorticity diffuses from the layers into the main body of the fluid, but is also generated in the boundary layer and added to the vortex-pair. At small Reynolds numbers a state of equilibrium is set up between the rates of generation and diffusion of vorticity, and the vortex-pair is stationary relative to the solid body. As the Reynolds number is raised the rate of generation of vorticity increases; in order to counteract this the vortex-layers become longer in the direction of flow, so as to provide a larger area from which vorticity can diffuse; at the same time the strength of the individual vortices in the vortex-pair increases. The vortex-layers are unstable; at some limiting Reynolds number they break away from the solid body and the vortices travel downstream.

It was mentioned in Chap. I, § 10, p. 39, that the motion of a concentrated vortex line-pair behind a circular cylinder in an inviscid fluid had been investigated mathematically by Föppl. If the vortex-pair is to remain stationary behind the cylinder the vortices must lie on the curve

$$2r \sin \theta = r - a^2/r$$

(Fig. 15, Chap. I); if the vortices are stationary at the points $(r, \pm\theta)$, then $\kappa$, the strength of the vortices, is given by the equation

$$\kappa = \pm 4\pi U_0 r(1 - a^4/r^4)\sin \theta,$$

† Homann, *loc. cit.* The photographs show the formation of the double row of vortices considered in § 242.

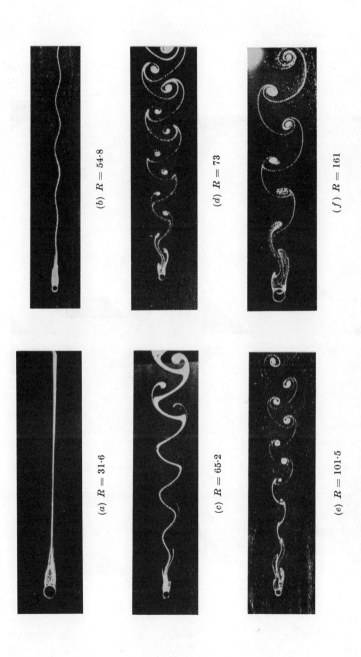

PLATE 32

(a) R = 31·6

(b) R = 54·8

(c) R = 65·2

(d) R = 73

(e) R = 101·5

(f) R = 161

PLATE 33

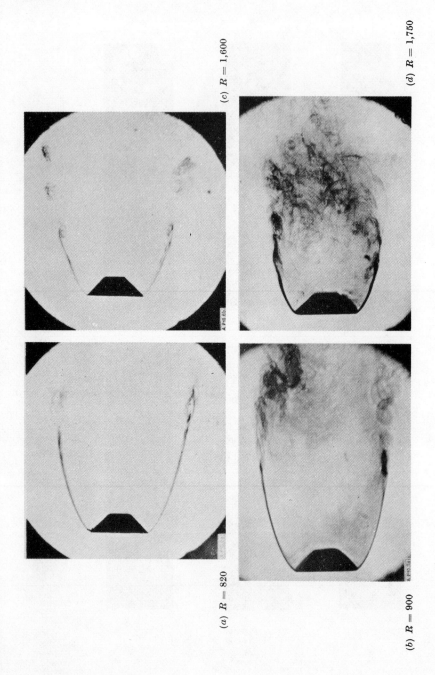

(a) $R = 820$

(b) $R = 900$

(c) $R = 1,600$

(d) $R = 1,750$

PLATE 34

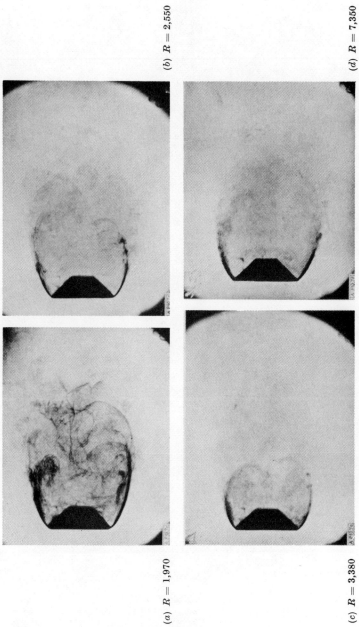

(a) R = 1,970

(b) R = 2,550

(c) R = 3,380

(d) R = 7,350

so that, if $\kappa$ increases, the vortices move away from the cylinder. This conclusion is verified experimentally by the elongation of the vortices. Moreover, the vortex-pair, although stable for disturbances symmetrical with respect to the axis of the wake, is unstable for anti-symmetrical disturbances. Except at quite low Reynolds numbers asymmetry sets in. Eventually the vortices move right away from the body.

## 241. Vortex-layers.

Smoke photographs of flow past a flat plate normal to the stream (two-dimensional motion) are shown in Pls. 33 and 34.† (The first and third were obtained with tin chloride, $SnCl_4$, and the others with titanium tetrachloride, $TiCl_4$.) The photographs show the thin vortex-layers which spring from the edges of the plate. (The Reynolds number $R$ is $U_0 b/\nu$, where $b$ is the breadth of the plate.) The vortex-layers break up nearer the plate as the Reynolds number increases. The shortening of the vortex-layers appears to take place gradually with increasing Reynolds number from $R = 1,600$ onwards; but it appears that a rather sudden shortening, from the forms shown in Pl. 33a and b ($R = 820$ and $900$) to that shown in Pl. 33c ($R = 1,600$) took place at about $R = 1,500$.

A comparison of the form of the layer as developed at $R = 820$ with the theoretical form of the surface of discontinuity on the Helmholtz-Kirchhoff-Rayleigh theory‡ is shown in Fig. 226.‖

In flow past such a body as a circular cylinder, 'free' vortex-layers, one on each side of the cylinder, originate in the boundary layer. Beyond each position of separation, the boundary layer is continued in the main body of the fluid as a 'free' vortex-layer. For plates, and also for wedges and ogival shapes having their greatest cross-sections at flat rear ends, separation takes place at the edges, from which the 'free' vortex-layers spring.

The first detailed experimental examination of the structure of these vortex-layers was made by Fage and Johansen,†† who, for two-dimensional motion, examined the layers from bodies of several shapes—flat plates, aerofoils, circular cylinders, wedges, and ogival

---

† Flachsbart, *Zeitschr. f. angew. Math. u. Mech.* **15** (1935), 32–37.

‡ Chap. I, § 10, p. 36; Lamb, *Hydrodynamics* (1932), pp. 94–105.

‖ Flachsbart, *loc. cit.* A comparison by Bryant and Williams (*A.R.C. Reports and Memoranda*, No. 962 (1925)), though not so clear cut, appears to present similar features.                    †† *Phil. Mag.* (7), **5** (1928), 417–441.

shapes. As is to be expected, the structures of the layers—for example, the velocity distribution across a section and the growth of the thickness of the layer—depend on the shape of the body, since, in particular, the initial characteristics of the layer depend on the characteristics of the boundary layer where it leaves the solid surface. Measurements of total head, and of velocity by means of a hot wire, were taken across sections of the layers, and the distribution of

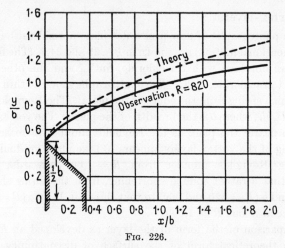

Fig. 226.

vorticity was deduced. The velocity distributions showed maxima at the outside edge of the sheet, and minima at the inside edge. The minimum is quite small just behind the body, and increases downstream. The maximum was in every case greater than the undisturbed stream velocity $U_0$, its ratio to $U_0$ just behind the body being 1·47, 1·42, 1·40, 1·13 for the flat plate at 90° incidence, the cylinder, the wedge, and the ogival model respectively. The maximum velocity decreased somewhat downstream in the case of the cylinder and the wedge; for the flat plate there did not appear to be any substantial change for half the plate breadth downstream, while the figures given for the ogival model show a slight initial increase. Somewhat similar results were found for the asymmetric models (aerofoil at 10°, 20°, 45° incidence, flat plate at 40° incidence).

The rates of discharge of vorticity are shown in the following table for the symmetrical models. Since the rate of discharge of vorticity may be affected by the constraint of the tunnel walls, the ratio of the width of the tunnel ($H$) to the breadth of the model ($b$) is also included.

|  | $H/b$ | Rate of discharge of vorticity $\div U_0^2$ |
|---|---|---|
| Flat plate .      .      .      . | 14 | 1·07 |
| Circular cylinder   .      .      . | 25 | 1·01 |
| Wedge .      .      .      .      . | 30 | 0·97 |
| Ogival model .      .      .      . | 30 | 0·64 |
| Long ogival model .      .      . | 30 | 0·71 |

Vorticity is shed from the two sides of an asymmetric model at the same rate. For the aerofoil at an incidence of 45°, this rate was just less than $U_0^2$; the rate decreased with incidence.

The flux of vorticity across a section of a layer decreases downstream.

If $q$ is the velocity along a layer and $\zeta$ the vorticity, the rate of discharge of vorticity is $\int q\zeta\, dy$ across a section of the layer. This is very nearly equal to half the difference of the squares of the maximum and minimum velocities, a relation which was verified by Fage and Johansen. On the Helmholtz-Kirchhoff-Rayleigh free stream-line theory for flow past a flat plate, the maximum velocity at the outside of the vortex-sheet would be $U_0$, and the rate of discharge of vorticity would be $\frac{1}{2}U_0^2$. It was actually about twice as much for the bluffer models.

It may be mentioned that the pressure in the outer boundary was slightly greater than that in the dead air region; and also that air was drawn into the layer through both boundaries, at a greater rate through the outer than through the inner boundary.

Fage and Johansen's measurements for the vortex-layers springing from the surface of a circular cylinder were performed at a Reynolds number between $2 \times 10^4$ and $3 \times 10^4$, and did not extend very far downstream. The variation with change of Reynolds number in the form of the layers from a circular cylinder has been studied by Schiller and Linke[†] in the range (roughly $3 \cdot 5 < \log R < 4 \cdot 5$) in which $C_D$ increases as $R$ increases (Chap. IX, Fig. 149). The results are shown in Fig. 227. If $d$ is the diameter of the cylinder and $\delta$ the thickness of the vortex-layer, it was found that, with $x$ measured downstream from a suitable origin (which was such that $x = 0$ for the points on the cylinder at 45° from the forward stagnation point), then at sufficiently

† *Zeitschr. f. Flugtechn. u. Motorluftschiffahrt*, **24** (1933), 193–198. See also Linke, *Physik. Zeitschr.* **32** (1931), 900–914.

low values of $U_0 d/\nu$ and $U_0 \delta/\nu$, $\delta$ was proportional to $x^{\frac{1}{2}}$. Since this law of growth is characteristic of laminar motion, it was assumed that the vortex-layers are laminar so long as this law holds, and the transition to turbulence was assumed to take place when this law broke

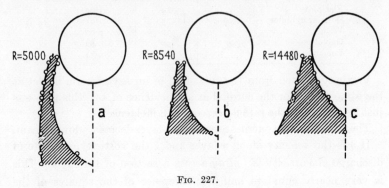

Fig. 227.

down. In this way, the following critical values of $U_0 \delta/\nu$ and $x/d$ were found:

$$\begin{aligned}
U_0 d/\nu &= 3{,}540 & 5{,}000 & & 6{,}090 & & 8{,}540 \\
(U_0 \delta/\nu)_{\text{crit}} &= 510 & 650 & & 730 & & 900 \\
(x/d)_{\text{crit}} &= 1\cdot41 & 1\cdot04 & & 0\cdot89 & & 0\cdot70
\end{aligned}$$

Thus the transition to turbulence approaches the cylinder as the Reynolds number increases. (The same feature is shown for a flat plate in Pls. 33 and 34. The effect on the drag coefficient was discussed in Chap. IX, § 187, p. 423.)

## 242. The double row of vortices.

For two-dimensional flow behind a bluff symmetrical obstacle at a Reynolds number above a certain critical value, the vortex-pair which is formed behind the obstacle at the beginning of the motion, after becoming more and more elongated in the direction of flow, takes up an asymmetrical position, and then moves away from the body. The asymmetrical arrangement alters the pressure distribution round the body, and vortices are discharged alternately from the two sides. In this way an eddying motion is set up which has a definite frequency for each Reynolds number.

Downstream the vortices assume what appears to be a regular pattern, which in most cases is evident at a distance of four or five diameters behind the solid body. The vortices arrange themselves in a double row, in which each vortex is opposite the mid-point of the interval between two vortices in the opposite row. In suitable circumstances the trail of vortices persists for a considerable distance behind the solid body.†

The vortices actually do not arrange themselves exactly on two parallel rows with a definite spacing ratio $h/a$ (that is, ratio of the lateral spacing $h$ between the rows to the distance $a$ between consecutive vortices in the same row), but the trail tends to widen out downstream, and the spacing ratio changes. For theoretical purposes the system far downstream was considered by Kármán‡ to be composed of isolated rectilinear vortices in two parallel rows. (The regularity shown in many photographs is often due, in part, to the effect of channel walls, which have a lateral compressing effect on the flow.)

At moderately high Reynolds numbers it is often impossible to speak of the formation of an initial vortex-pair. The vortex-layers which spring from the surface of the solid body are unstable, and roll up in such a way that vortices are formed at their ends and are shed alternately. An eddying motion of definite frequency is set up, and downstream the double row appears again.

At still higher Reynolds numbers the vortices diffuse so rapidly after their formation that it is no longer possible to speak of the formation of a double row. No definite value can be assigned to this upper limit, but from an investigation of the records of various experimenters it seems that no double row has been seen or photographed above $R = 2,500$. At the back of the solid body, however, vortices continue to be shed with unfailing regularity until $R = 4 \times 10^5$ or $5 \times 10^5$. Beyond this upper limit the flow is turbulent.

## 243. The double row of vortices. The calculation of the average drag.

If we consider the double vortex-trail behind a cylinder to extend without distortion to an infinite distance downstream, then by

---

† The double row of vortices behind an obstacle was first described by Mallock, *Proc. Roy. Soc.* A, **79** (1907), 262–265, and Bénard, *Comptes Rendus,* **147** (1908), 839–842. Bénard has returned to the subject in a number of papers in the *Comptes Rendus,*—**147** (1908), 970–972; **156** (1913), 1003–1005, 1225–1228; **182** (1926), 1375–1377, 1523–1525; **183** (1926), 20–22, 184–186, 379; **187** (1928), 1028–1030, 1123–1125, 1300.

‡ *Göttinger Nachrichten* (1911), pp. 509–517; (1912), pp. 547–556; Kármán and Rubach, *Physik. Zeitschr.* **13** (1912), 49–59.

considerations of momentum we can obtain a formula for the average drag force per unit length on the cylinder. We remark first that for a single infinite row of equidistant vortices, each of strength $\kappa$, at distances $a$ apart, with the origin at a vortex and the axis of $x$ along the row, the velocity components produced by the row are[†]

$$\left.\begin{aligned} u &= -\frac{\kappa}{2a}\frac{\sinh(2\pi y/a)}{\cosh(2\pi y/a)-\cos(2\pi x/a)}, \\ v &= \frac{\kappa}{2a}\frac{\sin(2\pi x/a)}{\cosh(2\pi y/a)-\cos(2\pi x/a)}. \end{aligned}\right\} \tag{1}$$

$u$ is an even periodic function of $x$, with period $a$. Its average value over a period is easily found to be $-\kappa/2a$ for $y$ positive and $+\kappa/2a$ for $y$ negative. The velocity components are exponentially small for large (positive or negative) values of $y$.

If now we have a double vortex-row which we consider to be infinite, then each vortex has no velocity due to the vortices in its own row, and has a velocity $(\kappa/2a)\tanh(\pi h/a)$ due to the vortices in the other row. Thus at a great distance downstream the whole vortex-street may be considered to be moving with a velocity[‡]

$$U_s = \frac{\kappa}{2a}\tanh\frac{\pi h}{a} \tag{2}$$

relative to the undisturbed fluid.

We now consider a cylinder fixed in a stream of velocity $U_0$. Then (Fig. 228) the vortex-street moves to the right with velocity $U_0-U_s$. The flow is periodic, with periodic time $a/(U_0-U_s)$ and frequency

$$N = \frac{U_0-U_s}{a}. \tag{3}$$

The average velocity at a fixed point over a time-period is the same as the average velocity at a given instant over an interval $a$ for $x$; since the average velocity for a single row is $\pm(\kappa/2a)$, the average velocity for the double row is zero outside the rows, and $\kappa/a$ between them. Thus (apart from the general streaming) there is an inward flux $\kappa h/a$ along the wake due to the vortex-street. The whole motion is assumed to be taking place in a fluid in which viscosity can be neglected, and, apart from the concentrated vortices, the flow is assumed to be a potential flow (except in the immediate neighbourhood of the cylinder). The flow at a great distance is composed of

[†] Lamb, *Hydrodynamics* (1932), § 156, p. 224.      [‡] Lamb, *op. cit.*, p. 225.

three constituents—the general streaming $U_0$, the velocity due to the vortex-street, and the remainder of the disturbance. There must be an average outflow $\kappa h/a$ to balance the inflow along the wake, and the leading part of the remainder of the disturbance must be a flow due to a source, and therefore of order $r^{-1}$ at a great distance, where $r$ is the distance from some point in the neighbourhood of the cylinder.

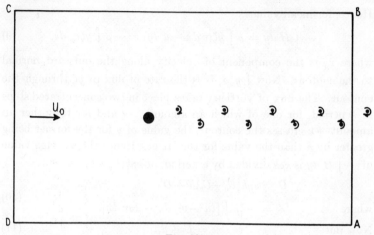

Fig. 228.

If now we consider a contour at a great distance from the cylinder and fixed relative to it, then the whole picture inside the contour is repeated after a period. We wish to calculate only the average drag, and the rate of change of the momentum inside the contour gives no contribution on the average. It may therefore be omitted from the momentum equation, which takes the form

$$D = -\int p \, dy + \rho \int u(v \, dx - u \, dy), \tag{4}$$

where it is assumed that average values are to be taken, the integrals are along the contour, $p$ is the pressure, $\rho$ the density, and $(u, v)$ the components of total velocity. Now put

$$p + \tfrac{1}{2}\rho(u^2 + v^2) = H. \tag{5}$$

Then $\qquad D = -\int H \, dy - \tfrac{1}{2}\rho \int [(u^2 - v^2) \, dy - 2uv \, dx]. \tag{6}$

If $\zeta$ is the vorticity, then from the equations of motion

$$\left.\begin{array}{l} \dfrac{\partial u}{\partial t}-v\zeta = -\dfrac{1}{\rho}\dfrac{\partial H}{\partial x}, \\[2mm] \dfrac{\partial v}{\partial t}+u\zeta = -\dfrac{1}{\rho}\dfrac{\partial H}{\partial y}. \end{array}\right\} \tag{7}$$

Also $\qquad -\displaystyle\int H\,dy = \int y\,dH = \int \left[y\dfrac{\partial H}{\partial x}\,dx + y\dfrac{\partial H}{\partial y}\,dy\right].$ $\qquad(8)$

Hence, for mean values,

$$-\int H\,dy = \rho\int y\zeta(v\,dx - u\,dy) = -\rho\int y\zeta v_n\,ds, \tag{9}$$

where $v_n$ is the component of velocity along the outward normal to the contour. Now $\int y\zeta v_n\,ds$ is the rate of flux of $y\zeta$ through the contour. The flux of vorticity takes place in two concentrated steps in a period, for one of which an amount $-\kappa$ and for the other an amount $+\kappa$ crosses the contour, the value of $y$ for the former being greater by $h$ than the value for the latter. Hence the average value of $-\int H\,dy$ is $\kappa\rho h$ divided by a period, or $\kappa\rho h(U_0-U_s)/a$; and[†]

$$\left.\begin{array}{l} D = \kappa\rho h(U_0-U_s)/a + D', \\[2mm] D' = -\tfrac{1}{2}\rho\displaystyle\int \left[(u^2-v^2)\,dy - 2uv\,dx\right]. \end{array}\right\} \tag{10}$$

where

Now put $\qquad\qquad\qquad u = U_0+u_1.$ $\qquad\qquad\qquad(11)$

Then $\qquad\qquad\qquad \displaystyle\int U_0^2\,dy = 0,$ $\qquad\qquad\qquad(12)$

and, since the total outflow out of the contour must be zero,

$$\int \left[(U_0+u_1)\,dy - v\,dx\right] = 0,$$

so that $\qquad\qquad\qquad \displaystyle\int [u_1\,dy - v\,dx] = 0.$ $\qquad\qquad(13)$

Hence $\qquad D' = -\tfrac{1}{2}\rho\displaystyle\int \left[(u_1^2-v^2)\,dy - 2u_1 v\,dx\right].$ $\qquad(14)$

The contour of integration must not pass through a vortex, but is otherwise at our disposal. We take it to be the rectangle $ABCD$ shown in Fig. 228. On $BC$, $CD$, and $DA$, $u_1$ and $v$ are of order $r^{-1}$; the integrand is of order $r^{-2}$, and when the contour is everywhere at an infinite distance the integral vanishes in the limit. On $AB$, put

$$u_1 = u_2+u_3, \qquad v = v_2+v_3, \tag{15}$$

where $(u_2, v_2)$ are the velocity components due to the vortex-street,

[†] J. Pérès, *Comptes Rendus*, **189** (1929), 1246–1248.

supposed the same as for an infinite street, and $(u_3, v_3)$ are the velocity components due to the rest of the disturbance, so that they are of order $r^{-1}$, and the integrals of their squares are zero in the limit and may be omitted. Hence

$$D' = -\tfrac{1}{2}\rho \int_A^B (u_2^2 - v_2^2)\, dy - \rho \int_A^B (u_2 u_3 - v_2 v_3)\, dy. \qquad (16)$$

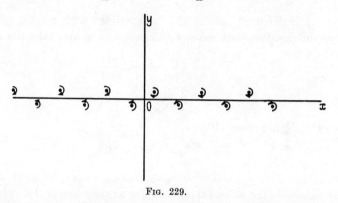

Fig. 229.

In the last integrand, $u_2, v_2$ are exponentially small when $|y|$ is large, and $u_3, v_3$ are of order $r^{-1}$. Hence, by correctly choosing a number $M$, the integrals along $AB$ from $y = M$ to $y = \infty$ and from $y = -M$ to $y = -\infty$ may be made as small as we like in absolute value, whilst between $y = +M$ and $y = -M$, the integrand is of order $r^{-1}$, and the integral from $-M$ to $+M$ may be made as small as we like by making $r$ large enough. The last integral therefore tends to zero, and

$$D' = -\tfrac{1}{2}\rho \int_A^B (u_2^2 - v_2^2)\, dy = -\tfrac{1}{2}\rho \Re \int_A^B (u_2 - iv_2)^2\, dy, \qquad (17)$$

where $\Re$ denotes 'the real part of'.

Now take a new origin half-way between a vortex in the bottom row and the nearest vortex to the right in the top row, as shown in Fig. 229. To begin with, take $AB$ along the new axis of $y$. With the new axes, it is easy to show (by adding the results for the two rows separately) that

$$\left.\begin{aligned}
u_2 - iv_2 &= \frac{i\kappa}{a}\, \frac{\sin(2\pi z_0/a)}{\cos(2\pi z_0/a) - \cos(2\pi z/a)}, \\[2mm]
\text{where}\qquad z &= x+iy, \qquad z_0 = \tfrac{1}{4}a + \tfrac{1}{2}ih.
\end{aligned}\right\} \qquad (18)$$

On the axis of $y$, $z = iy$, and hence

$$D' = \frac{\kappa^2 \rho}{2a^2} \mathfrak{R} \int\limits_{-\infty}^{\infty} \frac{\sin^2(2\pi z_0/a)}{\left[\cosh(2\pi y/a) - \cos(2\pi z_0/a)\right]^2} \, dy. \qquad (19)$$

Now

$$\int\limits_{-\infty}^{\infty} \frac{dy}{\cosh(2\pi y/a) - \cos(2\pi z_0/a)} = a \operatorname{cosec}(2\pi z_0/a)[1 - 2z_0/a]. \qquad (20)$$

We may differentiate under the integral sign with respect to $z_0$. If we differentiate both sides with respect to $z_0$ and take the real part, we find

$$D' = \frac{\kappa^2 \rho}{2a^2} \left\{ \frac{a}{\pi} - h \tanh \frac{\pi h}{a} \right\}$$

$$= \frac{\kappa^2 \rho}{2a\pi} - \frac{\kappa \rho h}{a} U_s \qquad (21)$$

from (2). Hence, from (10),

$$D = \frac{\kappa^2 \rho}{2a\pi} + \frac{\kappa \rho h}{a} (U_0 - 2U_s). \qquad (22)$$

Finally, since we are concerned with average values, we must show that we obtain the same value for $D'$ for all positions of $AB$ over an interval for $x$ of length $a$. This we do by writing $dz = i \, dy$, and substituting the value of $u_2 - iv_2$ from (18), in the integral in (17). If we have two positions of $AB$, we form a closed contour by drawing the lines parallel to the axis of $x$ which join the two positions of $AB$ at $y = \pm \infty$; the integrals taken along these two lines are zero, and by Cauchy's theorem in the theory of functions of a complex variable the difference in the values of the integrals along the two positions of $AB$ is zero if they do not enclose a vortex between them, and is $2\pi i$ times the residue of the integrand at the vortex if they do. This residue is found to be real; the difference in the integrals is purely imaginary, and by (17) the results found for $D'$ are the same, and any one of them is equal to the average value.[†]

† The formula for the average drag is due to Kármán, *Göttinger Nachrichten* (1912), pp. 547–556. See also Kármán and Rubach, *Physik. Zeitschr.* **13** (1912), 49–59. Reference may also be made to Villat, *Mécanique des Fluides* (Paris, 1930), pp. 131–140, and *Leçons sur les Tourbillons* (Paris, 1930), pp. 80–109. The part $\kappa \rho h(U_0 - U_s)/a$ of the formula for the average drag (see equation (10)) may also be obtained as the average rate of increase of momentum inside a contour fixed relative to the vortex-street, so that the body is advancing with velocity $U_0 - U_s$, and a new pair of vortices is added to the street in time $a/(U_0 - U_s)$. See also J. L. Synge, *Proc. Roy. Irish Acad.* **37** (1927), 95–109. The oscillating lift, with special reference to the amplitude of vibration of power lines in a steady wind, has been considered by R. Ruedy, *Canadian Journ. of Research*, **13** (1935), 82–92.

## 244. The double row of vortices. Stability. Drag.

It may be shown that, when subjected to a two-dimensional infinitesimal displacement, a double row of vortices of the character discussed above is unstable unless

$$\sinh(\pi h/a) = 1, \qquad h/a = 0.281. \tag{23}$$

In the calculation each vortex undergoes a small initial displacement parallel to itself.[†]

With the spacing ratio as in (23), $\tanh(\pi h/a) = 1/\sqrt{2}$, so that from equation (2)

$$\kappa = 2\sqrt{2}\,aU_s. \tag{24}$$

Hence with this spacing ratio we find from (22) that the drag coefficient $C_D$ is given by

$$C_D = \frac{D}{\frac{1}{2}\rho U_0^2 d} = \frac{a}{d}\left[1.59\frac{U_s}{U_0} - 0.63\left(\frac{U_s}{U_0}\right)^2\right], \tag{25}$$

$d$ being a representative length of the cylindrical obstacle. Now from (3)

$$\frac{U_s}{U_0} = 1 - \frac{Na}{U_0}, \tag{26}$$

and hence the theory provides a formula for the drag coefficient if the frequency, $N$, and the lateral spacing, $a$, are known. These two quantities (or two quantities from which they can be derived) must, however, be found by measurement.

In connexion with attempts to find further relations between the quantities involved we may note that the average rate at which vorticity passes downstream in one row of the vortex-street is $\kappa(U_0 - U_s)/a$. In § 241 we saw that for a very bluff obstacle the rate at which vorticity is discharged into the fluid along one vortex layer is nearly $U_0^2$, and also that the rate at which vorticity crosses a section of a vortex layer decreases downstream. The negative and positive vorticities mix to a considerable extent in the wake, and annihilate each other, and by the time the vortex-street is formed the rate of

---

[†] Kármán, *Göttinger Nachrichten* (1912), pp. 547–556; Lamb, *Hydrodynamics* (1932), pp. 225–229. In an earlier paper (*Göttinger Nachrichten* (1911), pp. 509–517) Kármán considered the stability when only one pair of vortices is displaced, the others being fixed, and found instability unless $\sinh(\pi h/a) = \sqrt{2}$, $h/a = 0.36$. A symmetrical double row, with the vortices in one row parallel to those in the other, will also move forward as a whole along the length, but is unstable. A double row of arbitrary stagger (with the vortices in one row neither exactly opposite, nor exactly half-way between, the vortices in the other row), although it will not move along its length, will keep its form unchanged. It is, however, unstable (Rosenhead, *Proc. Camb. Phil. Soc.* 25 (1929), 132–138).

passage of vorticity is, in some cases, reduced almost to $\frac{1}{2}U_0^2$.† For the case of a flat plate of breadth $b$ normal to the stream, Heisenberg‡ put simply $\kappa(U_0 - U_s)/a$ equal to $\frac{1}{2}U_0^2$ (which is the rate at which vorticity is shed from an edge of the plate according to the Helmholtz-Kirchhoff-Rayleigh free stream-line theory). From the preceding discussion we see that this may not be far from the truth in some cases, but the agreement is fortuitous. This equation and a second equation‖ also suggested by Heisenberg lead to values

$$U_s/U_0 = 0\cdot 2295, \quad a/b = 5\cdot 490, \quad Nb/U_0 = 0\cdot 140, \quad C_D = 1\cdot 82,$$

which are in fair agreement with representative values from observation. (The observed values differ considerably among themselves.) But (as pointed out by Prandtl in a note immediately following Heisenberg's paper) the assumptions underlying Heisenberg's theory are incorrect. Although the above values may be accepted as representative experimental values, there is in fact no theory from which we can deduce values for $U_s/U_0$ and $a/d$ (or $a/b$).

We now return to further considerations concerning the stability of a vortex-street. Even when the spacing ratio has the value given by (23), there will be a weak instability for certain finite disturbances— in particular, for a disturbance in which initially every alternate vortex in one row is moved the same distance perpendicular to the row, the positions of all the other vortices being unchanged.†† For certain particular initial displacements, in which alternate vortices in each row all have the same displacement (so that the whole street is divided into four sets of vortices), the later forms assumed by the street have been calculated.‡‡ These particular calculated forms of the street have not been observed.

A vortex-street is unstable for the two-dimensional infinitesimal disturbances considered by Kármán if the 'undisturbed' velocity is not uniform, but is greatest along the middle of the street.‖ Such

† Fage and Johansen, *Proc. Roy. Soc.* A, **116** (1927), 192, 193. (For hypotheses concerning the corrections to be applied for the constraint of the channel walls in these experiments see Glauert, *Proc. Roy. Soc.* A, **120** (1928), 34–46. Some of these corrections, in relation to the fully formed vortex-street, are considered below.)

‡ *Physik. Zeitschr.* **23** (1922), 363–366.

‖ Heisenberg puts the average inflow $\kappa h/a$ along the wake equal to $U_0 b$, on the assumption that when the plate is advancing in still fluid this is the inflow that would be required to maintain continuity. Actually the flow is too complicated to allow of any such deduction.

†† Schmieden, *Ingenieur-Archiv*, **7** (1936), 215–221.

‡‡ Schmieden, *ibid.*, pp. 337–341.

‖ Levy and Hooker, *Phil. Mag.* (7), **9** (1930), 489–502.

an effect might lead to instability if an obstacle is held stationary in a current in a channel; it will not enter if a model is towed through still fluid; and, if present, will be very weak for a wind-tunnel experiment in which the undisturbed velocity is nearly uniform across the working section.

The double row of vortices is unstable for a three-dimensional disturbance, the instability being caused by those components whose wave-length along the axis of the street is small.† (In considering the instability for a three-dimensional disturbance the effect of a vortex on itself enters, and in consequence the vortices cannot be taken to be concentrated along lines, but must have a certain thickness.) As a consequence of the instability the vortices become increasingly distorted. The distance between the rows either decreases or increases: in the first event the system always remains unstable and the double row disappears; in the second, a position of transitional stability is reached, the distance between the rows no longer tending to increase nor the vortices to become more distorted. The increase in the thickness of the vortices (arising in a real fluid from the action of viscosity) now becomes the dominating factor, and, as the vorticity spreads, the system, though always stable, tends to close up and disappear. In a general three-dimensional disturbance the pattern is always completely destroyed; it is only a matter of time before this occurs. Hence, on this account, at some distance behind the obstacle the vortices become increasingly distorted, and, if they do not at once come together and lose their identity, then the distance between the rows first increases and afterwards, as the vorticity diffuses, decreases so that the vortices come together.

Of the three instability effects mentioned the last is, in most cases, the most important.

In addition to the effect of instability the matter is further complicated, in the case of experiments in a channel, by the effect of the channel walls. In a real fluid, the effect of viscosity—in particular the diffusion of the vortices—may also lead to complications; whilst if experimental results are obtained from surface observations on water, surface tension will have some effect.

---

† Rosenhead, *Proc. Roy. Soc.* A, **127** (1930), 590–612. Three-dimensional disturbances have also been considered by Schlayer, *Zeitschr. f. angew. Math. u. Mech.* **8** (1928), 352–372.

### 245. The double row of vortices. The effect of channel walls.

The effect of the walls of a channel of breadth $H$ on the characteristics and stability of a symmetrically situated vortex-street have
been studied by Rosenhead.† We shall use dashed symbols to denote
the characteristics of the street in the channel, as distinct from those
in unlimited fluid. Then for the velocity of the street relative to
undisturbed fluid, Rosenhead finds

$$U_s' = \frac{\kappa'}{2a'}\left[ \tanh\frac{\pi h'}{a'} - \sum_{r=1}^{\infty} \frac{4e^{-2\pi rH/a'}}{1+(-1)^{r+1}e^{-2\pi rH/a'}} \sinh\frac{2\pi rh'}{a'} \right]. \quad (27)$$

Now the lateral spacing $a'$ may be about 5 or 6 times the breadth of
the obstacle, so that, unless the breadth of the obstacle is quite a
large fraction of the channel, this formula reduces very nearly to

$$U_s' = \frac{\kappa'}{2a'}\tanh\frac{\pi h'}{a'}. \quad (28)$$

As regards stability, if $a'/H < 0.815$, or $h'/H < 0.208$, the system
is unstable except for a definite value of $h'/a'$, which decreases from
$0.281$ to $0.256$ as $a'/H$ increases from 0 to $0.815$. For larger values of
$a'/H$ there is stability for a range of values of $h'/a'$, and if $a'/H > 1.419$
the system is always stable. If $a'/H < 0.815$, $h'/H < 0.208$, the
relevant value of $h'/a'$ for stability is given very nearly by

$$h'/a' = 0.281 - 0.090(a'/H)^6.$$

Thus, if the breadth of the obstacle is only a small fraction of the
width of the channel, the theoretical spacing ratio for stability is
almost unaltered.

If we now consider a vortex-street behind a cylindrical obstacle in
the channel, we first note that on account of the average inflow
$\kappa'h'/a'$ along the wake, if $U_0$ is the stream velocity far upstream, then
the general stream in the distant wake has a velocity $U_0+\kappa'h'/a'H$.
The vortices therefore pass downstream with a velocity

$$U_0+\kappa'h'/a'H-U_s',$$

and this is equal to $N'a'$, where $N'$ is the frequency.

Again, although the velocity of the street relative to the undisturbed fluid is given approximately by the same formula as in
unlimited fluid, and the spacing ratio is very nearly the same (unless

† *Phil. Trans.* A, **228** (1929), 275–329. Rosenhead has also shown that, if a double
row of vortices in a channel is to move forward as a whole, it must be symmetrically
situated with regard to the channel walls (*Proc. Camb. Phil. Soc.* **25** (1929), 277–281).

the breadth of the obstacle is a large fraction of $H$), yet the rate at which vorticity is discharged into the fluid will be different, and the spacing will be altered, so that $U'_s$, $h'$, and $a'$ will all be considerably affected by the value of the ratio of the breadth of the obstacle to $H$. By means of certain unproved hypotheses, Glauert† obtained estimates of these effects, but these estimates have not yet been sufficiently tested.

Finally, for the drag, if $a'/H < 0.815, h'/H < 0.208$, Rosenhead found the following approximate formula (representing to within $\frac{1}{2}$ per cent. the results of exact calculations):

$$D = \frac{\kappa'\rho h'}{a'}[U_0 - 2U'_s(1-c_1)] + \frac{\kappa'^2\rho}{2\pi a'}(1-c_2) + \frac{6\kappa'^2\rho h'^2}{4Ha'^2},$$

where

$$c_1 = \frac{H}{h'}\frac{\cosh(\pi h'/a')}{\sinh(\pi H/a') + \sinh(\pi h'/a')},$$

$$c_2 = \frac{4\pi H}{a'}\frac{\cosh^2(\pi h'/a')}{[\sinh(\pi H/a') + \sinh(\pi h'/a')]^2}$$
$$- \frac{\pi}{2}\frac{H}{a'}\left\{\frac{1}{\sinh^2[\pi(H+h')/2a']} - \frac{1}{\cosh^2[\pi(H-h')/2a']}\right\}.$$

(29)

When the width of the channel is large compared with the breadth of the obstacle, if we take $h'/a' = 0.281$ and the relation between $U'_s$ and $\kappa'$ to be given by (28), and neglect $c_1$ and $c_2$, we obtain the simpler approximate formula due to Glauert (loc. cit.):

$$C_D = \frac{a'}{d}\left[1.59\frac{U'_s}{U_0} - 0.63\left(\frac{U'_s}{U_0}\right)^2\right] + 24\left(\frac{U'_s}{U_0}\right)^2\frac{h'^2}{dH},$$

(30)

where $d$ is a representative length of the obstacle. (Cf. equation (25).) On the further unproved assumptions made by Glauert concerning the effect of the walls on $U'_s$, $a'$, and $h'$, it is found that the presence of the walls causes an increase in $C_D$ of approximately $32(U_s h/U_0 d)^2(d/H)$, or about $0.90(d/H)$ for a circular cylinder of radius $d$, and $4.00(b/H)$ for a flat plate of breadth $b$ normal to the stream.

## 246. The double row of vortices. Spacing and spacing ratio.

In order to estimate drag from (25) or (30), observations of frequency and lateral spacing are needed. In addition to drag, the most obvious observations to make to test the theory are those giving the spacing ratio $h/a$. In a real fluid the vortices are diffused, and the measurements are difficult, especially those of $h$.

† Proc. Roy. Soc. A, **120** (1928), 34–46.

The double row of vortices has been the subject of a large number of experiments. Generally it has been found that $h/a$ is larger than its theoretical value (0·28) and increases downstream, and that $a$ stays constant downstream, the increase in $h/a$ being due to an increase in $h$. It appears also that experiments carried out under apparently almost similar conditions sometimes do, but sometimes do not, lead to almost identical results. This fact is a strong indication of instability, probably rather weak, and probably also arising in the main from a three-dimensional disturbance. The suggestion of instability is supported by the graphs of the paths of the vortices, plotted by Rosenhead and Schwabe.†

It is also found that, as the Reynolds number increases, the longitudinal and lateral spacings decrease, while the ratio of the velocity of the vortex-street (relative to the fluid) to the velocity of the cylinder (relative to the fluid) increases—although, apart from the instability effects, these quantities probably reach constant values for sufficiently high Reynolds numbers.‡ Further, in Rosenhead and Schwabe's experiments on flow past a circular cylinder of diameter $d$ in a channel of breadth $H$, the ratios of the dimensions of the vortex-street to $d$ decreased, and the ratio of the velocity of the vortex-street to the velocity of the cylinder (relative to the fluid) increased (at a fixed Reynolds number), as $d/H$ increased.

Kármán and Rubach,‖ in their original experiments, found the values 0·28 and 0·30 for the ratio $h/a$ in the developed part of the street behind a circular cylinder and behind a flat plate respectively, corresponding to a Reynolds number ($U_0 d/\nu$ or $U_0 b/\nu$, where $d$ is the cylinder diameter and $b$ the plate width) of about 2,000. For the cylinder $a/d$ was about 4·27 and $U_s/U_0$ about 0·135 on the average, leading to $C_D = 0·87$ and $Nd/U_0 = 0·203$. The value of $C_D$ is rather low, and the value of $Nd/U_0$ rather high. (See Fig. 149, Chap. IX.) For the plate, $a/b$ was about 5·6, and $U_s/U_0$ about 0·203, leading to the value $C_D = 1·66$, which is also low. (See Chap. IX, footnote‖ on p. 423.) The average value of $Nb/U_0$ was 0·144.

In Fage and Johansen's experiments†† on a flat plate, on the other hand, when the plate was normal to the stream, $h/a$ increased

† *Proc. Roy. Soc.* A, **129** (1930), 115–135.

‡ Rosenhead and Schwabe, *loc. cit.*; Richards, *A.R.C. Reports and Memoranda*, No. 1590 (1934); *Phil. Trans.* A, **223** (1934), 279–301. The Reynolds numbers of these experiments were all well below 1,000.

‖ *Physik. Zeitschr.* **13** (1912), 49–59.  †† *Proc. Roy. Soc.* A, **116** (1927), 170–197.

from 0·25 at a distance $5b$ downstream to 0·38 at $10b$ and 0·52 at $20b$. The experiments were carried out in a tunnel only 14 times the breadth of the plate; the value of $Nb/U_0$ was 0·146, the measured value of $C_D$ was 2·13 and of $a/b$ was 5·25. If we assume that (29) or (30) is correct with the theoretical spacing ratio (0·281 very nearly), then $a'/b$ and $U_s'/U_0$ may be calculated from this drag equation and the equation

$$N'a' = U_0 + \kappa' h'/a' H - U_s'$$

together with the measured values of $C_D$ and $N'b/U_0$. The calculated value of $a'/b$ is definitely greater than the measured value 5·25, being about 5·7.†

The increase in the ratio $h/a$ found by Fage and Johansen is rather extreme, but other experimenters have found results in the same direction. Behind a circular cylinder, Tyler‡ found $h/a = 0·308$, with a gradual increase downstream; in the experiments of Rosenhead and Schwabe (*loc. cit.*) on a circular cylinder in a channel the spacing ratio in the regular part of the wake was 0·32 when $h$ was less than $\frac{1}{3}$ of the channel width $H$, but became 0·45 when the cylinder diameter was $\frac{2}{3}H$; whilst behind an elliptic cylinder of fineness ratio 6:1, according to the experiments of Richards,∥ the initial value of $h/a$ near the cylinder was 0·32, a typical value where the regular features were fully developed was 0·42, and the ratio increased slowly farther down the wake.

The wake behind a flat plate at incidences other than 90° has been examined by Fage and Johansen (*loc. cit.*); the wake behind plates and aerofoils at various incidences has also been studied by Tyler (*loc. cit.*). The ratio of the longitudinal spacing to the breadth facing the stream ($a/(b \sin \alpha)$, where $\alpha$ is the incidence) does not appear to vary much for incidences above about 30° or 40°.

An interesting suggestion has been made by Hooker.†† On account of viscosity, the vortices diffuse as they pass downstream. For a diffused vortex alone, the velocity at the centre falls to zero as soon as diffusion starts. But when account is taken of the velocity due to the other vortices, the velocity at the centre of any one vortex is not zero relative to the undisturbed fluid; the point of zero velocity

---

† For comparisons of the measured drag coefficients with calculated values obtained by means of additional assumptions, reference may be made to the papers of Rosenhead and Glauert cited in § 245.          ‡ *Phil. Mag.* (7), **11** (1931), 849–890.
∥ *A.R.C. Reports and Memoranda*, No. 1590 (1934); *Phil. Trans.* A, **223** (1934), 279–301.          †† *Proc. Roy. Soc.* A, **154** (1936), 67–89.

is farther away than the centre of the vortex from the axis of the vortex-street. Nor is the vorticity symmetrically distributed in any one vortex; it is concentrated in that portion of the vortex which is nearer the axis of the street than the point of zero velocity. Hooker suggests that the usual methods of determining the centres of the vortices seriously overestimate the distance between the rows; in particular, when measurements of photographs are made, the point of zero velocity is usually taken as the centre. Hooker estimated the increase in the apparent distance between the vortex-rows due to this cause for his own experiments, and found that after allowance had been made for this effect, the spacing ratio was very near the theoretical value for a very considerable distance downstream. Eventually, however, there was a further increase, not to be accounted for in this manner.

If the residual divergences, after allowance has been made for the effects mentioned by Hooker, are largely or wholly due to instability, we should expect the drag formula, obtained by momentum considerations, to be still valid. (Measured quantities have to be inserted therein, and difficulties of measurement, of course, remain.) There is not much really satisfactory evidence, but the formula seems to give a drag coefficient at any rate of the right order.

## 247. The frequency with which vortices are shed.

The note emitted by wires in a wind is connected with the frequency of the eddies discharged behind them. Relf and Ower[†] verified experimentally that the frequency of the note heard is the frequency of the eddies generated. The curve of $Nd/U_0$ against $R$ ($= U_0 d/\nu$) obtained by Relf and Simmons has been given in Chapter IX (Fig. 149). (They did not detect periodic motion below $R = 100$, but Richardson,[‡] using delicate apparatus, was able to detect it down to about $R = 34$, which is also about the value obtained by Thom[||] by visual means.) For the rising part of the curve at small Reynolds numbers, Tyler[††] suggests the formula

$$Nd/U_0 = 0{\cdot}198(1-19{\cdot}7/R),$$

† *A.R.C. Reports and Memoranda*, No. 825 (1923).
‡ *Proc. Phys. Soc.* **36** (1923), 153–165.
|| *A.R.C. Reports and Memoranda*, No. 1373 (1931). The critical Reynolds number depends both on the proximity of boundaries and on the turbulence in the main stream.
†† *Phil. Mag.* (7), **11** (1931), 849–890.

which agrees well with the formula suggested by Rayleigh,[†]

$$N d/U_0 = 0.195(1-20.1/R),$$

as a result of his analysis of the original observations of Strouhal.[‡]

Values of the frequency behind a flat plate at various incidences have been found by Fage and Johansen (loc. cit., § 246), and behind plates and aerofoils at various incidences by Tyler (loc. cit., § 246) and by Blenk, Fuchs, and Liebers.[||] All agree that if $b'$ is the breadth of the obstacle across the stream, $Nb'/U_0$ is nearly constant for incidences above about 30°. There are discrepancies between the actual values. For plates, Fage and Johansen found a mean value about 0.148, Tyler 0.158, and Blenk, Fuchs, and Liebers about 0.18. Tyler gives the average value 0.150 for his aerofoils (less than for plates), and Blenk, Fuchs, and Liebers the average value 0.21 for theirs (greater than for plates).[††]

## 248. Steady flow in a wake. The flat plate at zero incidence.

Very far downstream in a wake, if the motion were steady, the assumptions and approximations of the laminar boundary layer theory would be valid. The pressure would be nearly constant across a section of the wake, the transverse velocity small in comparison with the longitudinal velocity, and the rate of change of the longitudinal velocity along the axis of the wake small compared with the rate of change across a section. The effective Reynolds number, in fact, would be $u_0 x/\nu$, where $u_0$ is the undisturbed stream velocity and $x$ is measured downstream from an unspecified origin in the neighbourhood of the obstacle. Moreover, in an unlimited fluid, the pressure gradient along the axis of the wake would also be negligibly small, and the equation of motion would take the form

$$u\frac{\partial u}{\partial x} + v\frac{\partial u}{\partial y} = \nu\frac{\partial^2 u}{\partial y^2}. \qquad (31)$$

But at a very large distance downstream, $u$ would be nearly equal to $u_0$ and $v$ would be small. Thus, if we put

$$u = u_0(1-w), \qquad (32)$$

say, then a first approximation to $w$ satisfies the equation

$$u_0\frac{\partial w}{\partial x} = \nu\frac{\partial^2 w}{\partial y^2}. \qquad (33)$$

[†] Phil. Mag. (6), 29 (1915), 433–533; Scientific Papers, 6, 315–325.
[‡] Ann. der Phys. u. Chem. (Wiedemann's Ann.), 5 (1878), 216–251.
[||] Luftfahrtforschung, 12 (1935), 38–41.
[††] Their value for a circular cylinder (0.207) was also rather high.

Considerations of momentum show that the drag $D$ per unit breadth on the obstacle is given by

$$D = \rho u_0 \int (u_0 - u)\,dy = \rho u_0^2 \int w\,dy, \qquad (34)$$

where the integral is taken right across a section of the wake. Since the effective Reynolds number is $u_0 x/\nu$, according to boundary layer theory the breadth of the wake will increase as $x^{\frac{1}{2}}$. It follows from (34) that $w$ must be of order $x^{-\frac{1}{2}}$. In accordance with boundary layer theory we therefore introduce the variable

$$\eta = (u_0/2\nu x)^{\frac{1}{2}}y, \qquad (35)$$

and put

$$w = Ax^{-\frac{1}{2}}f(\eta). \qquad (36)$$

The boundary conditions are $w \to 0$ as $y \to \infty$, and, for a symmetrical wake, $\partial w/\partial \eta = 0$ when $\eta = 0$. The solution is

$$w = Ax^{-\frac{1}{2}}e^{-\frac{1}{2}\eta^2}, \qquad (37)$$

where $A$ is a constant. From (34) we find that

$$\frac{D}{\rho u_0^2} = A\left(\frac{2\nu}{u_0}\right)^{\frac{1}{2}} \int\limits_{-\infty}^{\infty} e^{-\frac{1}{2}\eta^2}\,d\eta = 2A\left(\frac{\pi\nu}{u_0}\right)^{\frac{1}{2}}. \qquad (38)$$

Consider now the wake behind a flat plate of length $l$ at zero incidence. From Chap. IV, § 53, equations (50) and (51),

$$D = 1\cdot328\rho u_0^2 l(u_0 l/\nu)^{-\frac{1}{2}},$$

and hence

$$\left.\begin{array}{l} A = \dfrac{0\cdot664}{\sqrt{\pi}}l^{\frac{1}{2}}, \\[2ex] u/u_0 = 1 - \dfrac{0\cdot664}{\sqrt{\pi}}\left(\dfrac{l}{x}\right)^{\frac{1}{2}}e^{-\frac{1}{2}\eta^2}, \end{array}\right\} \qquad (39)$$

where $\eta$ is given by (35).†

The first approximation to $v$ is found from the equation of continuity and the condition $v = 0$ at $\eta = 0$. It is of order $x^{-1}$. The second approximation to $w$, for which a term of order $x^{-1}$ must be added to (37), has been found by Goldstein,‡ who joined the solution so found (and valid for sufficiently large values of $x$) to a solution he had previously obtained|| for the velocity immediately behind the plate. This previous solution gives $u/x^{\frac{1}{2}}$ as a series of ascending powers of $x$, with coefficients which are functions of $y/x^{\frac{1}{2}}$ (suitably made non-dimensional), for small values of $y/x^{\frac{1}{2}}$; and gives $u$ as a series of ascending

† Tollmien, *Handbuch der Experimentalphysik*, **4**, part 1 (1931), 269.

‡ *Proc. Roy. Soc.* A, **142** (1933), 545–560.

|| *Proc. Camb. Phil. Soc.* **26** (1930), 18–30.

powers of $x^{\frac{1}{2}}$, with coefficients which are functions of $y$, for large values of $y/x^{\frac{1}{2}}$ (where $x$ is now measured downstream from the trailing edge). It was possible to calculate $u/u_0$ from this solution for a distance of $0\cdot 5\,l$ behind the trailing edge along, and near to, the axis, $y = 0$, and for a distance of $0\cdot 36\,l$ for other values of $y$. This

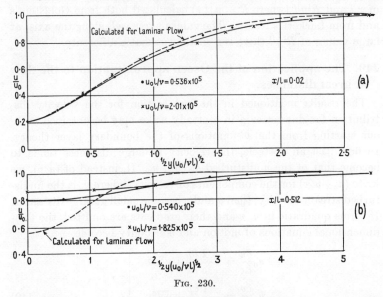

FIG. 230.

solution was joined smoothly to the solution for large $x$ along the axis, and it was found that in (37) the origin for $x$ must be taken at a distance $0\cdot 52\,l$ upstream from the trailing edge. For non-zero values of $y$, interpolation between the values of $u/u_0$ given by the two solutions was employed, and so a complete solution for $u/u_0$ behind a flat plate was found. In Fig. 230 this solution is compared with velocity distributions measured at two sections behind a plate by Fage and Falkner, the sections being at distances $0\cdot 02\,l$ and $0\cdot 512\,l$ behind the trailing edge.[†] There is fair agreement at the former section; at the latter the measurements show that the motion has become turbulent at both the Reynolds numbers of measurement.

The flow behind a flat plate at zero incidence, for $x/l \leqslant 1\cdot 0$ ($x$ measured from the trailing edge), has also been calculated by Luckert[‡]

† Fage, *Proc. Roy. Soc.* A, **142** (1933), 560–562.

‡ *Berlin Dissertation*, 1933; *Schriften des Math. Seminars und des Instituts f. angew. Math. der Universität Berlin*, **1** (1934), 245–274.

by the use of Mises's method. (Chap. IV, §§ 50 and 59, pp. 126, 127, 155.) Luckert tabulates $1-(u/u_0)^2$ against the (non-dimensional) stream-function for various values of $x/l$. The tables of Goldstein and of Luckert have been put into comparable form by Rosenhead and Simpson,[†] whose paper contains tables of $u/u_0$ at constant intervals of $x/l$ and $R^{\frac{1}{2}}y/l$ (where $R = u_0 l/\nu$) calculated both from Goldstein's and from Luckert's results. The values agree well along the axis; at large values of $R^{\frac{1}{2}}y/l$ there is a small systematic difference.

### 249. The application of the Oseen approximation to the flow at great distances.

The results mentioned in the last section for the velocity distribution far downstream in a steady wake may be obtained without starting from the assumptions of the boundary layer theory. In flow past an obstacle, the velocity at large distances tends to become that of the undisturbed stream $u_0$. If, instead of $(u, v)$, we write $(u_0+u, v)$ for the components of velocity, where $u_0$ is the undisturbed stream velocity, then $u$ and $v$ become small at large distances. If terms quadratic in $u$, $v$ and their gradients are omitted, the two-dimensional equations of motion and continuity become

$$\left.\begin{aligned}
u_0 \frac{\partial u}{\partial x} &= -\frac{1}{\rho}\frac{\partial p}{\partial x}+\nu\nabla^2 u, \\
u_0 \frac{\partial v}{\partial x} &= -\frac{1}{\rho}\frac{\partial p}{\partial x}+\nu\nabla^2 v, \\
\frac{\partial u}{\partial x}+\frac{\partial v}{\partial y} &= 0,
\end{aligned}\right\} \tag{40}$$

where $p$ is the difference of the pressure from the hydrostatic pressure, and $\nabla^2$ stands for the operator $\partial^2/\partial x^2+\partial^2/\partial y^2$. A general solution of these equations can be expressed in the form[‡]

$$\left.\begin{aligned}
p &= -\rho u_0 \frac{\partial \phi}{\partial x}, \\
u &= \frac{\partial \phi}{\partial x}+\frac{e^{kx}}{2k}\left(\frac{\partial \chi}{\partial x}-k\chi\right), \\
v &= \frac{\partial \phi}{\partial y}+\frac{e^{kx}}{2k}\frac{\partial \chi}{\partial y},
\end{aligned}\right\} \tag{41}$$

[†] *Proc. Camb. Phil. Soc.* **32** (1936), 385–391.

[‡] Lamb, *Hydrodynamics* (1932), pp. 610–615. It may be proved that this form gives the most general solution for two-dimensional motion.

where
$$k = u_0/2\nu, \tag{42}$$

and
$$\left. \begin{aligned} \nabla^2\phi &= 0, \\ (\nabla^2 - k^2)\chi &= 0. \end{aligned} \right\} \tag{43}$$

Consider now a symmetrical wake. Then the most general expressions for $\phi$ and $\chi$ are

$$\left. \begin{aligned} \phi &= A_0 \log r + \sum_{n=1}^{\infty} A_n \frac{\cos n\theta}{r^n}, \\ \chi &= \sum_{n=0}^{\infty} C_n K_n(kr)\cos n\theta, \end{aligned} \right\} \tag{44}$$

where $(r, \theta)$ are polar coordinates, $A_0$, $A_n$, $C_n$ are constants, and $K_n$ is the modified Bessel function which, for large values of $kr$, has the asymptotic expression†

$$K_n(kr) \sim \left(\frac{\pi}{2kr}\right)^{\frac{1}{2}} e^{-kr}. \tag{45}$$

Now the vorticity $\zeta$ is equal to $e^{kx} \partial\chi/\partial y$; and $\zeta$ and those terms in $u$, $v$ which depend on $\chi$ will, for large values of $kr$, all contain the factor $\exp[-k(r-x)]$ or $\exp[-kr(1-\cos\theta)]$. They are therefore insensible for large $r$ outside the region where $\theta$ is small, of order $(kr)^{-\frac{1}{2}}$, i.e. (since $\theta$ is approximately $y/r$, and $r/x$ approximately 1 in this region) outside the region where $y$ is of order $k^{-\frac{1}{2}}x^{\frac{1}{2}}$. This vaguely parabolic region is the wake.

If now we find the components of velocity in polar coordinates from (41), and approximate for large $kr$ and small $\theta$, we find that the part of the radial velocity $v_r$ which depends on $\chi$ is

$$-\left(\frac{\pi}{2kr}\right)^{\frac{1}{2}} e^{-kr(1-\cos\theta)} \sum_{n=0}^{\infty} C_n, \tag{46}$$

whilst the most important term arising from $\phi$ is $A_0/r$. Hence $v_r$ is of order $(kr)^{-\frac{1}{2}} \sum_{n=0}^{\infty} C_n$ inside the wake, and of order $A_0/r$ outside the wake. The breadth of the wake is of order $r^{\frac{1}{2}}$ and the velocity of order $r^{-\frac{1}{2}}$, so that, apart from the general streaming, there is a finite inflow along the wake, which balances the outflow $2\pi A_0$. This inflow is found to be $(\pi/k) \sum_{n=0}^{\infty} C_n$, so that there is the necessary relation

$$A_0 = \frac{1}{2k} \sum_{n=0}^{\infty} C_n \tag{47}$$

between the constants.

† Watson, *Theory of Bessel Functions* (Cambridge, 1922), p. 202.

Consideration of the outflow of momentum through a large circuit shows that the drag force per unit breadth on the obstacle is $2\pi\rho u_0 A_0$, i.e. $\rho u_0$ times the inflow along the wake.

A similar analysis may be carried through for the asymmetrical case. $\phi$ and $\chi$ now contain terms in $\sin n\theta$, and $\phi$ also contains a multiple of $\theta$, corresponding to a circulation. Considerations of momentum show that there is a lift force on the obstacle, whose magnitude per unit breadth is $\rho u_0$ times the circulation round a large circuit which is at right angles, where it cuts the wake, to the undisturbed velocity $u_0$.[†]

Along the wake, where $\theta$ is small and nearly equal to $y/x$, and $x/r$ very nearly equal to 1,

$$kr(1-\cos\theta) \doteqdot \tfrac{1}{2}kx\theta^2 \doteqdot \tfrac{1}{2}ky^2/x = u_0 y^2/(4\nu x) = \tfrac{1}{2}\eta^2, \qquad (48)$$

where $\eta$ is defined in equation (35). Hence (46) is a multiple of $x^{-\frac{1}{2}}e^{-\frac{1}{2}\eta^2}$, in agreement with (37), which was obtained by applying the approximations of the boundary layer theory in the wake.

A second approximation to the velocity distribution at large distances has been obtained, without the approximations of the boundary layer theory, by Filon,[‡] and it has been shown by Goldstein[||] that, for the symmetrical case, the results in the wake reduce to those derived by the aid of the boundary layer approximations if a term due to the pressure gradient $\partial p/\partial x$ is included. This term is of order of magnitude $(u_0 x/\nu)^{-\frac{1}{2}}$ times the first approximation (as given by (37), for example).

In the asymmetrical case the effect of the circulation is to displace the wake from the symmetrical position it occupies on the Oseen theory, and at an infinite distance downstream this displacement is logarithmically infinite. As a result, an attempt to calculate the moment on the obstacle in the asymmetrical case by using the second Oseen approximation and considering the outflow of angular momentum through a large contour leads to failure, giving a logarithmically infinite result.[††]

A similar analysis may be carried out for a three-dimensional wake. When there is symmetry about an axis, this is a comparatively

---

† Cf. Filon, *Proc. Roy. Soc.* A, **113** (1926), 7–27.

‡ *Phil. Trans.* A, **227** (1928), 93–135.

|| *Proc. Roy. Soc.* A, **142** (1933), 563–573.

†† Filon, *loc. cit.* (*Phil. Trans.*), and Goldstein, *loc. cit.*

simple matter, since the equations (41) generalize to

$$
\left.\begin{aligned}
p &= -\rho u_0 \frac{\partial \phi}{\partial x}, \\[6pt]
u &= \frac{\partial \phi}{\partial x} + \frac{e^{kx}}{2k}\left(\frac{\partial \chi}{\partial x} - k\chi\right), \\[6pt]
v &= \frac{\partial \phi}{\partial y} + \frac{e^{kx}}{2k}\frac{\partial \chi}{\partial y}, \\[6pt]
w &= \frac{\partial \phi}{\partial z} + \frac{e^{kx}}{2k}\frac{\partial \chi}{\partial z},
\end{aligned}\right\}
\tag{49}
$$

where
$$
\left.\begin{aligned}
\nabla^2\phi &= 0, \\
(\nabla^2 - k^2)\chi &= 0,
\end{aligned}\right\}
\tag{50}
$$

$k$ is given by (42), and $\nabla^2$ now stands for the operator
$$
\partial^2/\partial x^2 + \partial^2/\partial y^2 + \partial^2/\partial z^2.
$$

Then $\phi$ and $\chi$ can be expanded in Legendre polynomials. In $\chi$ the coefficients are multiples of $(kr)^{-\frac{1}{2}}K_{n+\frac{1}{2}}(kr)$, where $K$ is the same Bessel function as in (44) and (45). It is found that the breadth of the wake is of order $r^{\frac{1}{2}}$, and the inflow velocity along it of order $r^{-1}$, so that, apart from the general streaming, there is again a finite inflow along the wake, and a balancing outward flux in the irrotational motion outside the wake. The drag force on the obstacle is $\rho u_0$ times the inflow along the wake.[†] Moreover, the inflow velocity along the wake reduces to a multiple of $x^{-1}\exp(-u_0 y^2/4\nu x)$, and this result can be more simply obtained by applying the boundary layer approximations (compare Chap. IV, § 51, p. 130, equation (32)), as is done in § 248 for the two-dimensional case.

The general three-dimensional flow is naturally much more complicated, especially as, in the general case, the velocity arising from the potential of the irrotational part of the motion has a discontinuity in the wake which is cancelled by the rotational part. The drag result given above remains correct, and the lift force can be related to the circulation taken in distant circuits.[‡]

## 250. Three-dimensional wakes behind bluff obstacles.

The wake behind a solid body in three dimensions is in some ways similar to the two-dimensional wake, but there are striking differences.

[†] Goldstein, *Proc. Roy. Soc.* A, **123** (1929), 216–225.

[‡] Goldstein, *ibid.* **131** (1931), 198–208; Garstang, *Phil. Trans.* A, **236** (1936), 25–75. The latter paper contains a very full analysis by the method of expansion in harmonic functions.

At very small Reynolds numbers the stream-lines form a wake whose width ultimately varies as $r^{\frac{1}{2}}$ (§ 249). With increasing Reynolds numbers the stream-lines behind the body broaden out, and a stationary region of retarded fluid behind the body is separated from the main flow by a vortex-sheet. In this region of retarded fluid there is a permanent circulation, corresponding to a vortex-ring. Stanton and Marshall[†] found the vortex-ring at Reynolds numbers of about 5 behind circular disks in flowing water. When the Reynolds number $R$ is increased, the dimensions of the vortex-ring increase—especially longitudinally—until a critical value of $R$ is reached. At this stage the transverse diameter of the ring seems to be about one and a half times the diameter of the disk. Stanton and Marshall found that the vortex-ring behind a circular disk broke down at about $R = 195$. The critical value found by Simmons and Dewey[‡] for air flow past circular disks was about 100; the difference is probably due to different amounts of initial turbulence in the flows. When the Reynolds number exceeds the critical value, an oscillating disturbance of the surface of the vortex-ring becomes visible, with the result that successive portions of its substance are discharged downstream at regular intervals of time which depend on the rate of flow and the dimensions of the disk.

In investigating the wakes behind various obstacles experimenters have noticed some definite periodicity in the records of the velocity at points behind the obstacle, but have been unable to observe any uniformity or periodicity in the shape of the vorticity discharge. It was thought that the inability to make the vortices visible might be due to imperfection in experimental technique, and theoretical attempts were made to generalize the two-dimensional vortex-street to three dimensions.

A theoretical generalization of the two-dimensional double row is a series of vortex-rings, but a system of rings inclined to the wake axis would drift away from it. The only possible configuration of vortex-rings would be a row of rings with planes at right angles to the wake axis, and Levy and Forsdyke[||] have proved this configuration unstable.

Rosenhead[††] has suggested that the only possibility is a sequence

† *A.R.C. Reports and Memoranda*, No. 1358 (1932); *Proc. Roy. Soc.* A, **130** (1930), 295–301.                    ‡ *A.R.C. Reports and Memoranda*, No. 1334 (1931).
|| *Proc. Roy. Soc.* A, **114** (1927), 594–604.
†† Appendix to *A.R.C. Reports and Memoranda*, No. 1358 (1932).

PLATE 35

Loop discharge of vorticity behind a circular disk.
Simultaneous photographs in two perpendicular
directions

of irregularly shaped vortex-loops. Some photographs by Eden† show the gradual change in the form of the wake behind a flat plate inclined to the incident stream. (The introduction of a plate of any other shape, or of any other obstacle, altered only the degree of irregularity.) The photographs exhibit motion both in water and in air, and show clearly stream-line flow for small Reynolds numbers, and distorted surfaces of discontinuity with discharge of irregularly shaped vortex-rings at definite intervals at higher Reynolds numbers. Photographs by Stanton and Marshall‡ furnish evidence of a still more definite nature, since the wake was photographed simultaneously from two directions at right angles (Pl. 35). The oscillatory nature of the disturbance on the vortex-sheet was noted. There were two principal planes, one ($A$) defined by the axial line of flow and the line touching the crests of the vortex-loops, and the other ($B$) at right angles. The orientation of the plane ($A$) with respect to the obstacle producing the wake is purely a matter of chance; it depends on the position of the point in the vortex-sheet behind the body where the disturbance first becomes big enough to make part of the vortex-sheet break away, thus beginning the process of forming a vortex-loop. When this plane coincides with, or is at a small angle with, the photographic field of vision, the photographs show the shapes of a loop when seen end on and from above. When the principal planes make an appreciable angle with the field of vision, the shapes in the photographs are composite ones of the two main aspects of the loop. The fact that all the photographs belonged to one or other of these two groups supports the suggestion that the discharge consisted of vortex-loops with the orientation of the principal planes purely a matter of chance. Photographs by Simmons and Dewey (*loc. cit.*) of the flow in air provide additional confirmation of a succession of distorted vortex-rings moving along the axis in the wake. No systematic investigation of the periodicity of the discharge under different conditions appears to have been made.

At higher Reynolds numbers, the vortex-sheet and the vortex-loops diffuse very rapidly. A sheath of discontinuity along which corrugations travel in an irregular manner is usually seen immediately behind the body. At still greater Reynolds numbers the flow becomes completely turbulent.

† *Advisory Committee for Aeronautics, Reports and Memoranda*, No. 58 (1912), figs. 15–20.    ‡ *Loc. cit.*

### 251. The wake behind an aerofoil of finite span.

The Lanchester-Prandtl theory of flow past an aerofoil of finite span was sketched in Chap. I, § 12, where the rolling-up of the trailing vortex-sheet was also mentioned.

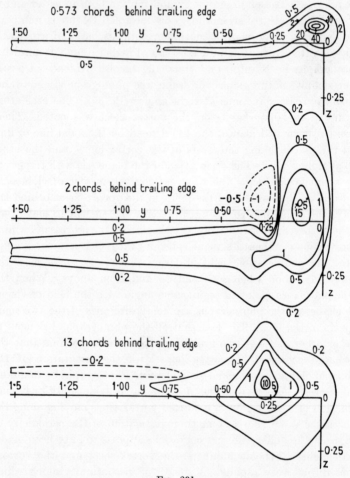

Fig. 231.

The flow in the wake behind an aerofoil of aspect ratio 6 has been investigated by Fage and Simmons.† Measurements of the speed and wind direction were made in three transverse planes behind the

† *Phil. Trans.* A, **225** (1925), 303–330.

aerofoil, at distances 0·573, 2·0, and 13·0 chords from the trailing edge, and in one plane in front of the aerofoil, 0·5 of a chord forward of the leading edge. It was verified that the total strength of the vorticity leaving a semi-span of an aerofoil, as obtained by integration over a transverse plane close behind it, was equal to the circulation round the median section; and that the variation of the component of vorticity in the direction of the stream was closely connected with the distribution of lift along the span. Within the limits of experimental error, the flow was irrotational in the plane of measurement in front of the aerofoil and in the regions beyond the wing tips behind the aerofoil. The velocity at right angles to the transverse planes was approximately constant in the planes behind the aerofoil, so that in each of the planes the flow relative to the longitudinal velocity was two-dimensional.

The diagrams in Fig. 231 give contour lines of constant vorticity component in the direction of the stream, and show the rolling-up of the trailing vortex-sheet. (The axis of $y$ is along the trailing edge of the aerofoil, with the origin at the tip and the median section at $y = 3$. The axis of $y$ is horizontal, and the undisturbed stream velocity is also horizontal, and at right angles to the axis of $y$. The axis of $z$ is vertically downwards.) A progressive inward displacement of the vorticity core is noticeable. Immediately behind the aerofoil the vorticity is concentrated in a band which increases in depth and intensity towards the aerofoil tips; at two chords behind, the total strength of the vorticity is unchanged, but the weak region at the centre has disappeared and the distribution at the tip has become less intense and is spread over a larger area. At thirteen chords behind, the rolling-up of the vortex band is almost complete; and the total measured vorticity is only about 82 per cent. of that leaving the aerofoil.

## 252. The turbulent wake behind a symmetrical cylinder.

When the boundary layer at the surface of a solid body becomes turbulent before it leaves the surface, the motion in the wake is turbulent. But even at lower Reynolds numbers, when a 'free' laminar layer of vorticity exists behind the solid body, the motion at some distance downstream becomes turbulent. (For the flow behind a flat plate at zero incidence, see § 248, especially Fig. 230. Even when a vortex-street is formed behind the obstacle, the motion sufficiently far downstream is completely turbulent.)

The velocity distribution at a considerable distance downstream in the wake behind a symmetrical cylinder (two-dimensional motion) may be calculated on either the momentum transfer or the vorticity transfer theory, with Prandtl's assumption for the apparent turbulent coefficient of diffusion and some further assumption concerning the mixture length $l$. (Chap. 5, §§ 81, 82, 83.) As in § 248 (where nonturbulent motion was considered), the assumptions and approximations of the boundary layer theory may be applied, and the pressure gradient $\partial \bar{p}/\partial x$ may be neglected, the axis of $x$ being taken along the axis of the wake, with the origin at some unspecified point in the neighbourhood of the obstacle. If $U_0$ is the undisturbed stream velocity and $(U_0 - U, V)$ the components of mean velocity in the wake, then for a first approximation terms of the second order in $U, V$ and their derivatives may be omitted. Hence

$$U_0 \frac{\partial U}{\partial x} = -\frac{1}{\rho} \frac{\partial \tau}{\partial y}, \qquad (51)$$

where $\tau$ is the shearing stress. (Cf. equation (33), § 248, for nonturbulent motion.) On the momentum transfer theory

$$\tau = -\rho \epsilon \, \partial U / \partial y, \qquad (52)$$

where $\epsilon \rho$ is the apparent turbulent coefficient of diffusion, and $\epsilon$ is denoted in Chapter V by $\overline{v(h_2 - h_1)}$ or $l'v$. (Chap. V, equations (20) and (22).) Prandtl's assumption for $\epsilon$ is

$$\epsilon = l^2 \left| \frac{\partial U}{\partial y} \right|. \qquad (53)$$

Prandtl now assumes† (i) that at sufficiently high Reynolds numbers or at a sufficient distance downstream there is geometrical and mechanical similarity in different sections of the wake, and that the values of $l$ at corresponding points of two different sections are proportional to the breadths, $2y_0$, of the sections, (ii) that $l$ is constant over any one section. If $l$ and $y_0$ are proportional to $x^m$, we put $\eta = y/x^m$ (so that $\eta$ is proportional to $y/y_0$), and $U/U_0 = f(\eta)/x^n$. Then $\epsilon$ is of order $x^{m-n}$, $\tau$ of order $x^{-2n}$, $\partial \tau/\partial y$ of order $x^{-(m+2n)}$, and $\partial U/\partial x$ of order $x^{-(n+1)}$. In order that both sides of (51) should be of the same order in $x$, we must have $n+1 = m+2n$. Moreover, if $D$ is the

† *Verhandlungen des 2. internationalen Kongresses für technische Mechanik, Zürich*, 1926, pp. 62–74.

drag per unit breadth on the obstacle, we find from considerations of momentum that

$$D = \rho U_0 \int U \, dy, \qquad (54)$$

where the integral is taken across a section of the wake far down-stream. Since the integral is of order $x^{m-n}$, we must have $m = n$, and hence $m = n = \frac{1}{2}$, so that

$$\eta = y/x^{\frac{1}{2}}, \qquad U/U_0 = f(\eta)/x^{\frac{1}{2}}, \qquad (55)$$

and

$$l^2 = a_1 x, \qquad (56)$$

where $a_1$ is a constant.† Then $U$ is greatest in the middle of the wake ($y = 0$), so that when $y$ is positive $\partial U/\partial y$ is negative, and

$$U_0 \frac{\partial U}{\partial x} = \frac{\partial}{\partial y}\left(\epsilon \frac{\partial U}{\partial y}\right) = -\frac{\partial}{\partial y}\left[l^2 \left(\frac{\partial U}{\partial y}\right)^2\right]$$

$$= -2a_1 x \frac{\partial U}{\partial y} \frac{\partial^2 U}{\partial y^2}. \qquad (57)$$

The equation for $f$ is

$$\frac{d}{d\eta}(\eta f) = 4a_1 f' f'', \qquad (58)$$

where dashes denote differentiation with respect to $\eta$. Since the wake is symmetrical, $\partial U/\partial y$ vanishes on the axis, and $f'(0) = 0$. Moreover $f$ is positive and $f'$ negative; and at the edge of the wake, where $U = 0$, so that $f = 0$, we should prefer $\partial U/\partial y = 0$, and there-fore $f' = 0$, in order that the velocity may pass smoothly over into that in the main stream. The first integral of (58) is

$$\eta f = 2a_1 f'^2, \qquad (59)$$

leading to

$$f = \frac{\eta_0^3}{18a_1}(1-\xi^{\frac{3}{2}})^2, \qquad (60)$$

where

$$\xi = \eta/\eta_0 = y/y_0, \qquad (61)$$

and $\eta = \eta_0$, $y = y_0$ at the edge of the wake, where $f = 0$, so that $y_0 = \eta_0 x^{\frac{1}{2}}$. The solution, which is based on (53) and makes $\epsilon$ vanish with $\partial U/\partial y$, therefore gives a finite breadth to the wake.‡

† The same results follow from taking $l \propto y_0 \propto \phi(x)$, $\eta \propto y/y_0$, $U/U_0 = f(\eta)/\psi(x)$, where $\phi$ and $\psi$ may be any two functions. It is found that $\phi \propto \psi \propto x^{\frac{1}{2}}$.

‡ Schlichting, *Ingenieur-Archiv*, **1** (1930), 533–571. Schlichting also calculated the second and the third approximations to $U$. For an attempt to find the velocity distribution in a wake with Kármán's form for $l$ $\left(\text{constant} \times \left|\frac{\partial U}{\partial y} \middle/ \frac{\partial^2 U}{\partial y^2}\right|\right.$; Chap. V, § 82$\left.\right)$, see Tollmien, *Ingenieur-Archiv*, **4** (1933), 1–15. The graph of $U$ against $y$ has a point of inflexion, so that difficulties are encountered both at the maximum value of $U$ (i.e. when $\partial U/\partial y = 0$) and at the inflexion (i.e. when $\partial^2 U/\partial y^2 = 0$).

Thus, according to this theory,

$$U/U_{\max} = (1-\xi^{\frac{3}{2}})^2, \tag{62}$$

and this result is compared with experiment in Fig. 232, which is due to Schlichting (*loc. cit.*).† The plotted points refer to experimental results at four sections, at distances downstream varying from about 100 cylinder diameters to 200 diameters. The agreement is satisfactory except at the edge of the wake, where the measured values

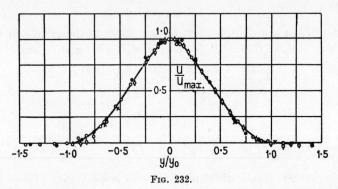

FIG. 232.

of $U$ approach zero more gradually than the calculated values. (This result is to be expected, since the turbulent interchange will not completely vanish with $\partial U/\partial y$.) Similar experimental results were found by Fage and Falkner.‡ Schlichting also verified that, when a suitable origin was chosen, the breadth of the wake sufficiently far downstream varied as $x^{\frac{1}{2}}$ and the maximum value of $U$ as $x^{-\frac{1}{2}}$. Again

$$\frac{U_{\max}}{U_0} = \frac{\eta_0^3}{18 a_1 x^{\frac{1}{2}}},$$

and it follows from (54) that

$$\frac{\eta_0^3}{2 a_1} = \frac{10 D}{\eta_0 \rho U_0^2},$$

so that $a_1$, and therefore the ratio of $l$ to the breadth $y_0$ of the wake, can be determined from measurements of $\eta_0$ and of either the drag coefficient or the maximum value of $U$. Since the edge of the wake is rather indefinite, the value, $Y$, of $y$ at which $U = \frac{1}{2}U_{\max}$ is inserted in the equations instead of $y_0$, this value being $0.441 y_0$ according to

† Measurements behind a pair of parallel cylinders were also made by Schlichting (*loc. cit.*).      ‡ *Proc. Roy. Soc.* A, **135** (1932), 702–705.

equation (62). Schlichting found, from the results he considered the most accurate, that $l/Y = 0.47$.

On the vorticity transfer theory, in place of equation (57) we get

$$U_0 \frac{\partial U}{\partial x} = \epsilon \frac{\partial^2 U}{\partial y^2}. \tag{63}$$

(Chap. V, equation (35)), which, with assumptions (53) and (56) for $\epsilon$ and $l$, becomes

$$U_0 \frac{\partial U}{\partial x} = -a_1 x \frac{\partial U}{\partial y} \frac{\partial^2 U}{\partial y^2}. \tag{64}$$

This equation is obtained either on the assumption of two-dimensional turbulence (Chap. V, § 83) or on the modified vorticity transfer theory (Chap. V, § 85). Apart from the fact that $a_1$ replaces $2a_1$ the equation is identical with (57), so the velocity distributions on the momentum transfer and vorticity transfer theories are the same. The value of $l/Y$ deduced from the experimental results would, however, be $\sqrt{2}$ times as big on the vorticity theory as on the momentum theory.

A criterion for judging between the two theories is provided by measurements of the temperature distribution in the wake behind a heated cylinder, since for a given distribution of velocity the apparent thermal conductivity due to turbulent mixing is twice as great on the vorticity transfer theory as on the momentum transfer theory, and thus, whereas the temperature and velocity distributions are the same on the momentum theory, the temperature distribution is the square root of the velocity distribution on the vorticity theory.[†] This matter is referred to again in Chap. XV, § 284, where it will be seen that the results on the vorticity theory are in fair agreement with experiment, whereas the results on the momentum theory disagree.

The components of the turbulent velocity in the wake behind a cylindrical obstacle have been examined with the ultramicroscope by Fage,[‡] who found that, though the turbulence is two-dimensional not too far downstream at low Reynolds numbers, it is three-dimensional at high Reynolds numbers, with the three components approximately equal.

We may note that, on the generalized vorticity transfer theory (Chap. V, § 84), (63) would be replaced by

$$U_0 \frac{\partial U}{\partial x} = \overline{L_2 v} \frac{\partial^2 U}{\partial y^2} - \overline{\left( v \frac{\partial L_3}{\partial z} - w \frac{\partial L_2}{\partial z} \right) \frac{\partial U}{\partial y}} \tag{65}$$

[†] Taylor, *Proc. Roy. Soc.* A, **135** (1932), 685–696.
[‡] *A.R.C. Reports and Memoranda*, No. 1510 (1933).

(Chap. V, equations (37) and (45)), where $\epsilon$ is $\overline{L_2 v}$. With two-dimensional turbulence, or on the modified vorticity transfer theory, the last term on the right goes out. The equation obtained by omitting the last term leads, in the case considered here, to fairly satisfactory agreement with experiment.

### 253. The turbulent wake behind a row of parallel rods.

The wake behind a row of parallel equally-spaced similar rods in a plane at right angles to the stream has been investigated by Anderlik[†] and by Gran Olsson.[‡] If $\lambda$ is the distance between two neighbouring rods, the velocity distribution in any section will be a periodic function of $y$, with period $\lambda$, the velocity being least behind the centre of a section of one of the rods and greatest behind a point half-way between two rods. The notation of the previous paragraph will be used, with the axis of $x$ through the centre of the section of one of the rods. Then $U$ has a maximum at $y = 0$ and a minimum at $y = \frac{1}{2}\lambda$. The undisturbed stream velocity in the wake is taken as the velocity at $y = \frac{1}{4}\lambda$, so that $U = 0$ there; and it is sufficient to find $U$ for $0 \leqslant y \leqslant \frac{1}{4}\lambda$, since it will be symmetrical about $y = 0$ and anti-symmetrical about $y = \frac{1}{4}\lambda$.

On the momentum transfer theory, the equation for $U$ may again be taken in the form (51), with $\tau$ and $\epsilon$ given by (52) and (53). It is assumed that $l$ is constant, so that in order that both sides of the equation may vary in the same way with $x$, $U$ must vary as $x^{-1}$; and we put

$$U/U_0 = AF(y)/x, \tag{66}$$

where $A$ is a constant, chosen so that $F = 1$ at $y = 0$, and $F$ vanishes at $y = \frac{1}{4}\lambda$. The equation for $U$ is

$$U_0 \frac{\partial U}{\partial x} = -2l^2 \frac{\partial U}{\partial y} \frac{\partial^2 U}{\partial y^2}, \tag{67}$$

so that

$$F = 2l^2 AF'F''. \tag{68}$$

Since $F = 1$ and $F' = 0$ at $y = 0$,

$$\tfrac{1}{2}F^2 = \tfrac{1}{2} + \tfrac{2}{3}l^2 AF'^3. \tag{69}$$

Hence

$$y = \left(\frac{4Al^2}{3}\right)^{\frac{1}{3}} \int_F^1 \frac{dF}{(1-F^2)^{\frac{1}{3}}}, \tag{70}$$

† *Magyar Tudományos Akadémia Matematikai és Természettudományi Értesítöje*, **52** (1934), 54–79.   ‡ *Zeitschr. f. angew. Math. u. Mech.* **16** (1936), 257–274.

and since $F = 0$ at $y = \frac{1}{4}\lambda$

$$\frac{\lambda}{4} = \left(\frac{4Al^2}{3}\right)^{\frac{1}{3}} \int\limits_0^1 \frac{dF}{(1-F^2)^{\frac{1}{3}}} = \frac{\Gamma(\frac{3}{2})\Gamma(\frac{2}{3})}{\Gamma(\frac{7}{6})}\left(\frac{4Al^2}{3}\right)^{\frac{1}{3}}, \tag{71}$$

so that

$$A = \frac{\lambda^3}{184 \cdot 7 l^2}. \tag{72}$$

On the vorticity transfer theory the equation (67), with $2l^2$ replaced by $l^2$, would be obtained, so the velocity distribution would be the same, only the experimentally determined value of $l$ being different.

We have previously pointed out that the turbulent interchange is not to be expected to vanish completely when $\partial U/\partial y = 0$. Prandtl (*loc. cit.*) has suggested that $|\partial U/\partial y|$ in (53) should be replaced by an average value throughout a neighbourhood, and that it is probably sufficient for this to take

$$\epsilon = l^2\left\{\left(\frac{\partial U}{\partial y}\right)^2 + l_1^2\left(\frac{\partial^2 U}{\partial y^2}\right)^2\right\}^{\frac{1}{2}}, \tag{73}$$

where $l_1$ is a second length, not too big. Gran Olsson noticed that if in (73) we take $l_1 = \lambda/2\pi$, then the equation

$$U_0\frac{\partial U}{\partial x} = \frac{\partial}{\partial y}\left(\epsilon\frac{\partial U}{\partial y}\right) \tag{74}$$

has the simply-harmonic solution

$$\frac{U}{U_0} = \frac{A}{x}\cos\frac{2\pi y}{\lambda}, \tag{75}$$

where

$$A = \frac{\lambda^3}{8\pi^3 l^2}. \tag{76}$$

Also

$$\epsilon = \frac{2\pi A U_0 l^2}{\lambda x}, \tag{77}$$

and is constant over a cross-section, so that the equations and solutions on the vorticity transfer and momentum transfer theories are identical.

Gran Olsson verified experimentally that $U \propto x^{-1}$. A comparison of the theoretical distribution over a section with his measured values at three sections sufficiently far downstream is shown in Fig. 233.†

The temperature distribution when the rods are heated is considered in Chap. XV, § 285.

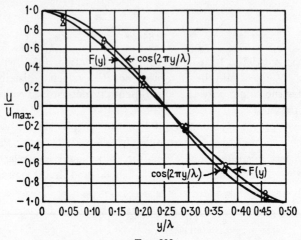

FIG. 233.

## 254. The turbulent wake behind a body of revolution.

In a wake behind a body of revolution, in which the mean motion is symmetrical about an axis, let $(r, \theta, x)$ be cylindrical polar coordinates, $x$ being measured along the axis of symmetry and $r$ being distance from it. If $U_0$ is the velocity of the undisturbed stream, $U_0 - U$ the velocity in the wake parallel to the axis, and $\tau$ the shearing stress, then far downstream the equation for $U$ is approximately

$$U_0 \frac{\partial U}{\partial x} = -\frac{1}{\rho} \frac{1}{r} \frac{\partial}{\partial r} (r\tau). \tag{78}$$

On the momentum transfer theory

$$\tau = -\rho \epsilon \, \partial U / \partial r, \tag{79}$$

and Prandtl's assumption for $\epsilon$ is

$$\epsilon = l^2 |\partial U / \partial r|. \tag{80}$$

Hence (since $\partial U / \partial r$ is negative)

$$U_0 \frac{\partial U}{\partial x} = -\frac{1}{r} \frac{\partial}{\partial r} \left[ l^2 r \left( \frac{\partial U}{\partial r} \right)^2 \right]. \tag{81}$$

† A comparison of (70) with experiment was also given by Anderlik (*loc. cit.*). In addition, he considered the equation with Kármán's form for $l$.

The drag force on the solid body of revolution is given by

$$D = 2\pi\rho U_0 \int Ur\,dr, \tag{82}$$

the integral being taken from the axis to the edge of the wake. With the same assumptions as in the two-dimensional case—similarity in different sections, $l$ proportional to the breadth of the wake at corresponding points of different sections, $l$ constant over any one section—we find, by proceeding as in the two-dimensional case (§ 252), that

$$U_{\max} \propto x^{-\frac{2}{3}}, \qquad l \propto r_0 \propto x^{\frac{1}{3}}, \tag{83}$$

where $r_0$ is the radius of the wake. Hence we put

$$\eta = r/x^{\frac{1}{3}}, \qquad U/U_0 = f(\eta)/x^{\frac{2}{3}}, \qquad l^2 = ax^{\frac{2}{3}}. \tag{84}$$

We then find, after a first integration of the equation for $f$, that $f'^2$ is proportional to $\eta f$, exactly as in (59), so that we once more have

$$U/U_{\max} = (1-\xi^{\frac{3}{2}})^2, \tag{85}$$

where

$$\xi = \eta/\eta_0 = r/r_0, \tag{86}$$

and $\eta = \eta_0$, $r = r_0$ at the edge of the wake.[†] $U/U_{\max}$ is shown plotted against $r/R$ as the full-line curve in Fig. 234, where $R$ is the value of $r$ at which $U = \frac{1}{2}U_{\max}$. The crosses in Fig. 234 represent mean experimental results obtained by Hall and Hislop, and the theoretical and experimental results have been made to agree at $r = R$.

On the generalized vorticity transfer theory we find from Chap. V, § 84, equations (38) and (44), that in cylindrical polar coordinates the equation for the first approximation to $U$ may be written[‡]

$$U_0\frac{\partial U}{\partial x} = \overline{L_r v_r}\left(\frac{\partial^2 U}{\partial r^2} - \frac{1}{r}\frac{\partial U}{\partial r}\right) - \overline{v_r\frac{\partial L_\theta}{\partial \theta} - v_\theta\frac{\partial L_r}{\partial \theta}}\frac{1}{r}\frac{\partial U}{\partial r}, \tag{87}$$

where $v_r$, $v_\theta$ are the components, in the directions of $r$ and $\theta$ increasing, of the turbulent velocity; $L_r$ and $L_\theta$ are the components in those directions of the vector 'mixing' displacement whose components are denoted by $L_1$, $L_2$, $L_3$ in Chap. V, § 84; and $\epsilon$ is $\overline{L_r v_r}$. There are two ways of simplifying this equation on the analogy of putting the last term in (65) zero. We may simply put the last term in (87) zero, so that

$$U_0\frac{\partial U}{\partial x} = \epsilon\left(\frac{\partial^2 U}{\partial r^2} - \frac{1}{r}\frac{\partial U}{\partial r}\right). \tag{88}$$

† Swain, *Proc. Roy. Soc.* A, **125** (1929), 647–659. A second approximation is found in this paper.
‡ Cf. Goldstein, *Proc. Camb. Phil. Soc.* **31** (1935), 358.

This is the equation that would be obtained on the assumption that the eddying motion, as well as the mean motion, is symmetrical about the axis.

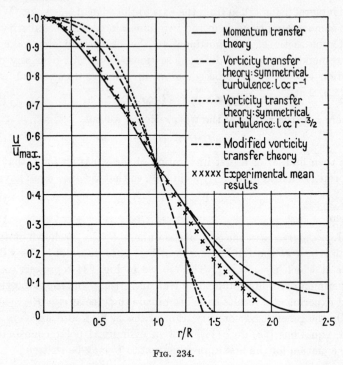

FIG. 234.

Alternatively we may use Cartesian coordinates $y$ and $z$ instead of $r$ and $\theta$ (with $y = r\cos\theta$, $z = r\sin\theta$), transform, and put

$$\overline{v\frac{\partial L_3}{\partial z} - w\frac{\partial L_2}{\partial z}} = 0 \qquad (89)$$

when $\theta = 0$, so that the axis of $y$ is then along the direction of $r$ increasing, and the axis of $z$ is along the direction of $\theta$ increasing, but is straight. The equations of the transformation are

$$v = v_r, \qquad w = v_\theta, \qquad \frac{\partial L_3}{\partial z} = \frac{\partial L_\theta}{\partial \theta} + L_r, \qquad \frac{\partial L_2}{\partial z} = \frac{\partial L_r}{\partial \theta} - L_\theta, \quad (90)$$

so

$$-\overline{\left(v_r\frac{\partial L_\theta}{\partial \theta} - v_\theta\frac{\partial L_r}{\partial \theta}\right)} = \overline{L_r v_r} + \overline{L_\theta v_\theta} \qquad (91)$$

when (89) holds, and (87) reduces to

$$U_0 \frac{\partial U}{\partial x} = \overline{L_r v_r} \frac{\partial^2 U}{\partial r^2} + \overline{L_\theta v_\theta} \frac{1}{r} \frac{\partial U}{\partial r}. \tag{92}$$

This may be shown to be the same equation as on the modified vorticity transfer theory (Chap. V, § 85). If we further assume that the turbulence is isotropic, so that

$$\overline{L_\theta v_\theta} = \overline{L_r v_r} = \epsilon,$$

then
$$U_0 \frac{\partial U}{\partial x} = \epsilon \left( \frac{\partial^2 U}{\partial r^2} + \frac{1}{r} \frac{\partial U}{\partial r} \right). \tag{93}$$

In two dimensions, when Prandtl's assumptions are made for $\epsilon$, the equations for $U$ on the vorticity transfer and momentum transfer theories become identical (apart from a constant to be determined from experiment). But this is not the case for motion symmetrical about an axis, since neither (88) nor (93) will reduce to the equation of the momentum theory.

For either (88) or (93) we take $\epsilon$ to be given by (80), and if we assume as before similarity in different sections, so that at corresponding points in different sections $l$ is proportional to the radius $r_0$ of the wake or of the value $R$ of $r$ at which $U = \frac{1}{2} U_{\max}$, then (without assuming $l$ constant over any one section) we find as before that $U_{\max} \propto x^{-\frac{2}{3}}$, $l \propto x^{\frac{1}{3}}$. We therefore put

$$\eta = r/x^{\frac{1}{3}}, \qquad U/U_0 = f(\eta)/x^{\frac{2}{3}}, \qquad l^2 = g(\eta)x^{\frac{2}{3}}. \tag{94}$$

Then from (88) we deduce the following equation for $f$:

$$2\eta f + \eta^2 f' = 3g f'(\eta f'' - f'). \tag{95}$$

At $\eta = 0$, $U$ is positive and $\partial U/\partial y$ zero, so that $f(0) > 0, f'(0) = 0$. With these conditions we find that, if $l$ is constant across a section ($g(\eta)$ constant), then (95) has no real solution at all, $f'$ being imaginary. If $g(\eta) \propto \eta^p$, then there is no real solution unless $p < -1$, so that $l$ becomes infinite on the axis. In spite of this, solutions have been worked out with $p = -2$ (so that $l \propto r^{-1}$ over a section) and $p = -3$ (so that $l \propto r^{-\frac{3}{2}}$). These are shown in Fig. 234, the former by the broken line - - -, and the latter by the dotted line ·····. The two run together over the middle of the range. In the first case, the solution was found by numerical integration; in the second

$$U/U_{\max} = (1 - \xi^3)^2,$$

where $\xi$ is given by (86). Moreover, $p = -3$ is the only value of

$p$ for which $U$ and $\partial U/\partial y$ vanish together at the edge of the wake. (In two dimensions, if $l^2 \propto y^p$, then $U$ and $\partial U/\partial y$ vanish together if, and only if, $p = 0$.)

In the case of (93) a solution for $f$ can be found with $l$ constant over a section. This solution never vanishes; it is always positive and tends to zero at infinity like a multiple of $\eta^{-2}$, so that there is no edge to the wake. The solution is certainly wrong in the outer parts of the wake. In the inner parts, it is very near the solution on the momentum transfer theory. In Fig. 234 it is shown by the chain-dotted line ·—·—· in the region in which it is distinct from the solution on the momentum transfer theory.†

The temperature distribution behind a heated body of revolution will be considered in Chap. XV, § 286, where it will appear that, except in the outer parts of the wake, the experimental evidence is in favour of the validity of the modified vorticity transfer theory.

## 255. The turbulent spreading of jets.

The methods of the last three sections may also be used to calculate the turbulent spreading of jets discharged into still fluid.‡ At some distance from the mouth of the nozzle from which the jet is discharged, the assumptions and approximations of the boundary layer theory may again be applied right across the section, and the pressure gradient along the axis of the jet may also be neglected. It appears that the breadth of the jet increases linearly downstream.

For two-dimensional motion, if $x$ is measured along the axis of the jet from some suitable origin, and $(U, V)$ are the components of mean velocity, the approximate equation of motion is

$$U\frac{\partial U}{\partial x} + V\frac{\partial U}{\partial y} = \frac{1}{\rho}\frac{\partial \tau}{\partial y}, \tag{96}$$

where $\tau$ is the shearing stress. For positive $y$, $\partial U/\partial y$ is negative, $U$ being greatest in the middle and zero at the edge of the jet; and on the momentum transfer theory, with Prandtl's assumption for $\tau$ (see eqns. (52) and (53)), the equation for $y$ positive is

$$U\frac{\partial U}{\partial x} + V\frac{\partial U}{\partial y} = -\frac{\partial}{\partial y}\left\{l^2\left(\frac{\partial U}{\partial y}\right)^2\right\}. \tag{97}$$

† For the solutions on the vorticity transfer theory see Goldstein, *Proc. Camb. Phil. Soc.* **34** (1938), 48–67. For the experimental results of Hall and Hislop see a forthcoming paper in the *Proc. Camb. Phil. Soc.* (July, 1938.)

‡ The calculations for laminar motion were set out in Chap. IV, § 57.

The rate $M$ at which momentum flows across a section of the jet is constant, and

$$M = 2\rho \int_0^\infty U^2 \, dy. \tag{98}$$

There is a stream-function $\psi$ such that

$$U = \frac{\partial\psi}{\partial y}, \qquad V = -\frac{\partial\psi}{\partial x}. \tag{99}$$

Then with the same assumptions as in § 252—similarity in different sections sufficiently far downstream, $l$ proportional to the breadth of a section and constant over any one section—$l$ and the breadth of a section are proportional to $x$ and the maximum velocity to $x^{-\frac{1}{2}}$. Hence we put

$$l = cx, \qquad \eta = y/x, \qquad \psi = Ax^{\frac{1}{2}}F(\eta), \tag{100}$$

so that

$$U = A\frac{F'(\eta)}{x^{\frac{1}{2}}}, \qquad V = \frac{A}{2x^{\frac{1}{2}}}(2\eta F' - F). \tag{101}$$

$A$ is a constant. The jet being symmetrical, $V$ and $\partial U/\partial y$ vanish on the axis, so that $F(0) = 0$, $F''(0) = 0$. The equation for $F$ is

$$F'^2 + FF'' = 2c^2\frac{d}{d\eta}(F''^2),$$

which integrates at once to

$$FF' = 2c^2 F''^2. \tag{102}$$

This equation also follows at once from the momentum equation for the portion of fluid on one side of a plane parallel to the axis and between two infinitesimally near sections, namely

$$\rho UV + \frac{\partial}{\partial x}\int^y \rho U^2 \, dy = \tau, \tag{103}$$

where the lower limit in the integral is taken at the edge of the jet. If we take $\eta/(2c^2)^{\frac{1}{3}}$ as independent variable instead of $\eta$, then $2c^2$ disappears from (102). We use the dash to denote differentiation with respect to the new independent variable. Then $A$ may be chosen so that $F'(0) = 1$, and the equation has to be integrated with the boundary conditions $F(0) = 0$, $F'(0) = 1$. At the edge of the jet $U = 0$, $\partial U/\partial y = 0$, so that $F' = 0$, $F'' = 0$. The equation was integrated by Tollmien,† who found that $\eta/(2c^2)^{\frac{1}{3}} = 4\pi/(3\sqrt{3}) = 2\cdot412$ at the edge of the jet. A comparison of the theoretical curve so

† *Zeitschr. f. angew. Math. u. Mech.* **6** (1926), 468–478.

obtained with measured values due to Förthmann[†] is shown in Fig.
235, where $U/U_{max}$ is plotted against $y/Y$, and $Y$ is the value of $y$ at
which $U = \frac{1}{2}U_{max}$. There is a small systematic discrepancy. Förthmann also verified that the breadth of the jet was proportional to $x$,
and the maximum velocity to $x^{-\frac{1}{2}}$, with the origin for $x$ suitably chosen.
He found $c = 0.0165$ and $l/Y = 0.17$ (in contrast to the value $0.47$
found by Schlichting on the same theory in a two-dimensional wake).

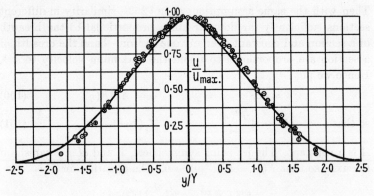

FIG. 235.

It follows from the value of $\psi$ in (100) that the flux of matter
across a section of the jet increases as $x^{\frac{1}{2}}$, so that the jet is drawing in
fluid from its surroundings.

On the vorticity transfer theory (97) would be replaced by

$$U\frac{\partial U}{\partial x} + V\frac{\partial U}{\partial y} = \epsilon\frac{\partial^2 U}{\partial y^2} = -l^2\frac{\partial U}{\partial y}\frac{\partial^2 U}{\partial y^2}, \tag{104}$$

which, with $l$ independent of $y$ as in (100), is the same as (97) with
$l^2$ instead of $2l^2$. Hence the same velocity distribution would be
obtained, only the value assigned to $l$ as a result of the comparison
with experiment being different. Whereas, however, the temperature
distribution for a jet of heated fluid would be the same as the velocity
distribution on the momentum transfer theory, it would be the square
root of the velocity distribution on the vorticity transfer theory. No
experiments appear to have been reported.

† *Ingenieur-Archiv*, **5** (1934), 42–54. Measurements have also been made by Peters
and Bicknell, for which see Rossby, *Papers in Physical Oceanography and Meteorology*,
*Massachusetts Institute of Technology and Woods Hole Oceanographic Institution*, **5**,
No. 1 (1936), 17.

For a jet symmetrical about an axis, if $x$ is measured along the axis of the jet and $r$ at right angles to it, and $U$, $V$ are the components of mean velocity in the directions of $x$ and $r$ increasing, then the approximate equation of motion on the momentum transfer theory, with Prandtl's assumption for $\epsilon$, is

$$U\frac{\partial U}{\partial x} + V\frac{\partial U}{\partial r} = -\frac{1}{r}\frac{\partial}{\partial r}\left\{l^2 r\left(\frac{\partial U}{\partial r}\right)^2\right\}, \qquad (105)$$

$U$ being greatest on the axis and zero at the edge of the jet, and $\partial U/\partial r$ negative.

The constant rate of flow of momentum across a section is

$$M = 2\pi\rho \int\limits_0^\infty U^2 r\, dr. \qquad (106)$$

From the equation of continuity there is a stream-function $\psi$ such that

$$U = \frac{1}{r}\frac{\partial\psi}{\partial r}, \qquad V = -\frac{1}{r}\frac{\partial\psi}{\partial x}. \qquad (107)$$

With the same assumptions as before, $l$ and the breadth of the wake are proportional to $x$, and the maximum velocity to $x^{-1}$. Hence we put

$$l = cx, \qquad \eta = r/x, \qquad \psi = AxF(\eta), \qquad (108)$$

so that

$$U = \frac{AF'}{x\eta}, \qquad V = \frac{A}{x}\left\{F' - \frac{F}{\eta}\right\}. \qquad (109)$$

The equation for $F$ is

$$\frac{d}{d\eta}\left(\frac{FF'}{\eta}\right) = \frac{d}{d\eta}\left\{\frac{c^2}{\eta}\left(F'' - \frac{F'}{\eta}\right)^2\right\}. \qquad (110)$$

Since $U$ is finite on the axis, $F'/\eta$ must tend to a finite value as $\eta$ tends to zero, and since $V$ and $\partial U/\partial r$ vanish on the axis, $F' - F/\eta$ and $F''/\eta - F'/\eta^2$ must tend to zero when $\eta$ tends to zero. Hence also $F(0) = 0$, and (110) integrates without the addition of any constant to

$$FF' = c^2(F'' - F'/\eta)^2. \qquad (111)$$

After changing the independent variable to $\eta/c^{\frac{2}{3}}$ in order to eliminate $c^2$ from the equation, we may choose $A$ so that $F'/\eta \to 1$ when $\eta \to 0$. The resulting equation was integrated by Tollmien (*loc. cit.*). At the edge of the wake $U = 0$ and $\partial U/\partial r = 0$, so that $F' = 0$, $F'' - F'/\eta = 0$. Tollmien found that $\eta/c^{\frac{2}{3}} = 3\cdot4$ at the edge of the wake. A comparison of the theoretical curve with the results of measurements by Ruden[†] and by Kuethe[‡] are shown in Fig. 236, where $U/U_{max}$ is plotted against $r/R$, and $r = R$ when $U = \frac{1}{2}U_{max}$. The full-line curve is the theoretical curve, the dotted curve was given by Ruden as representative of his experiments (no single results being given), and the circles and squares are due to Kuethe. It appears that,

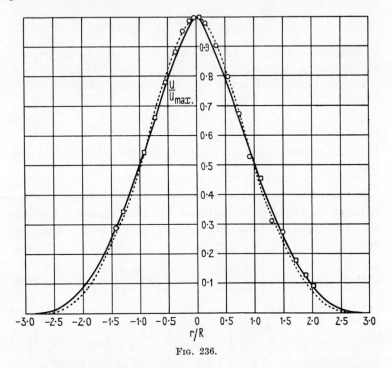

FIG. 236.

for a jet discharged from a nozzle, the theoretical results hold fairly accurately, at any rate over the middle portion of the jet, for distances downstream of the mouth of the nozzle greater than about eight diameters of the mouth.

The flux of matter across a section increases for the symmetrical jet in proportion to $x$.

Calculations for a symmetrical jet on the vorticity transfer theory have been made by Howarth[||] by the methods employed in § 254 for a symmetrical wake. Exactly analogous results are found, the equation

† *Die Naturwissenschaften*, **21** (1933), 375–378.
‡ *Journ. Applied Mechanics*, **2** (1935), 87–95.
|| *Proc. Camb. Phil. Soc.* **34** (1938), 185–194.

corresponding to (88) (symmetrical turbulence) having no real integral if $l$ is constant across a section. This equation was integrated both with $l$ proportional to $r^{-1}$ and to $r^{-\frac{3}{2}}$, as for wakes, and the equation corresponding to (93) (modified vorticity transfer theory) with $l$ constant over a section. The results found did not agree with the experimental results.

A two-dimensional problem on the spread of a mixing region, which is of some interest and which leads to analysis rather simpler than that for the motions discussed above, is the turbulent mixing of a stream of fluid with fluid at rest, the mixing taking place along a single boundary between them. Thus a stream of fluid, flowing parallel to the axis of $x$ with velocity $U_1$, is supposed to begin

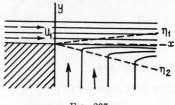

Fig. 237.

to mix with still fluid at some value of $x$ (Fig. 237). With the axes as shown in the figure, $\partial U/\partial y$ is positive, so the equation of motion on the momentum transfer theory is

$$U\frac{\partial U}{\partial x}+V\frac{\partial U}{\partial y} = \frac{\partial}{\partial y}\left\{l^2\left(\frac{\partial U}{\partial y}\right)^2\right\}, \tag{112}$$

the sign of the term on the right in (97) being changed. On the assumption of similarity in different sections, $l$ will be proportional to the breadth $b$ of the mixing region, and $U$ will be a function of $y/b$. It is easily found that in order that both sides of (112) should vary in the same way with $x$, $b$ must again be proportional to $x$. Hence we put

$$l = cx, \qquad \eta = y/x, \qquad \psi = AxF(\eta), \tag{113}$$

so that

$$U = AF'(\eta), \qquad V = A\{\eta F'(\eta)-F(\eta)\}, \tag{114}$$

where $A$ is a constant to be chosen later. The equation for $F$ is

$$F''(F+2c^2F''') = 0. \tag{115}$$

$F'' = 0$ corresponds to $U = $ constant. Inside the mixing region

$$F+2c^2F''' = 0. \tag{116}$$

Let $\eta = \eta_1$ and $\eta = \eta_2$ be the boundaries of the mixing region, as shown in the figure. Put

$$\eta' = \frac{1}{(2c^2)^{\frac{1}{3}}}(\eta-\eta_1). \tag{117}$$

Then
$$F + \frac{d^3F}{d\eta'^3} = 0, \tag{118}$$

and
$$F = d_1 e^{-\eta'} + e^{\frac{1}{2}\eta'}\left(d_2 \cos\frac{\sqrt{3}}{2}\eta' + d_3 \sin\frac{\sqrt{3}}{2}\eta'\right), \tag{119}$$

where $d_1$, $d_2$, $d_3$ are constants. These three constants, together with $\eta_1$ and $\eta_2$, must be determined from the boundary conditions. Four boundary conditions are

$$\left.\begin{array}{l} U = U_1, \quad \dfrac{\partial U}{\partial y} = 0 \quad \text{at} \quad \eta = \eta_1, \\[2mm] U = 0, \quad \dfrac{\partial U}{\partial y} = 0 \quad \text{at} \quad \eta = \eta_2. \end{array}\right\} \tag{120}$$

If then we take
$$A = (2c^2)^{\frac{1}{3}} U_1, \tag{121}$$

these four boundary conditions are

$$\left.\begin{array}{l} \dfrac{dF}{d\eta'} = 1, \quad \dfrac{d^2F}{d\eta'^2} = 0 \quad \text{at} \quad \eta' = 0 \\[3mm] \dfrac{dF}{d\eta'} = 0, \quad \dfrac{d^2F}{d\eta'^2} = 0 \quad \text{at} \quad \eta' = -\dfrac{1}{(2c^2)^{\frac{1}{3}}}(\eta_1 - \eta_2). \end{array}\right\} \tag{122}$$

These four equations determine $d_1$, $d_2$, $d_3$, and $(\eta_1 - \eta_2)/(2c^2)^{\frac{1}{3}}$. As fifth equation the condition $V = 0$ at $\eta = \eta_1$ is taken, which leads to $\eta_1/(2c^2)^{\frac{1}{3}} = F$ at $\eta' = 0$, and determines $\eta_1/(2c^2)^{\frac{1}{3}}$ and hence $\eta_2/(2c^2)^{\frac{1}{3}}$. Fluid is being drawn into the mixing zone from the stagnant region, but the flux of matter and momentum into the mixing zone from the stream is due only to the inclination of the boundary to the direction of the stream. The analysis is due to Tollmien (*loc. cit.*), who finds

$$d_1 = -0{\cdot}0062, \qquad d_2 = 0{\cdot}987, \qquad d_3 = 0{\cdot}577,$$
$$\eta_1/(2c^2)^{\frac{1}{3}} = 0{\cdot}981, \qquad \eta_2/(2c^2)^{\frac{1}{3}} = -2{\cdot}04.$$

The solution has been generalized by Kuethe (*loc. cit.*) to the case of the mixing of two streams with velocities $U_1$ and $U_2$, so that at $\eta = \eta_2$, $U = U_2$ instead of zero. For a discussion of the fifth boundary condition, and for numerical results, reference may be made to the original paper.

On the vorticity transfer theory the equation corresponding to (112) is the same with $l^2$ in place of $2l^2$, namely

$$U\frac{\partial U}{\partial x} + V\frac{\partial U}{\partial y} = l^2 \frac{\partial U}{\partial y}\frac{\partial^2 U}{\partial y^2}, \tag{123}$$

so that the velocity distribution is the same, only the experimentally obtained values of the constants being different. The temperature distribution (if one of the portions of fluid is hotter than the other) is, however, different (Chap. XV, § 287).

When a symmetrical jet is discharged from a large nozzle, with practically uniform velocity at the mouth of the nozzle, then near the mouth, where the breadth of the mixing zone at the edge of the jet is small compared with the jet radius, the curvature will have no appreciable effect. The solution above should then apply to the mixing zone. Tollmien has compared his solution with measurements at half a diameter from the mouth of the nozzle of the large Göttingen wind tunnel,† 224 cm. in diameter, and found satisfactory agreement with $c = 0.0174$ and the breadth of the mixing zone equal to $0.255x$. It is to be remarked, however, that the Göttingen measurements were made with a total-head and static-pressure tube, and that the velocity fluctuations in such a region are very large indeed (cf. Kuethe, *loc. cit.*), so that the measurements are difficult and, if no allowance is made for the velocity fluctuations, are subject to error (cf. Chap. VI, § 111, pp. 253, 254).

As we go downstream from the mouth along a symmetrical jet discharged from a large nozzle, as above, the annular mixing region at the edge will grow in breadth, and the core of fluid in which potential flow occurs will decrease in radius until it finally disappears, and the mixing region occupies the whole jet. Experiments and approximate calculations have been made by Kuethe (*loc. cit.*) to determine the velocity distribution between the mouth and a section where the core has disappeared, which is about 5 diameters downstream.

## ADDITIONAL REFERENCES

For an account of vortex-streets and three-dimensional wakes reference may be made to Falkenhagen, *Handbuch der Experimentalphysik*, **4**, part 1 (1931), 155–182, where other works are cited. To these may be added Lock, *Phil. Mag.* (6), **50** (1925), 1083–1089, and Carter, *Journ. Aero. Sciences*, **2** (1935), 159–161. For three-dimensional wakes see also Jeffreys, *Proc. Roy. Soc.* A, **128** (1930), 376–393, and (at low Reynolds numbers) Nisi and Porter, *Phil. Mag.* (6), **46** (1923), 754–768, and Schmiedel, *Physik. Zeitschr.* **29** (1928), 593–610. The wakes behind a strut and behind a sphere are considered by Ermisch, *Abhandl. Aero. Inst. Aachen*, No. 6 (1927), pp. 46–50. The wake behind a stream-line body of revolution at a non-zero

† *Ergebnisse der Aerodynamischen Versuchsanstalt zu Göttingen*, **2** (1923), 73.

angle of attack is discussed at some length by Harrington, *Guggenheim Airship Institute (Akron, Ohio)*, Publication No. 2 (1935), pp. 32–52.

The spread of turbulence is considered in the article by Prandtl, *Aerodynamic Theory* (edited by Durand), **3** (Berlin 1935), 162–178, and some further references are given in Bulletin No. 84 of the National Research Council, Washington: *Hydrodynamics*, by Dryden, Murnaghan, and Bateman (1932), pp. 410, 411.

Photographic and kinematographic investigations of the wake behind a sphere at Reynolds numbers between 150 and 10,000 have been reported by Möller, *Physik. Zeitschr.* **39** (1938), 57–66. This paper includes attempts by various methods to estimate the vortex frequency.

Calculations on the modified vorticity transfer theory of the velocity distributions in wakes and jets symmetrical about an axis have been published by Tomotika, *Proc. Roy. Soc.* A, **165** (1938), 53–72.

# HEAT TRANSFER (LAMINAR MOTION)

## 256. Introduction.

THE problem of heat transfer is difficult on account of the number of parameters involved. Moreover, the kinematic viscosity and the thermometric conductivity, two of the most important parameters, vary with temperature; and it follows that the velocity field depends upon the temperature differences present: in particular, the field for heat transfer from solid to fluid may be different from the field for heat transfer from fluid to solid.

Particular problems may conveniently be classified into problems involving laminar flow and those involving turbulent flow, and a further subdivision may be made into cases of free convection and of forced convection. By free convection is meant flow in which the motion is caused by the effect of gravity on the heated fluid of variable density; by forced convection, flow in which the velocities arising from the variable density distribution are negligible in comparison with the velocity of the main or forced flow.

In this chapter the governing equations are developed on the basis of certain simplifying assumptions, and are applied to the solution of several problems involving laminar flow. The assumptions are that the viscosity and the thermal conductivity are constant in the field,† that liquids are incompressible, and that gases obey the perfect gas law.

## 257. The equation of state.

The equation of state of a fluid relates the pressure $p$ to the density $\rho$ and the absolute temperature $T$. For liquids, which are assumed to be incompressible, this has the form

$$\rho = \rho_0,$$

while for gases, which are assumed to obey the perfect gas law,‡

$$\frac{p}{\rho T} = \frac{p_0}{\rho_0 T_0} = \frac{\bar{R}}{m}, \tag{1}$$

---

† This is strictly permissible only if the temperature differences involved are small, since the viscosity and thermometric conductivity vary with temperature. In an investigation by Busemann (*Zeitschr. f. angew. Math. u. Mech.* **15** (1935), 23–25) of a particular case of laminar flow along a flat plate, these factors are taken into consideration.　　　　　‡ Jeans, *Dynamical Theory of Gases* (1925), p. 114.

where $p_0$, $\rho_0$, $T_0$ are the pressure, density, and absolute temperature in some standard condition, $\bar{R}$ is the gas constant,[†] and $m$ is the molecular weight of the gas.

## 258. The equation of continuity.

For compressible fluids the equation of continuity is[‡]

$$\frac{\partial \rho}{\partial t}+\frac{\partial}{\partial x}(\rho u)+\frac{\partial}{\partial y}(\rho v)+\frac{\partial}{\partial z}(\rho w)=0, \tag{2}$$

which may also be written

$$\frac{1}{\rho}\frac{D\rho}{Dt}+\frac{\partial u}{\partial x}+\frac{\partial v}{\partial y}+\frac{\partial w}{\partial z}=0. \tag{3}$$

## 259. The equations of motion.

For compressible fluids the stress components are given by (17) of Chap. III with the addition of the term

$$-\tfrac{2}{3}\mu\left(\frac{\partial u}{\partial x}+\frac{\partial v}{\partial y}+\frac{\partial w}{\partial z}\right)$$

to each of $p_{xx}$, $p_{yy}$, $p_{zz}$.[||] The equations (19) of Chapter III for the accelerations $f_x$, $f_y$, $f_z$ of a fluid element are unchanged, and we obtain the equations of motion in the form[††]

$$\rho\frac{Du}{Dt}=-\frac{\partial p}{\partial x}+\rho X+\mu\left[\nabla^2 u+\frac{1}{3}\frac{\partial}{\partial x}\left(\frac{\partial u}{\partial x}+\frac{\partial v}{\partial y}+\frac{\partial w}{\partial z}\right)\right], \tag{4}$$

with two similar equations.

The external force per unit mass, which has the components $X, Y, Z$, and is denoted by $\mathbf{F}$, includes all the impressed forces. These forces will usually be conservative except for the force which arises from the effect of gravity on the fluid of variable density. We therefore put
$$\mathbf{F}=-\operatorname{grad}\Omega+\mathbf{G},$$

where $\mathbf{G}$ represents the excess force per unit mass due to the variable density. Now consider an element of volume $\delta\tau$, density $\rho$, pressure $p$, and absolute temperature $T$. By the principle of Archimedes the excess force on the element is $(\rho-\rho_0)\mathbf{g}\,\delta\tau$, where $\mathbf{g}$ is the vector of the earth's acceleration; and hence the excess force per unit mass is

$$\mathbf{G}=\mathbf{g}\left(\frac{\rho-\rho_0}{\rho}\right).$$

---

† $R$ is used for Reynolds number, and some other symbol must therefore be chosen for the gas constant.

‡ Lamb, *Hydrodynamics* (1932), p. 6.     || *Ibid.*, p. 544.     †† *Ibid.*, p. 546.

For liquids **G** vanishes, while for gases

$$\mathbf{G} = \mathfrak{g}\left(1 - \frac{p_0 T}{p T_0}\right)$$

from (1). This force is important only in problems of free convection: for these problems the pressure variations are small, so that

$$\mathbf{G} = -\mathfrak{g}\left(\frac{T - T_0}{T_0}\right), \tag{5}$$

and **G** is proportional to the excess temperature $(T - T_0)$. Hence

$$\mathbf{F} = -\operatorname{grad}\Omega - \mathfrak{g}\left(\frac{T - T_0}{T_0}\right). \tag{6}$$

## 260. The energy equation.

The rate at which work is being done by the surface stresses on the fluid inside any surface $S$ is

$$\int_S (u p_{nx} + v p_{ny} + w p_{nz})\, dS,$$

as in § 37 of Chapter III. Transformation by Green's theorem, as previously, may be used to show that this quantity is equal to

$$-\int \rho(uX + vY + wZ)\, d\tau + \frac{1}{2}\int \rho \frac{D}{Dt}(u^2 + v^2 + w^2)\, d\tau$$
$$-\int p\left(\frac{\partial u}{\partial x} + \frac{\partial v}{\partial y} + \frac{\partial w}{\partial z}\right) d\tau + \int \Phi\, d\tau, \tag{7}$$

which replaces (27) of Chapter III. The expression for the dissipation function $\Phi$ is[†]

$$\Phi = \tfrac{1}{2}\mu[e_{xx}^2 + e_{yy}^2 + e_{zz}^2 + 2e_{yz}^2 + 2e_{zx}^2 + 2e_{xy}^2] - \tfrac{1}{6}\mu[e_{xx} + e_{yy} + e_{zz}]^2$$
$$= \mu\left[ -\frac{2}{3}\left(\frac{\partial u}{\partial x} + \frac{\partial v}{\partial y} + \frac{\partial w}{\partial z}\right)^2 + 2\left(\frac{\partial u}{\partial x}\right)^2 + 2\left(\frac{\partial v}{\partial y}\right)^2 + 2\left(\frac{\partial w}{\partial z}\right)^2 \right.$$
$$\left. + \left(\frac{\partial w}{\partial y} + \frac{\partial v}{\partial z}\right)^2 + \left(\frac{\partial u}{\partial z} + \frac{\partial w}{\partial x}\right)^2 + \left(\frac{\partial v}{\partial x} + \frac{\partial u}{\partial y}\right)^2 \right], \tag{8}$$

which replaces (28) of Chapter III.

The first term in (7) is minus the rate at which work is being done by the external forces, and the second is the rate of increase of the kinetic energy of the fluid. Further, $\partial u/\partial x + \partial v/\partial y + \partial w/\partial z$ measures the rate of dilatation of the fluid at the point $(x, y, z)$,[‡] and hence the third term represents the rate at which work is done by the pressure $p$

[†] *Ibid.*, p. 548.                    [‡] *Ibid.*, p. 5.

in compressing the fluid inside the surface $S$.[†] The fourth term in (7) is proportional to the viscosity and represents the rate at which work is being done against viscosity; this work is no longer available as mechanical energy, and takes the form of heat developed in the fluid.[‡]

An equation for the temperature distribution can be obtained by considering the energy balance in an elementary parallelepiped, with centroid at $P$ $(x, y, z)$, of sides $\delta x$, $\delta y$, $\delta z$ and volume $\delta \tau$, through

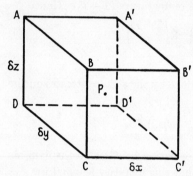

which the fluid is flowing. Let $E$ be the total kinetic energy, i.e. the sum of the kinetic and intrinsic energies, in this element at time $t$. Then

$$\frac{\partial E}{\partial t} + W = E_1 + E_2 + E_3, \quad (9)$$

where $W$ is the rate at which work is being done against the external forces, so that

$$W = -\rho(Xu + Yv + Zw)\,\delta\tau,$$

$E_1$ is the rate at which work is being done on the element by the surface stresses, $E_2$ is the rate of conduction of energy through the boundaries in the form of heat, and $E_3$ is the rate of convection of kinetic and intrinsic energy through the boundaries.

Let $\epsilon$ be the total kinetic energy per unit mass of the element. Then $E$ is equal to $\rho\epsilon\,\delta\tau$, and so

$$\frac{\partial E}{\partial t} = \left(\rho\frac{\partial\epsilon}{\partial t} + \epsilon\frac{\partial\rho}{\partial t}\right)\delta\tau.$$

For $E_1$ we have, from (7),

$$E_1 = \left[ -\rho(uX + vY + wZ) - p\left(\frac{\partial u}{\partial x} + \frac{\partial v}{\partial y} + \frac{\partial w}{\partial z}\right) \right.$$
$$\left. + \frac{1}{2}\rho\frac{D}{Dt}(u^2 + v^2 + w^2) + \Phi \right]\delta\tau.$$

To calculate $E_2$, the energy conducted into the element through the sides, we use the rule[||] that the rate of conduction of heat through an element of area $\delta S$ is $-k(\partial T/\partial n)\,\delta S$, where $k$ is the thermal conductivity and $n$ is the normal drawn in the direction considered.

† Lamb, *Hydrodynamics* (1932), p. 9.                           ‡ *Ibid.*, p. 549.
|| Carslaw, *Conduction of Heat* (1921), p. 4.

Now at $P$ $(x, y, z)$ the temperature gradient in the positive direction of the $x$-axis is $\partial T/\partial x$, and the temperature gradient at $ABCD$ in this direction is

$$\frac{\partial T}{\partial x} - \frac{\delta x}{2} \frac{\partial^2 T}{\partial x^2}.$$

Similarly at $A'B'C'D'$ it is

$$\frac{\partial T}{\partial x} + \frac{\delta x}{2} \frac{\partial^2 T}{\partial x^2}.$$

Hence the rate of conduction of heat inwards through $ABCD$ is

$$-k\,\delta y\delta z\left(\frac{\partial T}{\partial x} - \frac{\delta x}{2} \frac{\partial^2 T}{\partial x^2}\right),$$

and outwards through $A'B'C'D'$ it is

$$-k\,\delta y\delta z\left(\frac{\partial T}{\partial x} + \frac{\delta x}{2} \frac{\partial^2 T}{\partial x^2}\right);$$

therefore, for these two faces together, the rate of conduction of heat into the element is

$$k\,\delta x\delta y\delta z\frac{\partial^2 T}{\partial x^2};$$

and summing for all the faces we find

$$E_2 = Jk\nabla^2 T\,\delta\tau,$$

where $J$ is the mechanical equivalent of heat. In c.g.s. units $J$ has the value $4{\cdot}18 \times 10^7$ ergs per calorie.

The rate of convection of kinetic and intrinsic energy through an element of area $\delta S$ is $\rho v_n \epsilon\,\delta S$, where $v_n$ is the velocity normal to the element. It follows that the rate of convection of kinetic and intrinsic energy into the elementary parallelepiped through the face $ABCD$ is

$$\delta y\delta z\left[\rho u\epsilon - \frac{\delta x}{2} \frac{\partial}{\partial x}(\rho u\epsilon)\right]$$

and out of it through the face $A'B'C'D'$ it is

$$\delta y\delta z\left[\rho u\epsilon + \frac{\delta x}{2} \frac{\partial}{\partial x}(\rho u\epsilon)\right];$$

for these two faces together the rate of convection into the element is

$$-\delta x\delta y\delta z\frac{\partial}{\partial x}(\rho u\epsilon);$$

and summing for all the faces we find

$$E_3 = -\left[\frac{\partial}{\partial x}(\rho u\epsilon) + \frac{\partial}{\partial y}(\rho v\epsilon) + \frac{\partial}{\partial z}(\rho w\epsilon)\right]\delta\tau$$

$$= \left[\epsilon\frac{\partial\rho}{\partial t} - \rho\left(u\frac{\partial\epsilon}{\partial x} + v\frac{\partial\epsilon}{\partial y} + w\frac{\partial\epsilon}{\partial z}\right)\right]\delta\tau,$$

by the equation of continuity (2).

Substituting these expressions in (9) we obtain the equation

$$\rho\frac{D\epsilon}{Dt} = -p\left(\frac{\partial u}{\partial x} + \frac{\partial v}{\partial y} + \frac{\partial w}{\partial z}\right) + \frac{1}{2}\rho\frac{D}{Dt}(u^2+v^2+w^2) + Jk\nabla^2 T + \Phi. \quad (10)$$

Now for a perfect gas for which the specific heat at constant volume, $c_v$, is constant, the total kinetic energy per unit mass is given by

$$\epsilon = \tfrac{1}{2}(u^2+v^2+w^2) + Jc_v T. \quad (11)$$

The first term on the right-hand side of (11) represents the kinetic energy of the mass as a whole, and the second term represents its intrinsic energy.[†] With the expression (11) for $\epsilon$, equation (10) becomes

$$\rho Jc_v\frac{DT}{Dt} + p\left(\frac{\partial u}{\partial x} + \frac{\partial v}{\partial y} + \frac{\partial w}{\partial z}\right) = Jk\nabla^2 T + \Phi. \quad (12)$$

This energy equation determines the temperature distribution in any field of flow.

The same equation holds also for liquids, which are assumed to be incompressible, so that the second term on the left-hand side vanishes and the equation becomes

$$\rho Jc_v\frac{DT}{Dt} = Jk\nabla^2 T + \Phi. \quad (13)$$

For gases it is useful to transform equation (12) to another form. For perfect gases the relation

$$J(c_p-c_v) = \bar{R}/m$$

holds,[‡] where $c_p$ is the specific heat at constant pressure, and $\bar{R}$ and $m$ are the same as in equation (1). Hence

$$Jc_v T = Jc_p T - \bar{R}T/m = Jc_p T - p/\rho,$$

from the equation of state (1). This gives an alternative expression

---

† If $c_v$ is a function of $T$ this term must be replaced by $J\int c_v\,dT$.

‡ Jeans, *op. cit.*, p. 185.

for the intrinsic energy of a perfect gas; and for the total kinetic energy per unit mass we have

$$\epsilon = \tfrac{1}{2}(u^2+v^2+w^2)+(Jc_p\,T-p/\rho). \tag{14}$$

Then

$$\rho\frac{D\epsilon}{Dt} = \frac{1}{2}\rho\frac{D}{Dt}(u^2+v^2+w^2)+\rho Jc_p\frac{DT}{Dt}-\frac{Dp}{Dt}-p\!\left(\frac{\partial u}{\partial x}+\frac{\partial v}{\partial y}+\frac{\partial w}{\partial z}\right), \tag{15}$$

when we make use of the equation of continuity (3). Substituting in (10), we have an alternative form for the energy equation for a perfect gas:

$$\rho Jc_p\frac{DT}{Dt}-\frac{Dp}{Dt} = Jk\nabla^2 T+\Phi. \tag{16}$$

## 261. Dynamical similarity.

The isothermal flow of a viscous fluid at speeds which are small compared with the speed of sound is known to depend only on the Reynolds number, and the flow pattern is identical for two experiments provided that the Reynolds number is the same. When heat transfer is occurring, the temperature distribution may be shown to depend on three parameters:

(1) the Reynolds number $R = u_0 d/\nu$, where $u_0$ is a representative velocity and $d$ a representative length;

(2) the Prandtl number $\sigma = \nu/\kappa$, where $\kappa = k/(\rho c_p)$ and is called the thermometric conductivity; and

(3) the Grashoff number $\mathrm{Gr} = d^3 g(T_1-T_0)/(\nu^2 T_0)$, where $T_1$ and $T_0$ are two representative temperatures.

The Prandtl number $\sigma$ depends only on the material properties of the fluid; it is therefore preferable to the Péclet number $\mathrm{P\acute{e}} = u_0 d/\kappa$ which, like the Reynolds number, depends on the characteristics of the flow.† When $\sigma$ is unity the solution of a problem may be considerably simplified, because the temperature and velocity distributions are then, in some cases, identical.

The heat transfer coefficient or Nusselt number Nu is defined by the relation

$$Q = \mathrm{Nu}\,kS(T_1-T_0)/d, \tag{17}$$

where $Q$ is the quantity of heat transferred in unit time from an immersed body across an area $S$, $(T_1-T_0)$ is a representative tempera-

† Since $\mathrm{P\acute{e}} = \sigma R$, $\sigma$ or Pé may be adopted arbitrarily as a fundamental parameter.

ture difference, and $d$ a representative length.† It follows that the solution of the equations may be written

$$\text{Nu} = f(R, \sigma, \text{Gr}). \tag{18}$$

The analogy between heat transfer and skin-friction, described in Chap. XV, § 279, leads to the use of an alternative heat transfer number to replace the Nusselt number. This number is

$$\frac{\text{Nu}}{\sigma R} = \frac{Q}{\rho c_p\, Su_0(T_1 - T_0)},$$

and it may conveniently be denoted by the symbol $k_H$. Then (18) may be written

$$k_H = \frac{\text{Nu}}{\sigma R} = F(R, \sigma, \text{Gr}). \tag{19}$$

It has already been remarked that heat transfer problems may be divided into cases of free and of forced convection. For the former the Reynolds number is of no significance, since no representative velocity exists, and in such cases the solution of the equations has the form

$$\text{Nu} = f(\sigma, \text{Gr}). \tag{20}$$

For forced convection the forces due to the action of gravity on the fluid of variable density are assumed to be negligible in comparison with the impressed forces; the Grashoff number is then of no significance, and the solution of the equations is

$$\text{Nu} = f(R, \sigma), \tag{21}$$

or (alternatively)        $$k_H = \frac{\text{Nu}}{\sigma R} = F(R, \sigma). \tag{22}$$

Two other non-dimensional quantities have to be considered in some problems. These are $(T_1 - T_0)/T_0$, which is a measure of the variation of the physical properties of the fluid due to the temperature differences present,‡ and $u_0^2/[2Jc_p(T_1 - T_0)]$, which is the ratio of the kinetic energy to the change of intrinsic energy. The quantity $u_0^2/(2Jc_p)$ is a measure of the temperature differences which arise from the variation of pressure and from the dissipation due to viscosity,‖ and the number $u_0^2/[2Jc_p(T_1 - T_0)]$ will be unimportant

† In other books on heat transfer a heat transfer coefficient $q$ is defined as the rate of heat transfer per unit area per unit temperature difference. This quantity is convenient physically, but suffers from the disadvantage that it is not dimensionless.

‡ The experimental results for free convection from a heated horizontal cylinder in diatomic gases have been critically examined by Hermann (*Ver. deutsch. Ing., Forschungsheft* 379 (1936)) with special reference to the influence of this parameter, which was eliminated by extrapolation to small values.        ‖ Cf. § 263, p. 613.

only if these temperature differences are small compared with the imposed temperature difference $(T_1 - T_0)$. Thus there are definite limits to the applicability of the theory developed in this chapter. On the one hand, the temperature differences must be small enough to permit the assumption that the physical properties do not vary with temperature. On the other hand, the temperature differences must be large enough compared with the quantity $u_0^2/(2Jc_p)$ to permit the dissipation and the temperature changes due to pressure variations to be neglected, as they will be except in §§ 263, 264, 265, 269.

The above equations, which govern the convection of heat, apply also to the diffusion of other physical quantities—in particular, to the diffusion of solid particles in a moving fluid, provided that the heat convection parameters are replaced by corresponding mass diffusion parameters.

## 262. The convection of heat in potential flow.

The theory of the convection of heat in potential flow has been developed by Boussinesq† and by King.‡ Let $\phi$ and $\psi$ be the velocity potential and the stream-function for the potential flow of an incompressible fluid in two dimensions. Then the velocities are given by

$$\left. \begin{aligned} u &= \frac{\partial \phi}{\partial x} = \frac{\partial \psi}{\partial y}, \\ v &= \frac{\partial \phi}{\partial y} = -\frac{\partial \psi}{\partial x}, \end{aligned} \right\} \tag{23}$$

while the temperature distribution is given by (13):

$$u \frac{\partial T}{\partial x} + v \frac{\partial T}{\partial y} = \kappa \nabla^2 T, \tag{24}$$

where $\kappa = k/\rho c_v$; the dissipation function vanishes since viscosity is neglected. Changing coordinates from $(x, y)$ to $(\phi, \psi)$ and substituting from (23) for $u$ and $v$, we find from equation (24) that

$$\frac{\partial T}{\partial \phi} = \kappa \left( \frac{\partial^2 T}{\partial \phi^2} + \frac{\partial^2 T}{\partial \psi^2} \right). \tag{25}$$

Boussinesq neglects the term $\partial^2 T/\partial \phi^2$ in (25) and builds up solutions of the simplified equation based on the 'point-source solution'

$$T = \frac{1}{\phi^{\frac{1}{2}}} \exp\left( -\frac{\psi^2}{4\kappa\phi} \right), \tag{26}$$

† *Journ. d. Math.* 1 (1905), 285–332.          ‡ *Phil. Trans.* A, 214 (1914), 373–432.

while King retains the complete equation and builds up solutions from the following solution of (25):

$$T = \exp\left(\frac{\phi}{2\kappa}\right) K_0\left(\frac{(\phi^2+\psi^2)^{\frac{1}{2}}}{2\kappa}\right), \tag{27}$$

where $K_0$ is the modified Bessel function of the second kind.

For the circular cylinder, which King considered in detail, he chose a distribution of heat flux from the surface which gave results for the total heat flux in agreement with his experimental results. This distribution involves discontinuities of temperature between the surface and the fluid flowing over it; the theory is unsatisfactory, and it is clear that the effects of viscosity are of great importance. King's experimental results obtained with a whirling arm were used for the calibration of his hot wire anemometer, but Hilpert's† are probably more accurate.

## 263. Boundary layer equations.

When the Péclet number $u_0 d/\kappa$ is large, the conditions with respect to the temperature distribution near the surface of a heated body past which fluid is flowing are the same as those with respect to the velocity distribution when the Reynolds number is large. A boundary layer of small thickness exists near the surface, in which the temperature falls rapidly from its value at the surface to the value in the main body of the fluid, and the approximations of Chapter IV may be applied to the equations (4) and (16) for the velocity and temperature. The equations for the temperature layer will be developed on the assumption that the Prandtl number $\sigma$ is of order unity, so that the thickness of the temperature layer is of the same order of magnitude as that of the velocity layer.

Consider the flow near a cylindrical body held normal to a stream of undisturbed velocity $u_0$. The motion is two-dimensional, and if $x$ is measured along the surface, $y$ normal to it, and $z$ along the generators of the cylinder, the terms depending on $z$ disappear. Thus equations (4) become (eq. (8) of Chap. IV)

$$\left.\begin{aligned}\frac{\partial u}{\partial t}+u\frac{\partial u}{\partial x}+v\frac{\partial u}{\partial y} &= -\frac{1}{\rho}\frac{\partial p}{\partial x}+X+\nu\frac{\partial^2 u}{\partial y^2}, \\ -Ku^2 &= -\frac{1}{\rho}\frac{\partial p}{\partial y}+Y, \end{aligned}\right\} \tag{28}$$

† Cf. § 272.

where $K$ is the curvature of the surface. The components of $\mathbf{g}$ at the point $(x, y)$ are $(lg, mg)$, where $(l, m)$ are the direction cosines of the downward-drawn vertical and $g$ is the magnitude of the earth's acceleration. Hence for the components of the external force arising from the variable density, we have, from (5),

$$\left.\begin{aligned} X &= -lg\left(\frac{T-T_0}{T_0}\right), \\ Y &= -mg\left(\frac{T-T_0}{T_0}\right). \end{aligned}\right\} \tag{29}$$

Taking the equation of continuity in the form (3), and substituting for $\rho$ from (1), we have, for perfect gases,

$$\frac{\partial u}{\partial x} + \frac{\partial v}{\partial y} + \frac{1}{p}\frac{Dp}{Dt} - \frac{1}{T}\frac{DT}{Dt} = 0, \tag{30}$$

while for liquids the form

$$\frac{\partial u}{\partial x} + \frac{\partial v}{\partial y} = 0 \tag{31}$$

still holds. Now suppose that $\delta$ and $\delta'$ are the thicknesses of the velocity and temperature layers respectively, and further that $\delta$ and $\delta'$ are of the same order of magnitude.† Let $u_0$, the velocity at infinity, and $d$, a characteristic length of the body, be of standard order; then $\partial u/\partial x$ and $\partial v/\partial y$ are both $O(1)$. Also $p^{-1} Dp/Dt$ is small compared with unity if $\rho_0 u_0^2$ is small compared with the standard pressure $p_0$—that is, if the velocities in the field are small compared with the velocity of sound; $T^{-1} DT/Dt$ is small compared with unity if the temperature difference between the surface and the main body of the fluid is small compared with the standard absolute temperature $T_0$. Both these conditions will be assumed to hold, and the equation of continuity (30) for gases then reduces to the same form (31) as for incompressible fluids.‡

For two-dimensional motion the energy equation (16) becomes

$$\rho J c_p\left(\frac{\partial T}{\partial t} + u\frac{\partial T}{\partial x} + v\frac{\partial T}{\partial y}\right) - \left(\frac{\partial p}{\partial t} + u\frac{\partial p}{\partial x} + v\frac{\partial p}{\partial y}\right) = Jk\left(\frac{\partial^2 T}{\partial x^2} + \frac{\partial^2 T}{\partial y^2}\right) + \Phi. \tag{32}$$

It was shown in Chapter IV (equation (14), p. 122) that the dissipation function for incompressible fluids is given approximately by

$$\Phi = \mu\left(\frac{\partial u}{\partial y}\right)^2. \tag{33}$$

---

† This is equivalent to assuming that $\sigma^{\frac{1}{2}}$ is $O(1)$.

‡ It does not follow that the second term in (12) may be put equal to zero; $p$ is a large quantity and this term is of the same order of magnitude as the other terms.

For compressible fluids an additional term $-\frac{2}{3}\mu(\partial u/\partial x + \partial v/\partial y)^2$ is present (see equation (8)). The approximate form (31) of the equation of continuity shows that this is small, and $\Phi$ is therefore given as before by (33).

The velocity $u$ varies from zero at $y = 0$ to $u_1$ (say) at $y = \delta$, while the temperature $T$ varies from $T_1$ (the surface temperature) at $y = 0$ to $T_0$ (the temperature of the body of the fluid) at $y = \delta'$. Taking $u_1$ and $(T_1 - T_0)$ as quantities of standard order, $u$ is $O(1)$ and $v$ is $O(\delta)$, $\partial^2 T/\partial x^2$ is $O(1)$ and $\partial^2 T/\partial y^2$ is $O(\delta'^{-2})$ and hence $\partial^2 T/\partial x^2$ may be neglected. Also $u\,\partial T/\partial x$ is $O(1)$ and $v\,\partial T/\partial y$ is $O(\delta/\delta')$ which is $O(1)$; $u\,\partial p/\partial x$ is $O(1)$ while $v\,\partial p/\partial y$ is $O(\delta)$ (since, by (28), $\partial p/\partial y$ is at most $O(1)$) and therefore $v\,\partial p/\partial y$ may be neglected. Equation (32) then becomes

$$\rho J c_p\left(\frac{\partial T}{\partial t} + u\frac{\partial T}{\partial x} + v\frac{\partial T}{\partial y}\right) - \left(\frac{\partial p}{\partial t} + u\frac{\partial p}{\partial x}\right) = Jk\frac{\partial^2 T}{\partial y^2} + \mu\left(\frac{\partial u}{\partial y}\right)^2. \quad (34)$$

The corresponding equation for liquids obtained in the same way from (13) is

$$\rho J c_v\left(\frac{\partial T}{\partial t} + u\frac{\partial T}{\partial x} + v\frac{\partial T}{\partial y}\right) = Jk\frac{\partial^2 T}{\partial y^2} + \mu\left(\frac{\partial u}{\partial y}\right)^2. \quad (35)$$

An alternative form will now be derived for the steady motion of gases when no external force $X$ is present. Putting $X = 0$, and multiplying by $\rho u$, we see that the first equation of (28) may be written

$$\rho\frac{D}{Dt}\left(\frac{u^2}{2}\right) + u\frac{\partial p}{\partial x} = \mu u\frac{\partial^2 u}{\partial y^2}.$$

Adding this to (34) we have, since $\partial p/\partial t = 0$ for steady flow,

$$\rho J c_p\frac{DT}{Dt} + \rho\frac{D}{Dt}\left(\frac{u^2}{2}\right) = Jk\frac{\partial^2 T}{\partial y^2} + \mu\frac{\partial^2}{\partial y^2}\left(\frac{u^2}{2}\right),$$

whence†
$$\frac{D}{Dt}(J c_p T + \tfrac{1}{2}u^2) = \kappa\frac{\partial^2}{\partial y^2}(J c_p T + \tfrac{1}{2}\sigma u^2), \quad (36)$$

since $k/(\rho c_p) = \kappa$ and $\sigma = \nu/\kappa$. This form is used in § 269 to derive an equation for the reading of a thermometer in a moving fluid for the special case $\sigma = 1$.

To obtain the order of magnitude of $\delta'$, we note that $k/(\rho c_p\,\delta'^2)$ or $\kappa/\delta'^2$ must be $O(1)$ if the convection terms on the left-hand side of (34) are to be of the same order as the conduction term on the right-hand side. Hence $\delta'$ is $O(\kappa^{\frac{1}{2}})$. If the equations are reduced to non-dimen-

† Busemann, *Handbuch der Experimentalphysik*, **4**, part 1 (1931), 366.

sional form by replacing $u$ by $u/u_0$, $x$ by $x/d$, $T$ by $T/(T_1-T_0)$, etc. we find for the order of magnitude of the thickness of the temperature layer

$$\frac{\delta'}{d} = O\left(\sqrt{\frac{k}{\rho c_p u_0 d}}\right) = O\left(\sqrt{\frac{1}{\sigma R}}\right).$$

It is relevant to note the magnitudes of the terms $u\,\partial p/\partial x$ and $\mu(\partial u/\partial y)^2$ which occur in (34). Apart from the variations produced by gravity, the pressure variations in the field are $O(\rho u_0^2)$, and hence $u\,\partial p/\partial x$ is $O(\rho u_0^3/d)$. Also $\mu = \rho u_0 d/R$, and $\partial u/\partial y$ is

$$O(u_0/\delta) = O(u_0 R^{\frac{1}{2}}/d)$$

by (9) of Chapter IV; hence $\mu(\partial u/\partial y)^2$ is $O(\rho u_0^3/d)$, and therefore $u\,\partial p/\partial x$ and $\mu(\partial u/\partial y)^2$ are of the same order of magnitude. It follows that if account is taken of the dissipation term $\mu(\partial u/\partial y)^2$ it is necessary to take account also of the term $u\,\partial p/\partial x$ (except for the flat plate for which $\partial p/\partial x = 0$). Further, both these quantities may be neglected if they are small compared with the other terms in (34). Since the other terms are $O[\rho J c_p u_0(T_1-T_0)/d]$ this is permissible if

$$u_0^2/[J c_p(T_1-T_0)]$$

is a small quantity.

## 264. The momentum and energy equations for the boundary layer.

The momentum equation for the boundary layer is similar to that derived in § 52 of Chapter IV, except that it is necessary to add the term arising from the gravity forces (29). Equation (33) of Chapter IV is then replaced by

$$\frac{\partial}{\partial t}\int_0^\delta \rho u\,dy + \frac{\partial}{\partial x}\int_0^\delta \rho u^2\,dy - u_1\left[\frac{\partial}{\partial t}\int_0^\delta \rho\,dy + \frac{\partial}{\partial x}\int_0^\delta \rho u\,dy\right]$$

$$= -\delta\frac{\partial p}{\partial x} - lg\int_0^\delta \rho\left(\frac{T-T_0}{T_0}\right)dy - \mu\left(\frac{\partial u}{\partial y}\right)_{y=0}, \quad (37)$$

where $u_1$ is the velocity at the outside of the boundary layer.

The corresponding energy equation for the boundary layer may be derived either by integration of (34) across the layer or by considering the energy balance in an element of the layer. Only the former method will be given here. If $\delta'$ is the thickness of the temperature boundary layer, we obtain, on integrating (34) from $y = 0$ to $y = \delta'$,

$$
Jc_p \int_0^{\delta'} \left( \rho \frac{\partial T}{\partial t} + \rho u \frac{\partial T}{\partial x} + \rho v \frac{\partial T}{\partial y} \right) dy - \int_0^{\delta'} \left( \frac{\partial p}{\partial t} + u \frac{\partial p}{\partial x} \right) dy
$$

$$
= -Jk \left( \frac{\partial T}{\partial y} \right)_{y=0} + \mu \int_0^{\delta'} \left( \frac{\partial u}{\partial y} \right)^2 dy \quad (38)
$$

(since $\partial T/\partial y$ vanishes at $y = \delta'$, the outer edge of the temperature layer). Now

$$
\int_0^{\delta'} \rho u \frac{\partial T}{\partial x} \, dy = \frac{\partial}{\partial x} \int_0^{\delta'} \rho u T \, dy - \int_0^{\delta'} T \frac{\partial}{\partial x} (\rho u) \, dy - (\rho u T)_{y=\delta'} \frac{\partial \delta'}{\partial x},
$$

$$
\int_0^{\delta'} \rho v \frac{\partial T}{\partial y} \, dy = [\rho v T]_0^{\delta'} - \int_0^{\delta'} T \frac{\partial}{\partial y} (\rho v) \, dy,
$$

and $\quad \displaystyle \int_0^{\delta'} \rho \frac{\partial T}{\partial t} \, dy = \frac{\partial}{\partial t} \int_0^{\delta'} \rho T \, dy - (\rho T)_{y=\delta'} \frac{\partial \delta'}{\partial t} - \int_0^{\delta'} T \frac{\partial \rho}{\partial t} \, dy.$

Also $v$ vanishes for $y = 0$, and for $y = \delta'$ we have from (2)

$$
(\rho v)_{y=\delta'} = - \int_0^{\delta'} \left[ \frac{\partial \rho}{\partial t} + \frac{\partial}{\partial x} (\rho u) \right] dy
$$

$$
= -\frac{\partial}{\partial t} \int_0^{\delta'} \rho \, dy - \frac{\partial}{\partial x} \int_0^{\delta'} \rho u \, dy + (\rho)_{y=\delta'} \frac{\partial \delta'}{\partial t} + [\rho u]_{y=\delta'} \frac{\partial \delta'}{\partial x};
$$

hence, making use of (2),

$$
\int_0^{\delta'} \left( \rho \frac{\partial T}{\partial t} + \rho u \frac{\partial T}{\partial x} + \rho v \frac{\partial T}{\partial y} \right) dy
$$

$$
= \frac{\partial}{\partial t} \int_0^{\delta'} \rho T \, dy + \frac{\partial}{\partial x} \int_0^{\delta'} \rho u T \, dy - T' \left( \frac{\partial}{\partial t} \int_0^{\delta'} \rho \, dy + \frac{\partial}{\partial x} \int_0^{\delta'} \rho u \, dy \right),
$$

where $T'$ is the temperature at the outside of the layer. Substituting in (38) we obtain†

$$Jc_p\left[\frac{\partial}{\partial t}\int\limits_0^{\delta'}\rho T\,dy+\frac{\partial}{\partial x}\int\limits_0^{\delta'}\rho uT\,dy\right.$$

$$\left.-T'\left(\frac{\partial}{\partial t}\int\limits_0^{\delta'}\rho\,dy+\frac{\partial}{\partial x}\int\limits_0^{\delta'}\rho u\,dy\right)\right]-\int\limits_0^{\delta'}\left(\frac{\partial p}{\partial t}+u\frac{\partial p}{\partial x}\right)dy$$

$$=-Jk\left(\frac{\partial T}{\partial y}\right)_{y=0}+\mu\int\limits_0^{\delta'}\left(\frac{\partial u}{\partial y}\right)^2 dy. \tag{39}$$

This is the equation for perfect gases. The corresponding equation for liquids, obtained in the same way from (35), is

$$Jc_v\left[\int\limits_0^{\delta'}\rho\frac{\partial T}{\partial t}\,dy+\frac{\partial}{\partial x}\int\limits_0^{\delta'}\rho uT\,dy-T'\frac{\partial}{\partial x}\int\limits_0^{\delta'}\rho u\,dy\right]$$

$$=-Jk\left(\frac{\partial T}{\partial y}\right)_{y=0}+\mu\int\limits_0^{\delta'}\left(\frac{\partial u}{\partial y}\right)^2 dy. \tag{40}$$

## 265. Boundary conditions.

The conditions satisfied by the temperature require (i) that it shall have given values at the surface of the body or, alternatively, that there shall be no heat transferred, in which case the temperature gradient normal to the surface must vanish, and (ii) that the temperature shall have a given value at a large distance from the body. For flow through a pipe the condition at infinity is replaced by a corresponding condition at the entry. It is sometimes necessary to replace the condition at infinity by a condition at the outer edge of the boundary layer. Outside this layer the terms $Jk\nabla^2 T$ and $\Phi$ on the right-hand side of (13) and (16), due respectively to conductivity and dissipation, are negligible. For liquids (13) then becomes

$$\rho Jc_v\frac{DT}{Dt}=0,$$

and hence $\qquad\qquad T=T_0.$

For the flow of liquids the temperature is therefore constant except

---

† Frankl (*Trans. Centr. Aero-Hydrodyn. Inst., Moscow*, No. 176 (1934)) has derived an integral equation similar to (39) which is applied to calculate the effect of compressibility and dissipation on the drag of a flat plate.

in the regions in which conductivity and viscosity are important, and the condition at infinity may be replaced by an identical condition at the edge of the boundary layer. For gases, on the other hand, (16) becomes, for flow outside the boundary layer,

$$\rho J c_p \frac{DT}{Dt} - \frac{Dp}{Dt} = 0,$$

and the integral of this is

$$J c_p T - \int \frac{dp}{\rho} = \text{constant.} \tag{41}$$

Also outside the velocity layer the equation

$$\int \frac{dp}{\rho} + \tfrac{1}{2} q^2 + \Omega = \text{constant}$$

holds, where $\Omega$ is the potential of the external forces, i.e. the potential energy per unit mass. Hence from (41)

$$J c_p T + \tfrac{1}{2} q^2 + \Omega = \text{constant.} \tag{42}$$

It follows that the temperature $T'$ at the outer edge of the velocity or temperature boundary layer, whichever is the thicker, is given by

$$T' - T_0 = -\frac{1}{2 J c_p} (u_1^2 - u_0^2), \tag{43}$$

where $u_1$ is the velocity at the outer edge of the layer, and $u_0$, $T_0$ are the velocity and temperature in the undisturbed stream at the same value of $\Omega$. The equation (43) holds generally with $u_0$, $T_0$ denoting the velocity and temperature in the undisturbed stream, if the variation of $\Omega$ is neglected.

## 266. Laminar flow in a circular pipe. Wall at constant temperature.

The solution of the problem of heat transfer to a fluid flowing with the Poiseuille velocity distribution through a circular pipe of radius $a$, at a section of which the temperature of the surface changes discontinuously, has been obtained by Graetz† and Nusselt.‡ Take cylindrical coordinates $r$, $\phi$, $z$, with the origin at the centre of the section at which the wall temperature changes from $T_0$ to $T_1$, and the

---

† *Ann. d. Phys.* **18** (1883), 79–94; **25** (1885), 337–357.

‡ *Zeitschr. des Vereines deutscher Ingenieure*, **54** (1910), 1154–1158.

axis of $z$ along the axis of the pipe. The motion is steady, and equation (16) for the temperature becomes

$$\frac{D\theta}{Dt} = \kappa \nabla^2 \theta, \tag{44}$$

where
$$\theta = \frac{T - T_1}{T_0 - T_1},$$

and $\kappa$ is the thermometric conductivity; in this equation the terms due to the pressure gradient and the dissipation have been neglected.

The boundary conditions are

$$\theta = 0 \quad \text{for} \quad z > 0 \quad \text{and} \quad r = a, \tag{45}$$

$$\theta = 1 \quad \text{for} \quad z = 0 \quad \text{and} \quad r < a. \tag{46}$$

For the Poiseuille flow

$$\left. \begin{array}{l} u_r = 0, \qquad u_\theta = 0, \\ u_z = 2u_m[1 - (r/a)^2], \end{array} \right\} \tag{47}$$

where $u_m$ is the mean velocity. With these values for the velocity components equation (44) becomes

$$2u_m\left[1 - \left(\frac{r}{a}\right)^2\right]\frac{\partial \theta}{\partial z} = \kappa\left[\frac{\partial^2 \theta}{\partial r^2} + \frac{1}{r}\frac{\partial \theta}{\partial r} + \frac{\partial^2 \theta}{\partial z^2}\right]; \tag{48}$$

the term $\partial^2\theta/\partial\phi^2$ vanishes since there is symmetry about the axis of the pipe. It is assumed that $\partial^2\theta/\partial z^2$ may be neglected in comparison with $\partial^2\theta/\partial r^2 + r^{-1}\partial\theta/\partial r$. Solutions are then obtained by putting

$$\theta = A \exp\left(-\beta^2\frac{\kappa z}{2u_m a^2}\right) \cdot \psi(r),$$

where $\beta$ is an undetermined constant, so that equation (48) becomes

$$\frac{d^2\psi}{dr^2} + \frac{1}{r}\frac{d\psi}{dr} + \frac{\beta^2}{a^2}\left[1 - \left(\frac{r}{a}\right)^2\right]\psi = 0.$$

If we write
$$\beta r/a = r',$$
this equation is

$$\frac{d^2\psi}{dr'^2} + \frac{1}{r'}\frac{d\psi}{dr'} + \left[1 - \left(\frac{r'}{\beta}\right)^2\right]\psi = 0. \tag{49}$$

The series solution of this equation which is free from singularities at the origin is

$$\psi(r', \beta) = 1 - \frac{r'^2}{4} + \frac{3r'^4}{2 \cdot 4!}\left(\frac{1}{4} + \frac{1}{\beta^2}\right) + \cdots.$$

The boundary condition (45) requires that $\psi$ shall satisfy

$$\psi(\beta, \beta) = 0.$$

This equation has an infinite number of roots, of which the first three are

$$\beta_0 = 2.705, \qquad \beta_1 = 6.66, \qquad \beta_2 = 10.6.$$

Equation (49) has therefore an infinite number of partial solutions of the form

$$A_n \exp\left(-\beta_n^2 \frac{\kappa z}{2u_m a^2}\right) \chi_n\left(\frac{r}{a}\right),$$

where $\chi_n(r/a)$ stands for $\psi(\beta_n r/a, \beta_n)$. The constants $A_n$ are determined from the boundary condition (46), and Nusselt gives for the first three $A_0 = +1.477$, $A_1 = -0.810$, $A_2 = +0.385$. The temperature distribution is

$$\theta = A_0 \exp\left(-\beta_0^2 \frac{\kappa z}{2u_m a^2}\right) \chi_0\left(\frac{r}{a}\right)$$

$$+ A_1 \exp\left(-\beta_1^2 \frac{\kappa z}{2u_m a^2}\right) \chi_1\left(\frac{r}{a}\right)$$

$$+ A_2 \exp\left(-\beta_2^2 \frac{\kappa z}{2u_m a^2}\right) \chi_2\left(\frac{r}{a}\right) + \cdots,$$

or

$$\theta = f\left(\frac{\kappa z}{u_m d^2}, \frac{r}{a}\right), \tag{50}$$

where $d = 2a$ is the diameter of the pipe. Fig. 238† shows the temperature distribution given by (50). The outer layers of fluid take up the temperature of the walls very quickly and the radial temperature distribution soon becomes steady.

To obtain a comparison with experimental results it is necessary to calculate the mean temperature weighted with respect to the velocity, since this is the temperature which is measured when fluid which has passed through the pipe is mixed. This mean temperature $\theta_M$ is given by

$$\theta_M = \frac{\int_0^a \theta \cdot 2u_m[1-(r/a)^2]2\pi r \, dr}{\int_0^a 2u_m[1-(r/a)^2]2\pi r \, dr},$$

which becomes

$$\theta_M = \frac{\int_0^1 \theta(1-\xi^2)\xi \, d\xi}{\int_0^1 (1-\xi^2)\xi \, d\xi}$$

† Gröber and Erk, *Die Grundgesetze der Wärmeübertragung* (Berlin, 1933), 180.

on putting $\xi = r/a$. The value of the integral in the denominator is 0·25, while the value of the numerator has been calculated from (50) by Gröber[†] and by Jacob and Eck.[‡] They obtain

$$\theta_M = 0{\cdot}819 \exp\left(-14{\cdot}6272\,\frac{\kappa z}{u_m d^2}\right) + 0{\cdot}0976 \exp\left(-89{\cdot}22\,\frac{\kappa z}{u_m d^2}\right)$$

$$+\,0{\cdot}01896 \exp\left(-212\,\frac{\kappa z}{u_m d^2}\right) + \dots,$$

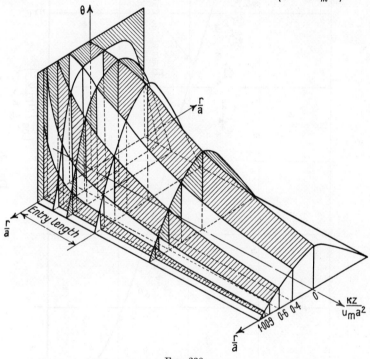

Fig. 238.

or
$$\theta_M = f_1\left(\frac{\kappa z}{u_m d^2}\right).$$

Putting
$$\theta_M = \frac{T_M - T_1}{T_0 - T_1},$$

we have for the mean absolute temperature $T_M$

$$T_M = T_1 + (T_0 - T_1)f_1.$$

† *Die Grundgesetze der Wärmeleitung und des Wärmeüberganges* (Berlin, 1921), p. 181.                                ‡ *Forsch. Ingwes.* **3** (1932), 121–126.

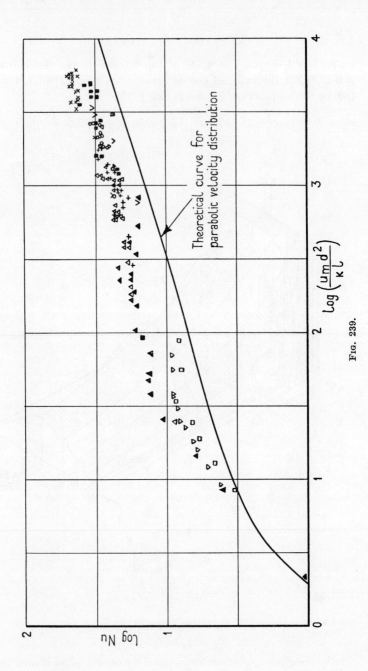

Fig. 239.

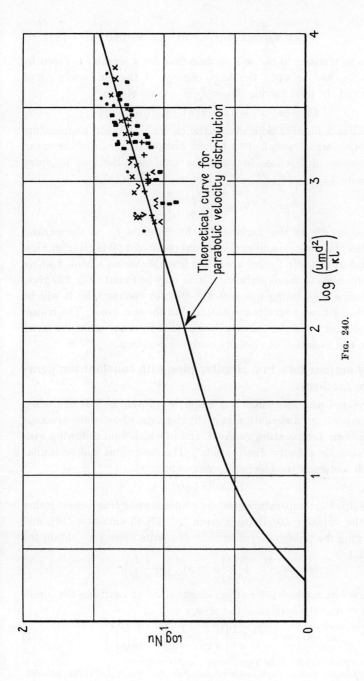

FIG. 240.

The heat transfer to the wall in unit time for a length $l$ is given by the difference between the heat content of the fluid entering at $z = 0$ and the fluid leaving at $z = l$. Hence we obtain

$$Q = \tfrac{1}{4}\pi d^2 \rho c_p \, u_m (T_0 - T_1)[1 - f_1(\kappa l/(u_m d^2))].$$

The Nusselt number depends on the choice of a mean temperature difference, which is taken to be the arithmetic mean of the mean temperature differences between the fluid and the wall at entry and exit—i.e. $\tfrac{1}{2}(1 + \theta_M)(T_0 - T_1)$. Hence (equation (17))

$$\text{Nu} = \frac{u_m d^2}{2\kappa l} \left( \frac{1 - f_1}{1 + f_1} \right),$$

with $\kappa l/(u_m d^2)$ as the variable in the function $f_1$. A comparison between this formula and experimental results for oil is given in Figs. 239 and 240. These figures are taken from McAdams's book,[†] where full references to the experimental data may be found. Fig. 239 gives the results for heating oils and Fig. 240 for cooling oils: it will be seen that different results are obtained in the two cases. The reason is the distortion of the velocity distribution from the parabolic form, due to the variation of viscosity with temperature.

### 267. Laminar flow in a circular pipe with constant temperature gradient.

A related problem which has a simple solution is that of finding the temperature distribution in a circular pipe whose walls are kept at a uniform temperature gradient, and in which fluid is flowing with the Poiseuille velocity distribution.[‡] The conditions will be similar at each section of the pipe, and we may put

$$T = Az + g(r),$$

where $A$ is the temperature gradient. Substituting this form together with the velocity distribution given by (47) in equation (16), and neglecting the pressure term and the dissipation term, we obtain the equation

$$2u_m\left[1 - \left(\frac{r}{a}\right)^2\right]A = \kappa\left(\frac{d^2g}{dr^2} + \frac{1}{r}\frac{dg}{dr}\right).$$

The solution which is free from singularities and satisfies the condition $g = 0$ at the surface of the pipe, $r = a$, is

$$g(r) = -\frac{2u_m A a^2}{\kappa}\left[\frac{3}{16} - \frac{1}{4}\left(\frac{r}{a}\right)^2 + \frac{1}{16}\left(\frac{r}{a}\right)^4\right]. \tag{51}$$

---

[†] *Heat Transmission* (New York, 1933), pp. 203–210.
[‡] Cf. Nusselt, *loc. cit.*; Eagle and Ferguson, *Proc. Roy. Soc.* A, **127** (1930), 540–566.

This gives
$$T_{\max} = Az - \frac{3}{8}\frac{u_m A a^2}{\kappa}$$

as the temperature at the centre of the pipe; for the mean temperatures $T_m$ and $T_M$ we obtain

$$T_m = \frac{2}{a^2}\int_0^a Tr\,dr = Az - \frac{1}{6}\frac{u_m A a^2}{\kappa}$$

and
$$T_M = \frac{2}{a^2 u_m}\int_0^a T u_z r\,dr = Az - \frac{11}{48}\frac{u_m A a^2}{\kappa},$$

where $T_m$ is the unweighted mean, and $T_M$ the mean weighted with respect to the velocity—i.e. the temperature which is measured in fluid which is mixed after passing through the pipe. The rate of heat transfer per unit area, $Q$, is equal to $k(\partial T/\partial r)_{r=a}$, which from (51) is $ku_m A a/2\kappa$; and the value of the Nusselt number Nu (equation (17) with $Q/S$ replaced by $Q$ and $d$ as the pipe diameter) is, when based on the mean temperature $T_m$,

$$\mathrm{Nu} = -\frac{2aQ}{k(T_m - Az)} = 6,$$

but when based on $T_M$ it is

$$\mathrm{Nu} = 48/11.$$

## 268. Forced convection in a laminar boundary layer at a flat plate along the stream.

For steady flow past a flat plate, at temperature $T_1$ and parallel to a stream of velocity $u_0$ and temperature $T_0$, the pressure $p$ is constant, so that equation (16) governing the temperature distribution is

$$\rho J c_p\left(u\frac{\partial T}{\partial x} + v\frac{\partial T}{\partial y}\right) = Jk\frac{\partial^2 T}{\partial y^2} + \mu\left(\frac{\partial u}{\partial y}\right)^2. \tag{52}$$

When the heat generated by dissipation is neglected, this becomes

$$u\frac{\partial T}{\partial x} + v\frac{\partial T}{\partial y} = \kappa\frac{\partial^2 T}{\partial y^2}, \tag{53}$$

and the boundary conditions are

$$T = T_1 \quad \text{at} \quad y = 0, \qquad T = T_0 \quad \text{at} \quad y = \infty.$$

For small temperature differences the velocity distribution is the same as for isothermal flow. The velocity $u_1$ outside the boundary layer is taken equal to the undisturbed velocity $u_0$, and the solution in Chap. IV, § 53 shows that, if

$$\eta = \tfrac{1}{2}(u_0/\nu x)^{\frac{1}{2}}y$$

and $\psi$ is the stream-function, then

$$\psi = (\nu u_0 x)^{\frac{1}{2}}f(\eta), \qquad u = \tfrac{1}{2}u_0 f', \qquad v = \tfrac{1}{2}(u_0 \nu/x)^{\frac{1}{2}}(\eta f' - f),$$

where $f(\eta)$ satisfies the equation

$$f''' + ff'' = 0.$$

With these values of $u$ and $v$ equation (53) can be satisfied if the temperature is a function of $\eta$ only. If we put

$$T = T_1 - (T_1 - T_0)\theta(\eta),$$

(53) becomes $\qquad\qquad \theta'' + \sigma f\theta' = 0,$ $\qquad\qquad\qquad$ (54)

where $\sigma = \nu/\kappa$, and the boundary conditions satisfied by $\theta$ are $\theta = 0$ at $\eta = 0$, $\theta = 1$ at $\eta = \infty$. The solution of (54) with these boundary conditions is†

$$\theta(\eta) = \alpha_1(\sigma) \int_0^\eta \exp\left\{-\sigma \int_0^\eta f \, d\eta\right\} d\eta,$$

where $\qquad\qquad \dfrac{1}{\alpha_1(\sigma)} = \displaystyle\int_0^\infty \exp\left\{-\sigma \int_0^\eta f \, d\eta\right\} d\eta.$

Substituting $f = -f'''/f''$, we obtain the alternative form

$$\theta(\eta) = \alpha_1(\sigma) \int_0^\eta \left[\frac{f''(\eta)}{f''(0)}\right]^\sigma d\eta.$$

Fig. 241 shows the variation of $\theta$ with $\eta$ for a number of values of $\sigma$. Pohlhausen (loc. cit.) tabulates $\alpha_1(\sigma)$, and remarks that

$$\alpha_1(\sigma) \doteqdot 0{\cdot}664\sigma^{\frac{1}{3}},$$

which is shown for comparison in Table 19.

TABLE 19

| $\sigma$ | $\alpha_1(\sigma)$ | $0{\cdot}664\sigma^{\frac{1}{3}}$ |
|---|---|---|
| 0·6 | 0·552 | 0·560 |
| 0·7 | 0·585 | 0·589 |
| 0·8 | 0·614 | 0·616 |
| 0·9 | 0·640 | 0·641 |
| 1·0 | 0·664 | 0·664 |
| 1·1 | 0·687 | 0·685 |
| 7·0 | 1·29 | 1·26 |
| 10·0 | 1·46 | 1·43 |
| 15·0 | 1·67 | 1·64 |

† Pohlhausen, Zeitschr. f. angew. Math. u. Mech. 1 (1921), 115–120.

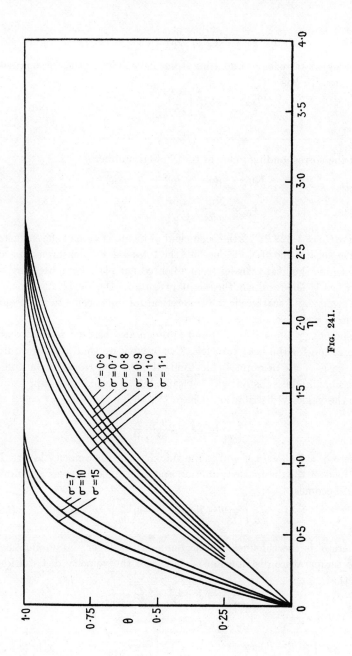

Fig. 241.

The heat transfer in unit time from one side of a plate of length $l$ and breadth $b$ is

$$Q = -kb \int\limits_0^l \left(\frac{\partial T}{\partial y}\right)_{y=0} dx$$

$$= \alpha_1(\sigma)kb(T_1-T_0)\sqrt{(u_0 l/\nu)},$$

and the corresponding value of the Nusselt number is

$$\mathrm{Nu} = \alpha_1(\sigma)\sqrt{(u_0 l/\nu)} = \alpha_1(\sigma)R^{\frac{1}{2}}.$$

Also
$$k_H = \frac{\mathrm{Nu}}{\sigma R} = \frac{\alpha_1(\sigma)}{\sigma R^{\frac{1}{2}}} \doteqdot \tfrac{1}{2}\sigma^{-\frac{2}{3}}C_f,$$

where $C_f (= 1\cdot328R^{-\frac{1}{2}})$ is the coefficient of mean friction of the surface.

The local value of Nu is one-half the total value. Measurements to determine the heat transfer from a heated flat plate have been made by Elias.† The results in the laminar region for the local heat transfer are scanty and scattered; the theoretical formula seems to represent them as well as could be expected.

The energy equations (39) and (40) can also be used to obtain an expression for the heat transfer. For brevity we shall consider only the case of an incompressible fluid, though the result for a compressible fluid is substantially the same. An approximate expression for the velocity distribution (Chap. IV, § 60, p. 157) is

$$\frac{u}{u_0} = 2\frac{y}{\delta} - 2\frac{y^3}{\delta^3} + \frac{y^4}{\delta^4},$$

where $\delta = 5\cdot83(\nu x/u_0)^{\frac{1}{2}}$ and is the thickness of the velocity layer. If we neglect the dissipation function and take the density as constant, (40) becomes

$$\frac{d}{dx} \int\limits_0^{\delta'} u(T-T_0)\,dy = -\kappa\left(\frac{\partial T}{\partial y}\right)_{y=0}, \tag{55}$$

where $\delta'$ is the thickness of the temperature layer. Assuming that the temperature distribution is similar to the velocity distribution, so that

$$\frac{T-T_0}{T_1-T_0} = 1 - 2\frac{y}{\delta'} + 2\frac{y^3}{\delta'^3} - \frac{y^4}{\delta'^4},$$

we obtain

$$\int\limits_0^{\delta'} u(T-T_0)\,dy = u_0(T_1-T_0)\,\delta'\left(\frac{2}{15}\chi - \frac{3}{140}\chi^3 + \frac{1}{180}\chi^4\right),$$

† *Zeitschr. f. angew. Math. u. Mech.* **9** (1929), 434–453; **10** (1930), 1–14.

where $\chi = \delta'/\delta$ and is the ratio of the thicknesses of the temperature and velocity layers, and is independent of $x$. Then (55) gives

$$u_0 \delta'^2 \left( \frac{2}{15}\chi - \frac{3}{140}\chi^3 + \frac{1}{180}\chi^4 \right) = 4\kappa x. \tag{56}$$

For the velocity layer

$$u_0 \delta^2 \left( \frac{2}{15} - \frac{3}{140} + \frac{1}{180} \right) = 4\nu x, \tag{57}$$

and dividing (56) by (57) we have

$$\chi^3 [1 + 0\cdot 182(1-\chi^2) - 0\cdot 047(1-\chi^3)] = 1/\sigma. \tag{58}$$

An approximate solution of (58) which, for values of $\sigma$ greater than unity, differs by less than 5 per cent. from the accurate solution is

$$\chi = \sigma^{-\frac{1}{3}}. \tag{59}$$

From (57) and (59) we obtain for the heat transferred in unit time from one side of a plate of length $l$ and breadth $b$

$$Q = 0\cdot 686\sigma^{\frac{1}{3}} k b(T_1 - T_0)\sqrt{(u_0 l/\nu)},$$

and the corresponding value of the Nusselt number is[†]

$$\mathrm{Nu} = 0\cdot 686\sigma^{\frac{1}{3}}\sqrt{(u_0 l/\nu)}.$$

This method of solution shows how the proportionality to $\sigma^{\frac{1}{3}}$, remarked by Pohlhausen, arises. The result applies only for $\delta'/\delta < 1$, that is for $\sigma > 1$, since the value of $u$ used in (55) holds only for $y < \delta$. For $\delta'/\delta > 1$ the algebra is more complicated, because two different expressions have to be used for $u$; but it gives an identical expression for the rate of heat transfer for values of $\sigma$ greater than 0·5.

## 269. The reading of a thermometer in a moving fluid.

Pohlhausen[‡] has also obtained the solution of the problem of the plate thermometer—i.e. the temperature which a thermometer in the form of a flat plate placed parallel to the stream will read when no heat is being transmitted to or from it by the fluid. In this problem the heat generated by dissipation cannot be neglected, and the temperature distribution is now given by (52) together with the

[†] Cf. Kroujiline, *Techn. Physics U.S.S.R.* **3** (1936), 183–194.
[‡] *Zeitschr. f. angew. Math. u. Mech.* **1** (1921), 120, 121.

condition of zero heat transfer from the surface. The velocity distribution is the same as before, but the boundary conditions for the temperature are now $\partial T/\partial y = 0$ at $y = 0$, $T = T_0$ at $y = \infty$. Changing to the new variable $\eta$ as in § 268, and putting

$$T = T_0 + \frac{1}{8}\frac{u_0^2}{Jc_p}\theta(\eta),$$

we obtain 

$$\theta'' + \sigma f\theta' + 2\sigma f''^2 = 0, \tag{60}$$

where $f(\eta)$ has the same meaning as in § 268. The boundary conditions are $\theta' = 0$ at $\eta = 0$, and $\theta = 0$ at $\eta = \infty$. The solution of (60) satisfying these conditions is

$$\theta(\eta) = 2\sigma \int_\eta^\infty \exp\left(-\sigma \int_0^\eta f\,d\eta\right)\left[\int_0^\eta f''^2 \exp\left(\sigma \int_0^\eta f\,d\eta\right) d\eta\right] d\eta. \tag{61}$$

Hence, if $\theta(0)$ is denoted by $\alpha_2(\sigma)$,

$$\alpha_2(\sigma) = 2\sigma \int_0^\infty \exp\left(-\sigma \int_0^\eta f\,d\eta\right)\left[\int_0^\eta f''^2 \exp\left(\sigma \int_0^\eta f\,d\eta\right) d\eta\right] d\eta. \tag{62}$$

Since $f = -f'''/f''$, and therefore

$$\exp\left(\sigma \int_0^\eta f\,d\eta\right) = [f''(\eta)/f''(0)]^{-\sigma},$$

(61) may be written

$$\theta(\eta) = 2\sigma \int_\eta^\infty (f'')^\sigma \left[\int_0^\eta (f'')^{2-\sigma}\,d\eta\right] d\eta,$$

with a corresponding expression for $\alpha_2$. For $\sigma = 1$ the expression for $\theta$ reduces to 

$$\theta(\eta) = 4 - f'^2$$

since $f'(0) = 0, f'(\infty) = 2$; and hence

$$T = T_0 + \frac{1}{2Jc_p}(u_0^2 - u^2). \tag{63}$$

This result, which is equivalent to the statement that the energy per unit mass $(Jc_p T + \frac{1}{2}u^2)$ is constant for $\sigma = 1$, may be verified directly from the original equation (52). (Cf. equation (36).)

Pohlhausen has tabulated the function (61) for a number of values of $\sigma$, and his results are reproduced in Fig. 242. He also gives values

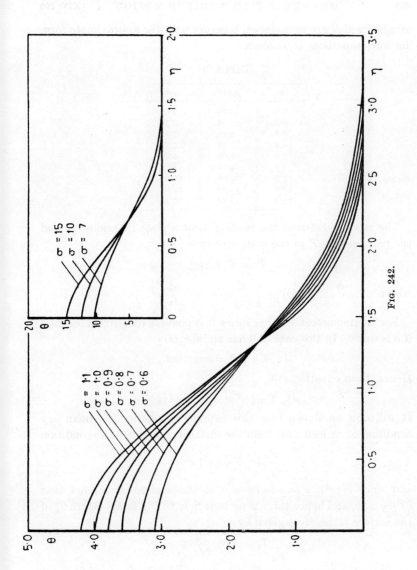

Fig. 242.

of $\alpha_2(\sigma)$, which are reproduced, together with the approximate form $4\sigma^{\frac{1}{4}}$ for comparison, in Table 20.

TABLE 20

| $\sigma$ | $\alpha_2(\sigma)$ | $4\sigma^{\frac{1}{4}}$ |
|---|---|---|
| 0·6 | 3·08 | 3·09 |
| 0·7 | 3·34 | 3·34 |
| 0·8 | 3·58 | 3·58 |
| 0·9 | 3·80 | 3·79 |
| 1·0 | 4·00 | 4·00 |
| 1·1 | 4·20 | 4·19 |
| 7·0 | 10·06 | 10·6 |
| 10·0 | 11·86 | 12·6 |
| 15·0 | 14·14 | 15·5 |

The relation between the reading $T_1$ of a plate thermometer and the temperature $T_0$ of the main stream is

$$T_1 = T_0 + \Delta(\sigma),$$

where
$$\Delta(\sigma) = \frac{1}{8}\frac{u_0^2}{Jc_p}\alpha_2(\sigma) \doteqdot \frac{u_0^2}{2Jc_p}\sigma^{\frac{1}{4}}.$$

For thermometers of other forms it is possible to derive a solution if $\sigma$ is unity. In this case (36) has an integral

$$Jc_p T + \tfrac{1}{2}u^2 = \text{constant}.$$

Hence, from equation (43),

$$Jc_p T + \tfrac{1}{2}u^2 = Jc_p T_0 + \tfrac{1}{2}u_0^2.$$

It will now be shown that this expression satisfies the boundary condition of no heat flow from the surface. It satisfies the condition

$$\frac{\partial}{\partial y}(Jc_p T + \tfrac{1}{2}u^2) = 0,$$

and since $\partial(\tfrac{1}{2}u^2)/\partial y = u\,\partial u/\partial y = 0$ at the surface, it follows that $\partial T/\partial y = 0$, and hence there is no heat flow.† The temperature $T_1$ of the surface is therefore given by

$$T_1 = T_0 + \frac{u_0^2}{2Jc_p};$$

to obtain the temperature $T_0$ of a stream of velocity $u_0$ of fluid for which $\sigma = 1$, the reading $T_1$ of a thermometer immersed in it must be reduced by an amount $u_0^2/(2Jc_p)$, which is the thermometric equivalent of the kinetic energy of the stream.

† Busemann, loc. cit. on p. 612.

For values of $\sigma$ of the order of, but not exactly equal to, unity, A. A. Griffith, in an unpublished communication, has suggested a formula

$$T_1 = T_0 + [u_0^2/(2Jc_p)][1 - (1-\sigma)m^2]$$

for either laminar or turbulent boundary layers, though the value of $m$ will depend on whether the boundary layer is laminar or turbulent, and on the shape of the body. Griffith suggests values of $m$ about 1 for a stream-line body (aerofoil) and about 1·4 for a circular cylinder. This formula has been experimentally verified by W. F. Hilton† for speeds below the 'compressibility' stall, with average values of $m$ over the surface about 1·1 or 1·2 for an aerofoil at small incidences and about 1·57 for a circular cylinder.

## 270. Forced convection from a cylinder near the forward stagnation point.

Near the forward stagnation point of a cylinder immersed in a moving fluid, the velocity $u_1$ outside the boundary layer increases linearly with distance from that point, so that $u_1 = \beta_1 x$. The velocities in the boundary layer are given in §54 of Chapter IV. If the body is heated to temperature $T_1$, equation (34) for the temperature distribution is

$$u\frac{\partial T}{\partial x} + v\frac{\partial T}{\partial y} = \kappa\frac{\partial^2 T}{\partial y^2}, \qquad (64)$$

if we neglect the term $u\,\partial p/\partial x$ and the dissipation function $\Phi$. The boundary conditions are $T = T_1$ at $y = 0$, $T = T_0$ at $y = \infty$.

The equation has a solution which is a function of $y$ only. Putting

$$\eta = (\beta_1/\nu)^{\frac{1}{2}}y, \quad u = \beta_1 x f'(\eta), \quad v = -(\nu\beta_1)^{\frac{1}{2}}f(\eta),$$

as in Chap. IV, §54, and further

$$T = T_1 - (T_1 - T_0)\theta(\eta),$$

we find that (64) becomes‡

$$\theta'' + \sigma f\theta' = 0. \qquad (65)$$

The boundary conditions for $\theta$ are $\theta(0) = 0$, $\theta(\infty) = 1$, and the relevant solution of (65) is

$$\theta(\eta) = \alpha_3(\sigma)\int_0^\eta \left[\exp\left(-\sigma\int_0^\eta f\,d\eta\right)\right]d\eta,$$

where

$$\frac{1}{\alpha_3(\sigma)} = \int_0^\infty \left[\exp\left(-\sigma\int_0^\eta f\,d\eta\right)\right]d\eta.$$

† See a forthcoming paper in the *Proc. Roy. Soc.*    ‡ Squire (unpublished).

The rate of heat transfer from a section of the cylinder of breadth $b$ and length $x$ measured from the stagnation point, is

$$Q = \alpha_3(\sigma)kbx(T_1-T_0)(\beta_1/\nu)^{\frac{1}{2}},$$

and the corresponding value of the Nusselt number is

$$\mathrm{Nu} = \alpha_3(\sigma)(\beta_1 d^2/\nu)^{\frac{1}{2}}, \tag{66}$$

where $d$ is a representative length. A set of values of $\alpha_3(\sigma)$, together with the values of an approximate form $0\cdot570\,\sigma^{0\cdot4}$ for comparison, is given in Table 21.

TABLE 21

| $\sigma$ | $\alpha_3(\sigma)$ | $0\cdot570\,\sigma^{0\cdot4}$ |
|---|---|---|
| 0·6 | 0·466 | 0·465 |
| 0·7 | 0·495 | 0·495 |
| 0·8 | 0·521 | 0·521 |
| 0·9 | 0·546 | 0·546 |
| 1·0 | 0·570 | 0·570 |
| 1·1 | 0·592 | 0·592 |
| 7·0 | 1·18 | 1·24 |
| 10·0 | 1·34 | 1·43 |
| 15·0 | 1·54 | 1·68 |

These results may be compared with experiments on the distribution of heat transfer from a heated cylinder in a fluid stream. Near the stagnation point of a circular cylinder the theoretical value for the velocity outside the boundary layer is approximately correct, and is

$$u_1 = 4u_0 x/d,$$

where $u_0$ is the velocity of the main stream and $d$ is the diameter of the cylinder. Hence $\beta_1 = 4u_0/d$; taking $\sigma$ for air as 0·733 we obtain from (66)

$$\mathrm{Nu} = 1\cdot01(u_0 d/\nu)^{\frac{1}{2}}.$$

For $u_0 d/\nu = 4 \times 10^4$ the theoretical value of Nu is 202. Values obtained experimentally are:

Lohrisch[†]     .     .     .     206
Klein[‡]     .     .     .     190
Drew and Ryan[||]     .     .     185 (approx.)

The agreement between theory and experiment is satisfactory.

† *Forschungsarbeiten des Ver. deutsch. Ing.*, No. 322 (1929), 46–67.
‡ *Archiv für Wärmewirtschaft und Dampfkesselwesen*, **15** (1934), 150.
|| *Trans. Amer. Inst. Chem. Eng.* **26** (1931), 118–147.

## 271. Further solutions for forced convection in a boundary layer.

Other problems of forced convection have been considered by Fage and Falkner.[†] They neglect the dissipation function and the term depending on the pressure, so that the energy equation for steady motion in the boundary layer has the form (64). It is assumed that the compressibility of the fluid may be neglected, so that the velocity distribution is unaffected by the heat transfer and may be supposed known.

Now put
$$\theta = \frac{T - T_1}{T_0 - T_1}$$

(as in § 270), where $T_1$ is the temperature of the surface, which is taken to be a function of $x$, and $T_0$ is the temperature at infinity. Then equation (64) becomes

$$u\frac{\partial\theta}{\partial x} + u\frac{\theta-1}{T_1-T_0}\frac{dT_1}{dx} + v\frac{\partial\theta}{\partial y} = \kappa\frac{\partial^2\theta}{\partial y^2}. \tag{67}$$

The boundary conditions are $\theta = 0$ at $y = 0$, $\theta = 1$ at $y = \infty$.

If the temperature at the surface is expressible in the form

$$T_1 = T_0 + Bx^\beta,$$

where $B$ and $\beta$ are constants, (67) becomes

$$u\left[\frac{\partial\theta}{\partial x} + \frac{\beta}{x}(\theta-1)\right] + v\frac{\partial\theta}{\partial y} = \kappa\frac{\partial^2\theta}{\partial y^2}. \tag{68}$$

Fage and Falkner consider cases for which it is sufficiently accurate to assume that the velocity at the outer edge of the layer is given by

$$u_1 = cx^m,$$

where $c$ and $m$ are constants, as in Chap. IV, § 54. Then (Chap. IV, equation (66)) the stream-function is $(u_1\nu x)^{\frac{1}{2}}f(\eta)$, where

$$\eta = (u_1/\nu x)^{\frac{1}{2}}y,$$

so that
$$u = u_1 f'(\eta),$$

$$v = -\tfrac{1}{2}(u_1\nu/x)^{\frac{1}{2}}[(m+1)f(\eta) + (m-1)\eta f'(\eta)].$$

There is a solution for which $\theta$ is a function of $\eta$ only, so that (68) reduces to

$$\theta'' + \sigma[\tfrac{1}{2}(m+1)f\theta' - \beta f'(\theta-1)] = 0.$$

Fage and Falkner now assume that it is sufficiently accurate to make

[†] *A.R.C. Reports and Memoranda*, No. 1408 (1931).

$u$ increase linearly in the boundary layer—i.e. they replace $f'(\eta)$ and $f(\eta)$ by $\alpha\eta$ and $\frac{1}{2}\alpha\eta^2$ respectively, where $\alpha = f''(0)$. The equation for $\theta$ then becomes†

$$\theta'' + \sigma\alpha[\tfrac{1}{4}(m+1)\eta^2\theta' - \beta\eta(\theta-1)] = 0. \qquad (69)$$

The boundary conditions are $\theta(0) = 0$, $\theta(\infty) = 1$. Equation (69) may be solved numerically and the rate of heat transfer per unit area at the point $x$ found in the form

$$Q = -k(\partial T/\partial y)_{y=0} = \alpha_4(\sigma)k(T_1-T_0)(u_1/\nu x)^{\frac{1}{2}},$$

where $\alpha_4(\sigma)$ is the value of $\theta'(\eta)$ at $\eta = 0$. Referred to $x$ as 'representative distance' the corresponding local heat transfer number is

$$\mathrm{Nu} = \alpha_4(\sigma)(u_1 x/\nu)^{\frac{1}{2}};$$

$\alpha_4(\sigma)$ depends on both $m$ and $\beta$. The total heat transfer is obtained by integration.

Taking the value of $\sigma$ for air as $0\cdot77$, Fage and Falkner calculated the value of $\alpha_4(0\cdot77)$ for a number of cases, and their values are given in Table 22.

TABLE 22

| $m$ | $\beta$ | $\alpha_4(0\cdot77)$ | Remarks |
|---|---|---|---|
| o | 0·5 | 0·425 | |
| | 0·4 | 0·406 | Flat plate |
| | 0·3 | 0·386 | $\alpha = 0\cdot332$ |
| | 0·2 | 0·363 | |
| | o | 0·310 | |
| $\frac{1}{3}$ | o | 0·450 | Uniform intensity of |
| | 1·0 | 0·674 | skin-friction |
| | −0·7 | o (approx.) | $\alpha = 0\cdot757$ |
| 1 | o | 0·600 | Uniform gradient of |
| | 1 | 0·830 | skin-friction |
| | −1 | o (approx.) | $\alpha = 1\cdot233$ |

For a flat plate kept at uniform temperature ($m = 0$, $\beta = 0$) Fage and Falkner's method gives $\alpha_4 = 0\cdot310$, whereas the corresponding exact value obtained by Pohlhausen is $0\cdot303$.‡ This difference arises from the assumption that in the heated layer the velocity increases linearly with distance from the surface.

The calculated results were compared with experimental results

---

† $\alpha$ depends on $m$, and numerical values may be found from a short table given in Chap. IV, § 64, p. 180, wherein $(d^2F/dY^2)_{Y=0}$ would be here denoted by $[2/(m+1)]^{\frac{1}{2}}\alpha$, and what is there denoted by $\beta$ is equal to $2m/(m+1)$.

‡ See p. 624. For the flat plate the local value of Nu is one-half the total value.

obtained by measuring the heat transfer (i) from a piece of platinum foil in the form of a flat plate, and (ii) from a nickel strip embedded in a circular cylinder, which in both cases were heated by passing an electric current through them; reasonably good agreement was found.

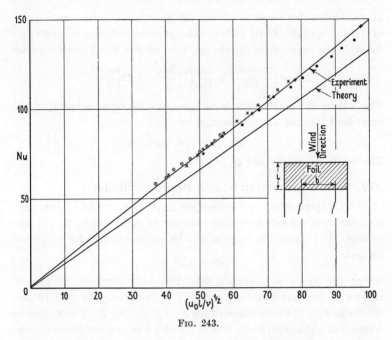

FIG. 243.

In the case of the platinum foil ($m = 0$), experiments were made on the middle half of the foil, this half being of breadth $b$ (say). The average temperature elevation was estimated to be 150° C. from the mean electrical resistance. The value (0·4) of $\beta$ was chosen to make the distribution of heat transfer as determined from the solution above in as close agreement as possible with that determined from the physical properties of the foil on the assumption that the heat transferred to the air from any strip was equal to the heat generated in the strip. The theoretical temperature change $Bl^{\beta}$ (where $l$ is the length of the foil) was found by making the measured resistance of the foil equal to that calculated from the assumed temperature law and the physical properties of the foil; and for $\beta = 0·4$ was found to be 220° C. If we integrate the expression obtained above for the rate of heat transfer per unit area at the point $x$, after replacing $T_1 - T_0$ by $Bx^\beta$ and putting $u_1$ equal to the undisturbed velocity $u_0$ of the stream, we find that the heat transfer from one side of the plate is

$$\frac{2\alpha_4}{2\beta+1} kbBl^\beta \left(\frac{u_0 l}{\nu}\right)^{\frac{1}{2}}.$$

If we double this so as to obtain the heat transfer from both sides of the foil, and divide by $kb$ times the average temperature elevation found in the experiment (as above), then we find the Nusselt number

$$\text{Nu} = \frac{4 \times 0 \cdot 406}{1 \cdot 8} \times \frac{220}{150} \left(\frac{u_0 l}{\nu}\right)^{\frac{1}{2}} = 1 \cdot 32 \left(\frac{u_0 l}{\nu}\right)^{\frac{1}{2}}.$$

The values found for this same quantity by measurement of the total heat transfer are represented by

$$\text{Nu} = 1 \cdot 50 (u_0 l / \nu)^{\frac{1}{2}}.$$

The results are shown in Fig. 243.

## 272. Forced convection from a circular cylinder.†

Many experimental investigations have been made of the heat transfer from a heated circular cylinder at right angles to the air-stream. For gases the results can be expressed by the empirical formula
$$\text{Nu} = CR^n,$$
where the index $n$ increases with Reynolds number. Hilpert's‡ results for laminar flow for a range of $R$ from unity to $2 \cdot 4 \times 10^5$ are shown in Fig. 244 as a curve of $\log \text{Nu}$ against $\log R$. The cylinders were all at a temperature of $100°$ C.; the air was at room temperature. Hilpert also investigated the influence of the temperature difference between the cylinder and the air, and found that it could be represented by the formula
$$\text{Nu} = C \left[ R \left(\frac{T_{\text{surface}}}{T_{\text{air}}}\right)^{\frac{1}{4}} \right]^n.$$

For gases the dependence of the heat transfer coefficient on the Prandtl number $\sigma$ is unimportant, since $\sigma$ does not differ greatly from unity. For liquids the Nusselt number will, however, vary with $\sigma$, and from the experiments of Davis‖ this dependence can be expressed in the form
$$\text{Nu} = C\sigma^{0 \cdot 31} R^n.$$

† The subject-matter of this section is from a lecture by Schmidt, *Proc. 4th Internat. Congress for Applied Mechanics, Cambridge*, 1934, pp. 97, 98.

‡ *Forsch. Ingwes.* **4** (1933), 215–224; *Ver. deutsch. Ing., Forschungsheft* 355 (1932).

‖ *Phil. Mag.* (6), **44** (1922), 920–940.

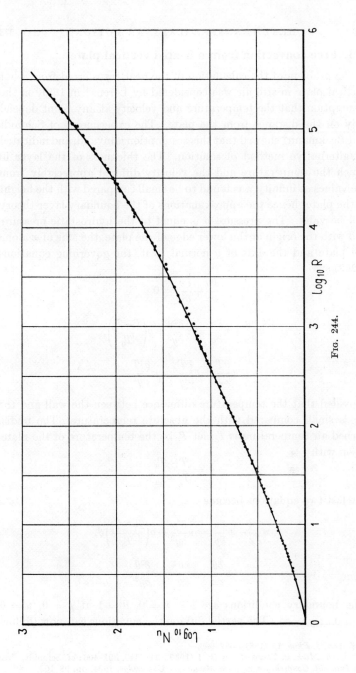

FIG. 244.

## 273. Free convection from a heated vertical plate.

The problem of the calculation of the heat transferred from a hot vertical plate in still air was considered by Lorenz† in 1881, on the assumption that the temperature and velocity at any point depend only on the distance from the plate. The experiments of Schmidt and Beckmann‡ showed that this assumption is invalid, and indicated an alternative method of solution. The thickness of the layer in which the temperature and the velocity differed appreciably from the values at infinity was found to be small compared with the height of the plate: hence the approximations of the boundary layer theory will be valid. The pressure $p$ is equal to the hydrostatic pressure, and with the origin at the lower edge of the plate, the axis of $x$ along the plate and the axis of $y$ normal to it, the governing equations (§ 263) are

$$\frac{\partial u}{\partial x} + \frac{\partial v}{\partial y} = 0,$$

$$u\frac{\partial u}{\partial x} + v\frac{\partial u}{\partial y} = \nu\frac{\partial^2 u}{\partial y^2} + g\left(\frac{T-T_0}{T_0}\right),$$

$$u\frac{\partial T}{\partial x} + v\frac{\partial T}{\partial y} = \kappa\frac{\partial^2 T}{\partial y^2},$$

provided that the temperature difference between the wall and the air is small compared with the absolute temperature. The undisturbed air temperature is $T_0$; let $T_1$ be the temperature of the plate. Then with

$$\theta = \frac{T-T_0}{T_1-T_0},$$

the last two equations become

$$u\frac{\partial u}{\partial x} + v\frac{\partial u}{\partial y} = \nu\frac{\partial^2 u}{\partial y^2} + g\left(\frac{T_1-T_0}{T_0}\right)\theta,$$

$$u\frac{\partial \theta}{\partial x} + v\frac{\partial \theta}{\partial y} = \kappa\frac{\partial^2 \theta}{\partial y^2}.$$

The boundary conditions are $u = v = 0$, $\theta = 1$ at $y = 0$, $u = 0$, $\theta = 0$ at $y = \infty$. The partial differential equations can now (follow-

† *Ann. d. Phys.* **13** (1881), 582–606.

‡ *Tech. Mech. u. Thermodynamik*, **1** (1930), 341–349, 391–406; cf. Schmidt, *Proc. 4th Internat. Congress for Applied Mechanics, Cambridge*, 1934, pp. 98–102.

ing Pohlhausen) be transformed into ordinary differential equations by the substitutions

$$\eta = \left[\frac{g(T_1-T_0)}{4\nu^2 T_0}\right]^{\frac{1}{4}} \frac{y}{x^{\frac{1}{4}}} = C\frac{y}{x^{\frac{1}{4}}} \quad \text{(say)},$$

$$\psi(x,y) = 4\nu C x^{\frac{3}{4}} f(\eta),$$

$$\theta(x,y) = g(\eta),$$

where $\psi$ is the stream-function defined by

$$u = \frac{\partial \psi}{\partial y}, \qquad v = -\frac{\partial \psi}{\partial x},$$

so that          $u = 4\nu C^2 x^{\frac{1}{2}} f'(\eta), \qquad v = \nu C x^{-\frac{1}{4}}(\eta f' - 3f).$

The equations for $f$ and $g$ are

$$f''' + 3ff'' - 2f'^2 + g = 0,$$

$$g'' + 3\sigma fg' = 0,$$

and the boundary conditions are $f(0) = 0$, $f'(0) = 0$, $g(0) = 1$, $f'(\infty) = 0$, $g(\infty) = 0$. These boundary conditions also suffice to make $u = 0$, $\theta = 0$ at $x = 0$, and the transformations indicate that $\theta$ and the quantity $u/(4\nu C^2 x^{\frac{1}{2}})$ should be functions of $Cy/x^{\frac{1}{4}}$ only, where

$$C = \left[\frac{g(T_1-T_0)}{4\nu^2 T_0}\right]^{\frac{1}{4}}.$$

The equations have been solved in series by Pohlhausen for air ($\sigma = 0.733$), and the theoretical local value of the Nusselt number at a point distant $h$ from the lower edge of the plate was found to be

$$\text{Nu} = \frac{\alpha h}{k} = 0.508\left[\frac{gh^3(T_1-T_0)}{4\nu^2 T_0}\right]^{\frac{1}{4}} = 0.359(\text{Gr})^{\frac{1}{4}},$$

where $\alpha$ is the heat transfer in unit time from unit area for unit difference of temperature, and Gr is the Grashoff number (p. 607):

$$\text{Gr} = gh^3(T_1-T_0)/(\nu^2 T_0).$$

Schmidt and Beckmann, in their experiments on heat transfer from vertical plates, found good agreement with the results of this theory.[†] Figs. 245 and 246 show comparisons between the theoretical and the measured temperature and velocity, respectively, for a flat plate of height 12·5 cm. at a temperature of 65° C. in air at 15° C.

† Measurements of heat transfer from plates in a pressure chamber over a range of pressures from 0·043 to 65 atmospheres by Saunders (*Proc. Roy. Soc.* A, **157** (1936), 278–291) gave results in agreement with those of Schmidt and Beckmann.

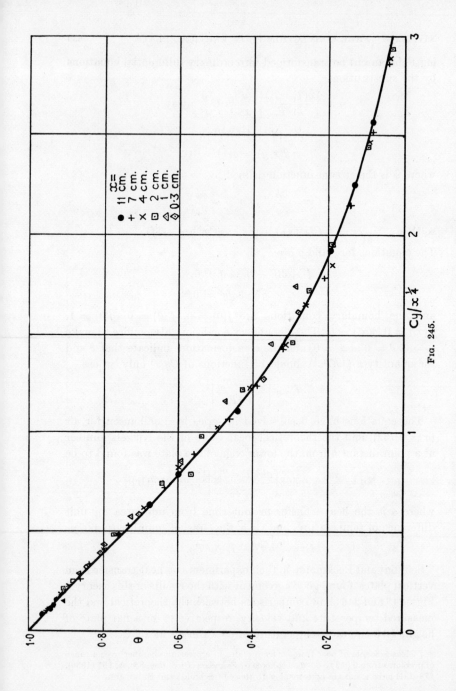

Fig. 245.

An alternative method of solution may be obtained by use of equations (37) and (39).† For steady flow, with constant pressure and with dissipation neglected, these equations simplify to

$$\frac{d}{dx}\int_0^\delta \rho u^2 \, dy = g\int_0^\delta \rho\left(\frac{T-T_0}{T_0}\right)dy - \mu\left(\frac{\partial u}{\partial y}\right)_{y=0};\tag{70}$$

$$Jc_p\left[\frac{d}{dx}\int_0^\delta \rho uT \, dy - T_0\frac{d}{dx}\int_0^\delta \rho u \, dy\right] = -Jk\left(\frac{\partial T}{\partial y}\right)_{y=0},\tag{71}$$

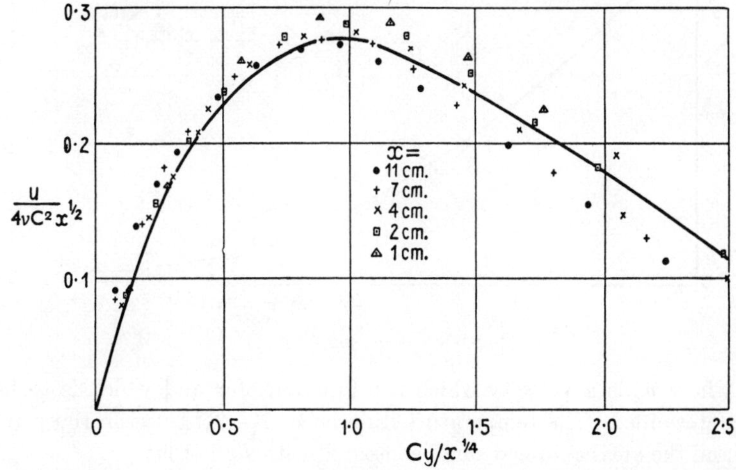

FIG. 246.

where $\delta'$ has been put equal to $\delta$, since the fluid will tend to ascend only in the region where the temperature of the fluid is raised. Also, since $p$ is constant (the variation of pressure with height being neglected), equation (1) gives

$$\rho T = \rho_0 T_0,$$

where $\rho_0$ and $T_0$ are the density and temperature at infinity. Hence (70) and (71) become

$$\frac{d}{dx}\int_0^\delta \frac{T_0}{T}u^2 \, dy = g\int_0^\delta\left(1-\frac{T_0}{T}\right)dy - \nu\left(\frac{\partial u}{\partial y}\right)_{y=0},\tag{72}$$

$$\frac{d}{dx}\int_0^\delta u\left(1-\frac{T_0}{T}\right)dy = -\kappa\left[\frac{\partial}{\partial y}\left(\frac{T}{T_0}\right)\right]_{y=0}.\tag{73}$$

† Squire (unpublished).

Suitable approximate expressions for $u$ and $T$ which satisfy the boundary conditions are

$$u = u_x \frac{y}{\delta}\left(1 - \frac{y}{\delta}\right)^2,$$

$$\frac{T}{T_0} = 1 + \left(\frac{T_1 - T_0}{T_0}\right)\left(1 - \frac{y}{\delta}\right)^2,$$

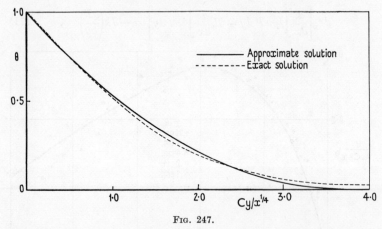

FIG. 247.

where $u_x$ is a velocity which is a function of $x$ and which is to be determined. The temperature difference $(T_1 - T_0)$ between the plate and the air is assumed small compared with $T_0$, so that

$$\frac{T_0}{T} = 1 - \left(\frac{T_1 - T_0}{T_0}\right)\left(1 - \frac{y}{\delta}\right)^2,$$

to a sufficient approximation. With these expressions for $u$ and $T$ (72) and (73) give approximately

$$\frac{d}{dx}\left(\frac{u_x^2 \delta}{105}\right) = g\left(\frac{T_1 - T_0}{T_0}\right)\frac{\delta}{3} - \frac{\nu u_x}{\delta}$$

and

$$\frac{d}{dx}\left(\frac{u_x \delta}{30}\right) = \frac{2\kappa}{\delta},$$

where, in the first equation, a term has been omitted which is of higher order in $(T_1 - T_0)/T_0$ than those included. The solution of these equations is

$$u_x = 5 \cdot 17\nu\left(\sigma + \frac{20}{21}\right)^{-\frac{1}{2}}\left[\frac{g(T_1 - T_0)}{\nu^2 T_0}\right]^{\frac{1}{2}}x^{\frac{1}{2}},$$

$$\delta = 3 \cdot 93\sigma^{-\frac{1}{2}}\left(\sigma + \frac{20}{21}\right)^{\frac{1}{4}}\left[\frac{g(T_1 - T_0)}{\nu^2 T_0}\right]^{-\frac{1}{4}}x^{\frac{1}{4}},$$

and the corresponding value for the local heat transfer number at a point distant $h$ from the lower edge of the plate is

$$\text{Nu} = \frac{\alpha h}{k} = 0.508\sigma^{\frac{1}{2}}\left(\sigma+\frac{20}{21}\right)^{-\frac{1}{4}}\left[\frac{gh^3(T_1-T_0)}{\nu^2 T_0}\right]^{\frac{1}{4}}.$$

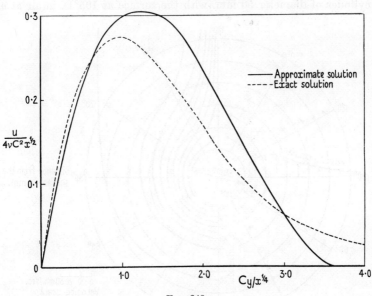

FIG. 248.

For air $\sigma = 0.733$, and this expression gives

$$\text{Nu} = 0.382\left[\frac{gh^3(T_1-T_0)}{\nu^2 T_0}\right]^{\frac{1}{4}} = 0.382(\text{Gr})^{\frac{1}{4}},$$

which is 6 per cent. higher than the accurate solution. (Closer approximations could be obtained by taking more complicated expressions for $u$ and $T$.) Figs. 247 and 248 show the relation between the approximate and the exact solutions for the temperature and the velocity, respectively, for air, $\theta$ again denoting $(T-T_0)/(T_1-T_0)$ in Fig. 247.

## 274. Other problems of free convection.

Hermann† has extended the solution for free convection from a vertical plate (§ 273) to the case of free convection from a horizontal cylinder in the range $(10^4 < \text{Gr} < 3.10^8)$ in which the laminar

† *Ver. deutsch. Ing., Forschungsheft* 379 (1936), 1–24. See also the footnote on p. 608.

boundary layer equations may be used; good agreement was obtained with the experimental results of Jodlbauer.[†] As in the case of the plate, the Nusselt number is proportional to $(Gr)^{\frac{1}{4}}$.[‡] Fig. 249 shows the measured temperature and velocity fields for a cylinder of diameter 50 mm. with the surface at 105° C. in air at a

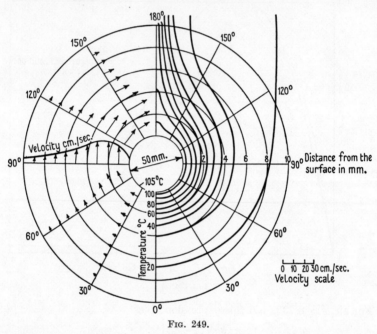

Fig. 249.

temperature of 18° C. (The radial distances outside the cylinder are enlarged 10 times in comparison with the scale for the cylinder diameter.)

If the Grashoff number is sufficiently high the flow in the boundary layer associated with free convection may become turbulent. Hermann‖ has investigated the transition to turbulence for vertical plates and circular cylinders. It was found that $R_\delta$, the Reynolds number formed from the maximum velocity in the boundary layer and the boundary layer thickness, was about 300 at the transition

† *Forsch. Ingwes.* **4** (1933), 157–172. See also Schmidt, *Proc. 4th Internat. Congress for Applied Mechanics, Cambridge,* 1934, pp. 102, 103.

‡ The proportionality to $(Gr)^{\frac{1}{4}}$ holds for all cylinders, as may be deduced from the governing equations.

‖ *Loc. cit.* on p. 643.

point for both plates and cylinders; this is associated with a value of about $10^9$ for the Grashoff number for a vertical plate and $3.5 \times 10^8$ for a horizontal cylinder.

Griffiths and Davis[†] and Saunders[‡] have made experiments on heat transfer from plates with a turbulent boundary layer. These gave the result that the Nusselt number is proportional to $Gr^{\frac{1}{3}}$, as compared with $Gr^{\frac{1}{4}}$ for a laminar boundary layer.

The complete measurement of the temperature and velocity fields for free convection from a body is a lengthy process, but the heat transfer can be found by Töpler's 'Schlieren' method. This depends on the variation of the refractive index of air with density, and involves only the measurement of the deviation of light rays passing through the air. A full account of the 'Schlieren' method has been given by Schardin,[||] and its application to problems of free convection has been described by Schmidt.[††]

[†] 'Transmission of Heat by Radiation and Convection' (Special Report of the Food Investigation Board, No. 9, London, 1922; revised edition, 1931).

[‡] *Loc. cit.* on p. 639.

[||] *Ver. deutsch. Ing., Forschungsheft* 367 (1934). See also Chap. VI, § 134.

[††] *Proc. 4th Internat. Congress for Applied Mechanics, Cambridge,* 1934, pp. 103–110.

## ADDITIONAL REFERENCES

A list of additional references for Chapters XIV and XV will be found at the end of Chapter XV.

# HEAT TRANSFER (TURBULENT FLOW)

## 275. Introduction.

THE exact analysis of heat transfer in turbulent flow is dependent on a knowledge of the distributions of mean velocity and of the velocity fluctuations, concerning which theories have been formulated only in recent years. Before these developments the only important step taken was due to Reynolds,† who suggested that heat and momentum were transferred in the same way, and deduced that heat transfer was proportional to skin-friction. The momentum transfer theory of turbulent flow (Chap. V, § 81) neglects the effect of fluctuating pressure gradients, and since these pressure fluctuations do not influence the heat transfer the assumptions of the momentum transfer theory and of Reynolds's theory are equivalent (§ 279 *infra*).

In this chapter the heating due to pressure changes and to the dissipation of energy by viscosity will be neglected, and consideration will be limited to incompressible fluids; the distributions of mean velocity and of the velocity fluctuations will be assumed to be the same as for isothermal flow; and the variations in conductivity, viscosity, and specific heat with temperature will be neglected. The application of the results to heat transfer in compressible fluids (provided that $c_p$, the specific heat at constant pressure, is used in the formulae) may be justified for small temperature differences under the same conditions as in Chapter XIV (pp. 608, 609).

The temperature at any point in a field of turbulent flow has a mean value denoted by $T$ and a fluctuation of magnitude $T'$ (the mean value of which is zero), while the velocity has mean components $U$, $V$, $W$ and fluctuating components $u$, $v$, $w$.‡

## 276. The equation of eddy heat transfer.

The equation (13) of Chapter XIV for the temperature $T$ in an incompressible fluid, with dissipation neglected, is

$$\rho c \frac{DT}{Dt} = k\nabla^2 T, \tag{1}$$

---

† *Proc. Manchester Lit. and Phil. Soc.* **14** (1874), 7–12; *Collected Papers*, **1**, 81–85.

‡ The same remarks with regard to taking means apply to the temperature as to the velocity. (See Chap. V, § 68.)

where $c$ is the specific heat. If $U+u$ is substituted for $u$, $V+v$ for $v$, $W+w$ for $w$, and $T+T'$ for $T$, and mean values are taken, (1) becomes (when use is made of the equation of continuity, as in Chap. V, §69)

$$\rho c\left(\frac{\partial T}{\partial t}+U\frac{\partial T}{\partial x}+V\frac{\partial T}{\partial y}+W\frac{\partial T}{\partial z}\right)$$

$$=\frac{\partial}{\partial x}\left(k\frac{\partial T}{\partial x}-\rho c\overline{uT'}\right)+\frac{\partial}{\partial y}\left(k\frac{\partial T}{\partial y}-\rho c\overline{vT'}\right)+\frac{\partial}{\partial z}\left(k\frac{\partial T}{\partial z}-\rho c\overline{wT'}\right). \quad (2)$$

The rates of heat transfer by molecular conductivity across unit areas normal to the coordinate axes in the positive directions of these axes are $-k\,\partial T/\partial x$, $-k\,\partial T/\partial y$, $-k\,\partial T/\partial z$, respectively. Equation (2) shows that the effect of the fluctuations is to add to these rates of heat transfer by molecular conductivity the eddy heat transfer terms $\rho c\overline{uT'}$, $\rho c\overline{vT'}$, $\rho c\overline{wT'}$, so that the resultant rates of heat transfer are $(-k\,\partial T/\partial x+\rho c\overline{uT'})$, etc. In turbulent flow the effect of the eddy heat transfer terms is large compared with the effect of the molecular conductivity terms except where the temperature gradient is very large. (This is analogous to the largeness of the Reynolds stresses compared with the viscous stresses except where the rate of deformation is very large.) When the molecular conductivity terms are neglected, (2) reduces to

$$\frac{\partial T}{\partial t}+U\frac{\partial T}{\partial x}+V\frac{\partial T}{\partial y}+W\frac{\partial T}{\partial z}=-\frac{\partial}{\partial x}(\overline{uT'})-\frac{\partial}{\partial y}(\overline{vT'})-\frac{\partial}{\partial z}(\overline{wT'}). \quad (3)$$

In cylindrical polar coordinates $(r,\theta,z)$ (3) becomes

$$\frac{\partial T}{\partial t}+V_r\frac{\partial T}{\partial r}+\frac{V_\theta}{r}\frac{\partial T}{\partial\theta}+V_z\frac{\partial T}{\partial z}$$

$$=-\frac{1}{r}\frac{\partial}{\partial r}(r\overline{v_rT'})-\frac{1}{r}\frac{\partial}{\partial\theta}(\overline{v_\theta T'})-\frac{\partial}{\partial z}(\overline{v_zT'}). \quad (4)$$

## 277. Example. Flow between parallel walls with a constant temperature gradient in the direction of flow.

Consider the heat transfer for steady flow between two parallel fixed walls each of which is kept at the same constant temperature gradient. The origin being taken midway between the walls, the axis of $x$ parallel to the mean flow and the axis of $y$ normal to it, the mean temperature may be written $T=Ax+\theta(y)$, while for the mean velocity $U=U(y)$, $V=0$, $W=0$. Of the eddy heat transfer terms

$\overline{uT'}$ is independent of $x$ and $\overline{wT'}$ is zero, so (2) becomes

$$\frac{\partial}{\partial y}\left(k\frac{\partial T}{\partial y} - \rho c\overline{vT'}\right) = \rho c UA;$$

and this may be integrated to give

$$\rho c\overline{vT'} = k\frac{\partial T}{\partial y} - \rho cA \int_0^y U\, dy, \tag{5}$$

the constant of integration vanishing since $\overline{vT'}$ and $\partial T/\partial y$ vanish by symmetry at the centre of the channel, $y = 0$.

The corresponding result for flow through a circular pipe with a constant temperature gradient in the direction of flow is easily derived in the same way, and is

$$\rho c\overline{v_r T'} = k\frac{\partial T}{\partial r} - \frac{\rho cA}{r} \int_0^r Ur\, dr, \tag{6}$$

where $U$ now stands for the axial velocity at a distance $r$ from the axis of the pipe.

## 278. The mixture length theory.

It has been shown (Chap. V, eq. (19), p. 206) that, with the conception of mixture as in the kinetic theory of gases, the rate of transfer of a property $\theta$ across unit area normal to the $y$-axis and in the positive direction of this axis is

$$Q = -\overline{Lv}\, d\theta/dy, \tag{7}$$

where $L$ is written for $(h_2 - h_1)$. The two 'mixture length' theories of turbulent flow, the momentum transfer theory and the vorticity transfer theory, differ in their assumptions that momentum and vorticity are transferable in this sense, the effect of the fluctuating pressure gradients being neglected in the former theory. These fluctuating pressure gradients, however, cannot affect the heat transfer, and we may therefore identify $Q$ with the rate of heat transfer across unit area normal to the $y$-axis in the positive direction, so that $Q = \rho c\overline{vT'}$. The property $\theta$ which is being transferred is the heat content per unit volume, equal to $\rho cT$. Then (7) becomes†

$$\rho c\overline{vT'} = -\rho c\overline{Lv}\, dT/dy. \tag{8}$$

† Eq. (8) holds if $T$ is a function of $y$ only. In (10) and (11) no allowance is made for the effect of the axial temperature gradient on $\overline{vT'}$, so a term of the form $A\overline{L'v}$ is neglected: this term is small compared with the terms retained.

With Prandtl's further hypothesis that

$$\overline{Lv} = l^2 \left| \frac{dU}{dy} \right|,$$

(8) becomes        $$\overline{vT'} = -l^2 \left| \frac{dU}{dy} \right| \frac{dT}{dy}.$$       (9)

It is reasonable to suppose that the quantity $\overline{Lv}$ and the mixture length $l$ are the same for heat transfer as for momentum or vorticity transfer, since the same mechanism is at work, and it will be assumed that this is the case.

For the flow in a channel considered in § 277, (5) gives

$$\frac{\partial T}{\partial y}(\kappa + \overline{Lv}) = A \int_0^y U\, dy,$$       (10)

where $\kappa = k/\rho c$ and is the thermometric conductivity of the fluid; for the corresponding flow in a pipe (6) gives

$$\frac{\partial T}{\partial r}(\kappa + \overline{L_r v_r}) = \frac{A}{r} \int_0^r Ur\, dr,$$       (11)

where $L_r$ is measured in the direction of $r$ increasing and $v_r$ is the turbulent velocity component in that direction.

## 279. Reynolds's analogy between heat transfer and skin-friction.

Reynolds[†] suggested that in turbulent flow momentum and heat were transferred in the same way. If we now consider flow parallel to the $x$-axis with heat transfer in the positive direction of the $y$-axis, this statement (which requires that the velocity fluctuations parallel to the mean flow should be proportional to the temperature fluctuations) is represented analytically by the equation

$$\frac{\overline{vT'}}{(U_m - U_1)(T_m - T_1)} = \frac{\overline{uv}}{(U_m - U_1)^2},$$       (12)

where $U_m$, $T_m$ are the mean velocity and mean temperature for the region of flow considered, and $U_1$, $T_1$ are the velocity and temperature at some reference plane. It follows that the rate of heat transfer $Q \;(= \rho c \overline{vT'})$ in the positive direction of the $y$-axis across unit area

† *Loc. cit.* on p. 646.

normal to this axis is correlated with the Reynolds shear stress $\tau \ (= -\rho\overline{uv})$ in such a way that

$$\frac{Q}{\rho c(U_m-U_1)(T_m-T_1)} = -\frac{\tau}{\rho(U_m-U_1)^2}. \tag{13}$$

Between the heat transfer from a fixed wall at temperature $T_1$, for which $U_1 = 0$, and the skin-friction at this wall we obtain from (13) the relation

$$k_H = \tfrac{1}{2}c_f, \tag{14}$$

where $k_H$ is the heat transfer coefficient defined by $Q_0/[\rho c U_m(T_1-T_m)]$ (cf. Chap. XIV, equation (19)), and $c_f$ is the skin-friction coefficient $\tau_0/(\tfrac{1}{2}\rho U_m^2)$; the zero suffix indicates that values are taken at the boundary wall.

To examine more closely the conditions for which Reynolds's theory is applicable, consider the flow of a fluid in which the mean velocity $U$ is a function of $y$ only, or alternatively is large compared with the component $V$ (as for flow in channels, wakes, and boundary layers). The equations for the velocity distribution on the momentum and vorticity transfer theories are then

$$U\frac{\partial U}{\partial x} + V\frac{\partial U}{\partial y} = -\frac{1}{\rho}\frac{\partial \bar{p}}{\partial x} + \frac{\partial}{\partial y}\left(\nu\frac{\partial U}{\partial y} + \overline{Lv}\frac{\partial U}{\partial y}\right), \tag{15}$$

and

$$U\frac{\partial U}{\partial x} + V\frac{\partial U}{\partial y} = -\frac{1}{\rho}\frac{\partial \bar{p}}{\partial x} + (\nu+\overline{Lv})\frac{\partial^2 U}{\partial y^2}, \tag{16}$$

respectively. When heat is being transferred the equation for the temperature, obtained from (2) and (8), is

$$U\frac{\partial T}{\partial x} + V\frac{\partial T}{\partial y} = \frac{\partial}{\partial y}\left(\kappa\frac{\partial T}{\partial y} + \overline{Lv}\frac{\partial T}{\partial y}\right). \tag{17}$$

Inspection of these equations shows that there is no direct correlation between velocity and temperature if the velocity distribution is controlled by the vorticity transfer theory.† If, however, the momentum transfer theory is applied, so that $U$ satisfies (15), equation (17) has a solution for which $(T-T_0)$ is proportional to $U$ provided that the conditions $\partial\bar{p}/\partial x = 0$ and $\sigma = \nu/\kappa = 1$ are satisfied,

---

† The quantity $\overline{Lv}$ can be eliminated between (16) and (17), so that the temperature distribution can be calculated from the observed velocity distribution without making any assumption about $\overline{Lv}$. This calculation, however, involves a differentiation of the experimental velocity curve, which is not a very accurate process.

and provided also that the boundary conditions for $T$ and $U$ are analogous. In these circumstances the relations

$$\tau = \mu \frac{\partial U}{\partial y} + \rho \overline{Lv} \frac{\partial U}{\partial y}, \tag{18}$$

$$Q = -k \frac{\partial T}{\partial y} - \rho c \overline{Lv} \frac{\partial T}{\partial y}, \tag{19}$$

for the shear stress and the rate of heat transfer per unit area, lead to the results (13) and (14).[†]

It follows that Reynolds's theory holds only under special conditions. The condition $\sigma = 1$ is of importance only when a region of laminar flow is present or a region in which the viscous stresses and molecular heat transfer are comparable with the Reynolds stresses and eddy heat transfer—for example, in the neighbourhood of a wall. An extension of the theory in such cases for values of $\sigma$ not equal to unity has been given by Taylor, Prandtl, and Kármán.[‡]

The importance of the condition $\partial \bar{p}/\partial x = 0$ is rather uncertain. It is satisfied for flow in the boundary layer along a flat plate at zero incidence, and in wakes and jets. Prandtl[||] has pointed out that for flow in pipes, where a pressure gradient is present, the correct value for the heat transfer would be obtained in this way if heat sources and sinks were present in the fluid. If these sources are of strength $q$ per unit volume, equation (17) becomes

$$U \frac{\partial T}{\partial x} + V \frac{\partial T}{\partial y} = \frac{q}{\rho c} + \frac{\partial}{\partial y} \left( \kappa \frac{\partial T}{\partial y} + \overline{Lv} \frac{\partial T}{\partial y} \right);$$

then $(T-T_0)$ is proportional to $U$, the velocity distribution given by (15), provided that

$$\frac{q U_m}{c(T_m - T_0)} = -\frac{\partial \bar{p}}{\partial x},$$

where $U_m$ and $T_m$ are the average values of $U$ and $T$.[††] The sources could be realized experimentally by passing an electric current

---

[†] The above discussion is concerned with turbulent flow. The corresponding results for laminar flow are obtained simply by omitting the eddy transfer terms from the equations for the velocity and the temperature. The equations for the velocity and temperature are again analogous if the conditions $\partial p/\partial x = 0$ and $\sigma = 1$ are satisfied, and if the boundary conditions for $T$ and $U$ are analogous.

[‡] See § 280 below.

[||] *Physik. Zeitschr.* **29** (1928), 487–489.

[††] The equations have been written out to cover the case of two-dimensional flow, but exactly the same conclusions hold for the flow in a circular pipe.

through a conducting liquid, such as mercury, flowing in a channel or pipe having non-conducting walls which are kept at a constant temperature. The conclusions of the theory are usually compared, however, with the results of experiments on heat transfer from fluids in pipes for which no sources are present.

For the case in which the walls of a pipe are kept at a constant temperature gradient of magnitude $A$ per unit length, the temperature distribution has been calculated on the basis of the momentum transfer theory by G. I. Taylor.[†] From equation (11)

$$(\kappa + \overline{L_r v_r})\frac{\partial T}{\partial r} = \frac{A}{r} \int_0^r Ur\, dr;$$

the rate of heat transfer $Q$ per unit area for any section, given by (8), is

$$Q = -\rho c(\kappa + \overline{L_r v_r})\frac{\partial T}{\partial r} = -\rho c\frac{A}{r} \int_0^r Ur\, dr. \qquad (20)$$

The equation for the shear stress on the momentum transfer theory is

$$\tau = \rho(\nu + \overline{L_r v_r})\frac{\partial U}{\partial r} = \frac{r}{2}\frac{\partial \bar{p}}{\partial x}, \qquad (21)$$

where, since $r$ is measured from the axis of the pipe and $\partial U/\partial r$ is negative, $\tau$ is negative.[‡] (If $A$ is positive, then $\partial T/\partial r$ is positive and $Q$ is negative.) Eliminating $\overline{L_r v_r}$ between (20) and (21), we obtain for the temperature at any point, when $\sigma = 1$, the equation

$$T - T_c = -\frac{2}{c(\partial \bar{p}/\partial x)} \int_0^r \frac{Q}{r}\frac{\partial U}{\partial r}\, dr$$

$$= -\frac{2}{c(\partial \bar{p}/\partial x)} \int_{U_c}^U \frac{Q}{r}\, dU, \qquad (22)$$

where $T_c$ and $U_c$ are the temperature and velocity, respectively, at the centre of the pipe. From Stanton's measurements of the velocity distribution at about $R = 10^5$, Taylor calculated the values of $-Q/(\rho c A U_c r)$ from (20) and the temperature distribution from (22).

---

† *Proc. Roy. Soc.* A, **129** (1930), 25–30.

‡ In the corresponding equation in Chap. VIII, equation (31), $\tau$ denotes the positive shear stress.

The results, which apply for $\sigma = 1$ and $R = 10^5$, are partly reproduced in Table 23, where $T_0$ is the temperature at the wall.

TABLE 23

| $\dfrac{r}{a}$ | $\dfrac{U}{U_c}$ | $\dfrac{T_0-T}{T_0-T_c}$ |
|:---:|:---:|:---:|
| 0·0 | 1·000 | 1·000 |
| 0·1 | 0·995 | 0·9941 |
| 0·2 | 0·982 | 0·9790 |
| 0·3 | 0·966 | 0·9614 |
| 0·4 | 0·944 | 0·9353 |
| 0·5 | 0·916 | 0·9040 |
| 0·6 | 0·882 | 0·8665 |
| 0·7 | 0·844 | 0·8255 |
| 0·8 | 0·794 | 0·7730 |
| 0·85 | 0·764 | 0·7419 |
| 0·9 | 0·722 | 0·6990 |
| 0·95 | 0·664 | 0·6435 |
| 0·98 | 0·596 | 0·5750 |
| 0·99 | 0·548 | 0·5285 |
| 1·00 | 0 | 0 |

The value of $T_0 - T_c$ was

$$T_0 - T_c = -0.4264 \frac{2\rho A U_c^2}{\partial \bar{p}/\partial x}, \tag{23}$$

and if $Q_0$ and $\tau_0$ are the rate of heat transfer per unit area and the shear stress at the wall respectively, so that $\tau_0 = \frac{1}{2}a(\partial\bar{p}/\partial x)$, it was found that

$$\frac{Q_0}{\tau_0} = -0.411 \frac{2\rho c A U_c}{\partial \bar{p}/\partial x}. \tag{24}$$

The mean velocity $U_m$ over a section, the mean unweighted temperature difference $T_0 - T_m$, and the mean temperature difference $T_0 - T_M$ weighted with respect to the velocity—i.e. the temperature difference measured in mixed fluid which has passed through the pipe— were found by graphical integration to have the values

$$U_m = \frac{2}{a^2} \int_0^a Ur \, dr = 0.822 U_c,$$

$$T_0 - T_m = \frac{2}{a^2} \int_0^a (T_0-T)\, r \, dr = 0.801(T_0-T_c),$$

$$T_0 - T_M = \frac{2}{a^2 U_m} \int_0^a (T_0-T)Ur \, dr = 0.823(T_0-T_c).$$

From these and from (23) and (24) we obtain the relationship

$$\frac{1}{k_H} = \frac{0\cdot4264\times0\cdot823}{0\cdot411\times0\cdot822}\times\frac{2}{c_f} = 1\cdot04\left(\frac{2}{c_f}\right)$$

between the heat transfer coefficient, defined by

$$k_H = |Q_0/[\rho c U_m(T_M-T_0)]|,$$

and the skin-friction coefficient defined by $c_f = |\tau_0/(\frac{1}{2}\rho U_m^2)|$. Thus the correct application of Reynolds's analogy for a particular case leads to a result different from (14) even for $\sigma = 1$.

An investigation of the effect of inlet length on heat transfer has been made by Latzko,[†] for fluids for which $\sigma = 1$, on the basis of the 1/7th power law for the velocity distribution near a wall. The results obtained depend on the velocity and its derivative being given accurately by this power law; no sufficiently accurate experimental data are available to check the results.

As remarked above, an important limitation to Reynolds's theory arises in connexion with the validity of the momentum transfer theory: when this does not apply, Reynolds's theory also fails. It is shown later that this is certainly the case for wakes. It seems, however, that the extended theory (for $\sigma$ not equal to unity) gives good agreement with the experimental results for heat transfer from pipes to fluids flowing through them. The momentum transfer theory gives the correct velocity distribution near a wall, and it may be that the agreement for heat transfer from pipes is due to the correct formulation of conditions near the wall.

### 280. The extension of Reynolds's analogy.

Reynolds's analogy fails for fluids for which $\sigma$ is not equal to unity, owing to the existence of a layer near the wall in which the transfer of heat and momentum by conductivity and viscosity are of importance. Extensions to allow for the effect of this layer have been made by Prandtl,[‡] Taylor,[||] and Kármán.[††]

Taylor and Prandtl assume that a sharp boundary exists between a laminar wall layer and the turbulent core. In this core the conditions postulated by Reynolds are assumed to hold, so that equation

---

[†] *Zeitschr. f. angew. Math. u. Mech.* **1** (1921), 268–290.

[‡] *Loc. cit.* in footnote || on p. 651.

[||] *A.R.C. Reports and Memoranda*, No. 272 (1919).

[††] *Proc. 4th Internat. Congress for Applied Mechanics*, Cambridge, 1934, pp. 77–83.

(13) is satisfied, $U_1$, $T_1$ now representing the velocity and temperature at the boundary between the wall layer and the core. In the wall layer the velocity and temperature distributions are assumed to be linear, with the Reynolds stresses and the eddy heat transfer negligible, so that the shear stress and heat flow are given by

$$\left.\begin{array}{l}
\tau = \tau_0 = \mu\dfrac{\partial U}{\partial y} = \dfrac{\mu U_1}{y_1}, \\[2mm]
Q = Q_0 = -k\dfrac{\partial T}{\partial y} = -k\dfrac{T_1 - T_0}{y_1},
\end{array}\right\} \tag{25}$$

where $\tau_0$, $Q_0$ are the shear stress and rate of heat transfer per unit area at the wall, $T_0$ is the wall temperature, and $y_1$ is the thickness of the wall layer. Eliminating $y_1$ and $T_1$ from (13) and (25) we obtain the result

$$\frac{Q_0}{\rho c(T_0 - T_m)} = \frac{\tau_0}{\rho U_m[1 + (\sigma - 1)U_1/U_m]},$$

or

$$\frac{1}{k_H} = \frac{2}{c_f}\left[1 + (\sigma - 1)\frac{U_1}{U_m}\right]. \tag{26}$$

In this equation $U_1/U_m$ must be determined from some theory of turbulent flow. If the thickness of the wall layer can be defined by $U_\tau y_1/\nu = $ constant, where $U_\tau$ is the 'friction velocity', $(\tau_0/\rho)^{\frac{1}{2}}$, then with $U/U_\tau$ a function of $yU_\tau/\nu$ (as with either a power law (Chap. VIII, equation (23), p. 340) or a logarithmic law (Chap. VIII, equation (8), p. 333)) for the velocity distribution near the wall but outside the viscous layer, we shall have $U_1/U_\tau = $ constant. Further, since $U_\tau/U_m = \sqrt{(\frac{1}{2}c_f)}$,† $U_1/U_m = A\sqrt{(\frac{1}{2}c_f)}$, where $A$ is a constant. Hence (26) becomes

$$\frac{1}{k_H} = \frac{2}{c_f} + A(\sigma - 1)\sqrt{\left(\frac{2}{c_f}\right)}. \tag{27}$$

With Blasius's power law (Chap. VIII, equation (20), p. 339) for flow in pipes, $c_f$ is proportional to $R^{-\frac{1}{4}}$, where $R$ is $2aU_m/\nu$ and $a$ is the radius of the pipe. Prandtl found that the assumption $U_1/U_m = 1\cdot 1R^{-\frac{1}{8}}$ best fits the experimental results, and this corresponds to a value of $5\cdot 6$ for $A$. The formula (27), with this value of $A$, is in satisfactory agreement with the rather scattered experimental results for heat transfer to fluids for which the value of $\sigma$ does not differ greatly from unity. For other values of $\sigma$ the agree-

† The more general $c_f$ is used here for pipe flow also, instead of the $\gamma$ of Chaps. VII and VIII.

ment is not satisfactory. Kármán† pointed out that this is probably due to the assumption that a sharp boundary exists between the wall layer (in which the velocity and temperature distributions are assumed linear, and the eddy stresses and eddy heat transfer negligible) and the turbulent core. He suggested that the field of flow should be divided into three regions: (1) a turbulent core in which heat and momentum are transferred by turbulent mixing and in which Reynolds's analogy applies; (2) a transition layer in which heat and momentum are transferred both by turbulent mixing and by molecular processes (conductivity or viscosity, respectively); (3) a laminar layer at the wall in which no turbulent mixing occurs, and in which the velocity and temperature distributions may be taken as linear.

In both the laminar and transition layers we take $Q = Q_0$ and $\tau = \tau_0$, where the zero suffix denotes values at the wall. Measuring $y$ normal to the wall, we put $U = U_1$, $T = T_1$, $y = y_1$, at the edge of the laminar layer and $U = U_2$, $T = T_2$, $y = y_2$, at the edge of the transition layer. In the laminar layer the relations (25) hold, where $T_0$ is the temperature of the wall. In the turbulent core Reynolds's analogy is assumed to hold, so that

$$\frac{Q}{\rho c (U_m - U_2)(T_m - T_2)} = -\frac{\tau}{\rho (U_m - U_2)^2}, \tag{28}$$

where $U_m$, $T_m$ are the mean velocity and temperature respectively. This relation, with $Q = Q_0$ and $\tau = \tau_0$, will hold at the junction of the core with the transition layer. In the transition layer a velocity distribution of the form

$$U/U_\tau = a + b \log_e U_\tau y/\nu$$

is assumed, as in the fully turbulent region, but with different values for $a$ and $b$. The values of $a$ and $b$ in the transition layer are determined so that $U$ and $dU/dy$ shall be continuous at $y = y_1$. For the laminar layer (25) gives $dU/dy = \tau_0/\mu = U_\tau^2/\nu$ and $U_1/U_\tau = y_1 U_\tau/\nu$. For the transition layer $dU/dy = bU_\tau/y$. If $U$ and $dU/dy$ are continuous at $y = y_1$, we obtain

$$b = y_1 U_\tau/\nu = U_1/U_\tau,$$

and

$$a = -\frac{y_1 U_\tau}{\nu}\left(\log_e \frac{y_1 U_\tau}{\nu} - 1\right) = -b(\log_e b - 1).$$

† *Loc. cit.* in footnote †† on p. 654. The derivation of the resulting equation (equation (30) below) was not given.

An examination of the experimental values of $U/U_\tau$ in the transition layer leads to the choice of $b = 5$. This makes $a = -3\cdot05$, and the curve

$$U/U_\tau = -3\cdot05 + 5\log_e yU_\tau/\nu$$

passes fairly well through the experimental points in the transition layer (cf. Chap. VIII, Figs. 92 and 93). This curve meets the curve of $U/U_\tau$ in the fully turbulent region, whose equation is given by (9) of Chapter VIII, where $yU_\tau/\nu = 30$; accordingly we take $y_2 U_\tau/\nu = 30$. Then in the transition layer

$$\tau_0/\rho = U_\tau^2 = (\nu + \overline{Lv})\, dU/dy = (\nu + \overline{Lv})bU_\tau/y,$$

so that
$$\overline{Lv} = yU_\tau/b - \nu.$$

Further

$$Q_0/\rho c = -(\kappa + \overline{Lv})\, dT/dy = -(\kappa - \nu + yU_\tau/b)\, dT/dy,$$

and the integral of this equation is

$$T_2 - T_1 = -\frac{Q_0 b}{\rho c U_\tau} \log_e\left(\frac{\kappa - \nu + y_2 U_\tau/b}{\kappa}\right),$$

since $y_1 U_\tau/b = \nu$. Hence, since $U_\tau^2 = \tau_0/\rho$ and $\nu/\kappa = \sigma$,

$$T_1 - T_2 = \frac{Q_0}{c\tau_0} bU_\tau \log_e(1 - \sigma + \sigma y_2/y_1). \tag{29}$$

Elimination of $T_1$ and $T_2$ from (25), (28), and (29) leads to the equation

$$T_0 - T_m = \frac{Q_0}{c\tau_0}[U_m - U_2 + \sigma U_1 + bU_\tau \log_e(1 - \sigma + \sigma y_2/y_1)].$$

We put
$$U_2/U_\tau = U_1/U_\tau + b\log_e y_2/y_1$$

(by the condition of continuity at $y = y_2$) and $U_\tau/U_m = \sqrt{(\tfrac{1}{2}c_f)}$ in this equation, and obtain

$$\frac{1}{k_H} = \frac{\rho c U_m(T_0 - T_m)}{Q_0}$$

$$= \frac{2}{c_f} + b\sqrt{\left(\frac{2}{c_f}\right)}\left[(\sigma - 1) + \log_e\left\{(1 - \sigma)\frac{y_1}{y_2} + \sigma\right\}\right].$$

With $b = 5$ and $y_2 U_\tau/\nu = 30$, this becomes

$$\frac{1}{k_H} = \frac{2}{c_f} + 5\sqrt{\left(\frac{2}{c_f}\right)}[(\sigma - 1) + \log_e\{1 + 0\cdot83(\sigma - 1)\}]. \tag{30}$$

## 281. Comparison between theory and experiment for flow in pipes.

The experimental values obtained for heat transfer from pipes to fluids flowing through them are mostly rather scattered owing to difficulties of measurement. These difficulties arise (1) from the effect of the entry length, in which the temperature distribution has not settled down and in which the rate of heat transfer is higher than the final value, and (2) from the variation of the properties of fluids with temperature, which results in the heat transfer coefficient for a finite rate of heat flow being different from the coefficient for an infinitesimal rate. The theoretical formulae apply only to the coefficient for an infinitesimal rate of heat flow. For gases the variation with temperature of the Prandtl number $\sigma$ is not large; but further errors may be caused by the convection currents which arise from the variation of density with temperature. For water the variation of $\sigma$ with temperature is important; Table 24 gives a set of values of $\sigma$ for water at various temperatures, and it will be seen that $\sigma$ changes from 12·45 at 0° C. to 1·02 at 160° C.

TABLE 24

| °C. | 0 | 5 | 10 | 15 |
|---|---|---|---|---|
| $\sigma$ | 12·45 | 10·46 | 8·93 | 7·73 |
| °C. | 20 | 25 | 30 | 40 |
| $\sigma$ | 6·75 | 5·96 | 5·30 | 4·29 |
| °C. | 50 | 60 | 70 | 80 |
| $\sigma$ | 3·55 | 2·99 | 2·56 | 2·22 |
| °C. | 90 | 100 | 110 | 120 |
| $\sigma$ | 1·95 | 1·74 | 1·56 | 1·40 |
| °C. | 130 | 140 | 150 | 160 |
| $\sigma$ | 1·27 | 1·17 | 1·09 | 1·02 |

Eagle and Ferguson† have made a complete series of experiments on the heat transfer from pipes to water flowing through them, in which the difficulty of the variation of $\sigma$ with temperature was surmounted by measuring the heat transfer at three different rates of heat flow and extrapolating the results to obtain the heat transfer coefficient for an infinitesimal temperature difference. The effect of the entry length was eliminated by heating the tube electrically, so that a linear distribution of temperature along the pipe was

† *Proc. Roy. Soc.* A, **127** (1930), 540–566.

obtained. The section for which the heat flow was measured was at such a distance from the beginning of the heated section that the entry effects had completely died away. The experiments extended over a range of values of $\sigma$ from 3 to 10, and a range of Reynolds numbers from $5.10^3$ to $2.10^5$. Eagle and Ferguson proposed an empirical formula to fit their experimental results, of the form

$$\frac{1}{k_H} = A + B(\sigma-1) - C(\sigma-1)^2$$

$$= \frac{2}{c_f}[\alpha + \beta(\sigma-1) - \gamma(\sigma-1)^2],$$

where $A$, $B$, $C$, $\alpha$, $\beta$, $\gamma$, are functions of the Reynolds number and $c_f$ is the skin-friction coefficient $\tau_0/(\frac{1}{2}\rho U_m^2)$. The experimental values of these quantities are given in Table 25, together with the values of $2/c_f$ obtained from the experiments of Stanton and Pannell.†

TABLE 25

| $\log R$ | $A$ | $B$ | $100C$ | $2/c_f$ | $\alpha$ | $\beta$ | $100\gamma$ |
|---|---|---|---|---|---|---|---|
| 3·7 | 307 | 99·5 | 123 | 207 | 1·483 | 0·480 | 0·59⁵ |
| 3·8 | 311 | 97 | 124·5 | 220 | 1·414 | 0·439 | 0·56⁵ |
| 3·9 | 315 | 96 | 126 | 234 | 1·343 | 0·411 | 0·53⁵ |
| 4·0 | 321 | 97·5 | 127·5 | 250 | 1·284 | 0·390 | 0·51 |
| 4·1 | 329 | 99 | 129 | 266 | 1·237 | 0·372 | 0·48⁵ |
| 4·2 | 341 | 101·5 | 130·5 | 284 | 1·202 | 0·358 | 0·46 |
| 4·3 | 356 | 104·5 | 132 | 303 | 1·175 | 0·345 | 0·43⁵ |
| 4·4 | 372 | 108 | 133 | 323 | 1·152 | 0·333 | 0·41 |
| 4·5 | 390 | 111·5 | 134 | 344 | 1·133 | 0·324 | 0·39 |
| 4·6 | 409 | 116 | 135·5 | 366 | 1·118 | 0·316 | 0·37 |
| 4·7 | 429·5 | 120·5 | 137·5 | 388 | 1·105 | 0·311 | 0·35⁵ |
| 4·8 | 450 | 126 | 140 | 411 | 1·094 | 0·307 | 0·34 |
| 4·9 | 470·5 | 132 | 143 | 434 | 1·084 | 0·305 | 0·33 |
| 5·0 | 491·5 | 138·5 | 146·5 | 457 | 1·075 | 0·303 | 0·32 |
| 5·1 | 512·5 | 145 | 150·5 | 480 | 1·069 | 0·302 | 0·31⁵ |
| 5·2 | 535 | 151·5 | 155·5 | 503 | 1·064 | 0·302 | 0·31 |
| 5·3 | 555 | 158·5 | 160·5 | 524 | 1·060 | 0·301 | 0·30⁵ |

The values of $k_H = Q/[\rho c(T_M - T_0)U_m]$ are based on the mean water temperature

$$(T_M - T_0) = \frac{2}{a^2 U_m} \int_0^a (T - T_0) U r \, dr,$$

as obtained by the 'mixing cup' method; hence the theoretical formulae developed in the previous section are not strictly comparable with them. To reduce them to comparable forms an analysis

† *Phil. Trans.* A, **214** (1914), 199–224.

similar to that given by Taylor[†] for $\sigma = 1$ and $R = 10^5$ is needed. However, Kármán has compared the results given by (30) directly with Eagle and Ferguson's experimental results; the comparison is shown in Fig. 250, and there is good agreement. The results of other

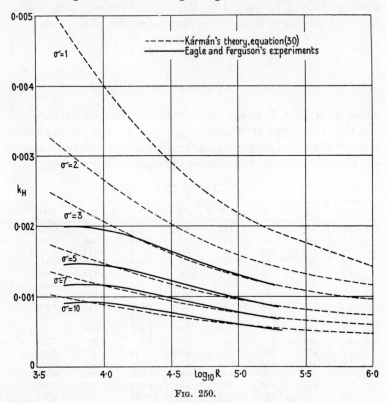

FIG. 250.

experimenters are rather scattered, and are not reproduced here. A collected account of them has been given by Lorenz.[‡]

## 282. Flow along a heated flat plate.

The calculation on mixture length theories of the velocity distribution for flow in a turbulent boundary layer along a flat plate was considered in Chap. VIII, §165, where a comparison with experiment was also shown (Chap. VIII, Fig. 112). If the plate is heated,

† See pp. 652–654.

‡ *Zeitschr. f. techn. Physik*, **15** (1934), 155–162, 201–206.

and maintained at a constant temperature, then on the momentum transfer theory the temperature and velocity distributions are similar. To find the temperature distribution on the vorticity transfer theory, we note first that the equation for the temperature is

$$U\frac{\partial T}{\partial x} + V\frac{\partial T}{\partial y} = \frac{\partial}{\partial y}\left(l^2\left|\frac{\partial U}{\partial y}\right|\frac{\partial T}{\partial y}\right),$$

if $\overline{Lv}$ is replaced by $l^2|\partial U/\partial y|$, on Prandtl's assumption (9). The temperature distributions at different sections are assumed to be similar, and the same assumption for $l$ is made as in the calculation of the velocity distribution (Chap. VIII, eqns. (81) and (82)). Putting $\eta = y/\delta$, where $\delta$ is the boundary layer thickness, and $(T_1-T)/(T_1-T_0) = g(\eta)$, where $T_1$ is the temperature of the plate and $T_0$ the temperature of the main stream, we find† that, on the vorticity transfer theory,

$$g = 1-C(1/\eta-1).$$

This formula satisfies the condition $g = 1$ at $\eta = 1$; but neither of the conditions, (1) that the fluid in contact with the plate should have the same temperature as the plate, or (2) that the temperature difference between the fluid in the main stream and the fluid in the boundary layer should pass smoothly to zero, can be satisfied. The constant $C$ is therefore determined by making the theoretical and experimental values of $g$ agree near $\eta = 0.25$. The resulting curve for the temperature distribution is shown in Fig. 112 of Chapter VIII (where $\theta/\theta_1$ is written for $(T_1-T)/(T_1-T_0)$), for comparison with the curve obtained from the momentum transfer theory and with the experimental results of Elias.

The calculation of the temperature distribution from the observed velocity distribution on the vorticity transfer theory, without any assumption concerning $\overline{Lv}$ (by eliminating this quantity between the equations for the velocity and the temperature), involves differentiation of the observed velocity distribution, and has not yet been carried out in view of the scatter of the experimental velocity observations, and also of the doubtful validity of the assumption of similarity in the velocity and temperature distributions (the calculations being very difficult without this assumption).

## 283. The temperature distribution between rotating cylinders.

The velocity and temperature distributions between rotating cylinders at different temperatures have been investigated by

† Howarth, *Proc. Roy. Soc.* A, **154** (1936), 373–376.

Taylor.† It was found that the velocity distribution over the central region could be explained by the vorticity transfer theory but not by the momentum transfer theory (see Chap. VIII, § 170). If the

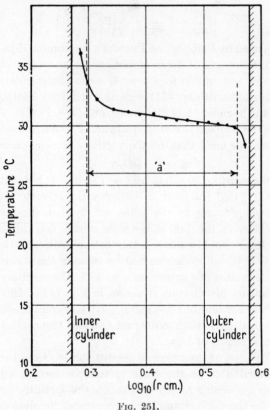

Fɪɢ. 251.

cylinders are kept at a constant temperature difference, the rate of heat transfer $Q$ per unit length is constant, and therefore

$$Q = 2\pi\rho c \overline{L_r v_r} \, r \, dT/dr,$$

where $r$ is measured radially from the axis of the cylinders. The results of Taylor's experiments are shown in Fig. 251, where the temperature is plotted against $\log_{10} r$. A linear distribution over the central region is obtained, showing that $\overline{L_r v_r}$ is constant in this

† Proc. Roy. Soc. A, **151** (1935), 494–512.

region. On the momentum transfer theory we should have

$$\frac{dT}{dr} \Big/ \frac{d}{dr}(Ur) = \text{constant},$$

but, whereas $Ur$ varied by only 0·4 per cent. over a region (marked 'a' in Fig. 251) covering 83·5 per cent. of the space between the cylinders, when the rotating inner cylinder was maintained at a higher temperature than the outer one the temperature decreased continually outwards, the change in the same region being 22 per cent. of the temperature difference between the cylinders. Near the walls the velocity and temperature distributions could not be defined closely owing to difficulties of measurement, but they were similar to those which would be predicted by the momentum transfer theory.

### 284. The temperature distribution in the wake of a heated cylinder.†

It has been shown (Chap. XIII, equations (55), (60), and (61), p. 583) that on certain assumptions the velocity distribution in the wake of a symmetrical cylinder, on either the momentum transfer or the vorticity transfer theory, tends far downstream to

$$\begin{aligned} U/U_0 &= Kx^{-\frac{1}{2}}f(\eta), \\ \text{where} \qquad f(\eta) &= [1-(\eta/\eta_0)^{\frac{3}{2}}]^2, \end{aligned} \tag{31}$$

$(U_0-U)$ is the velocity in the wake, $U_0$ is the undisturbed velocity of the stream, $K$ is a constant, $x$ is measured along the axis of the wake from an origin near the body, $\eta = yx^{-\frac{1}{2}}$, and $\eta_0$ is the value of $\eta$ at the edge of the wake. The result (31) is in good agreement with the experiments of Schlichting and of Fage and Falkner.

If the body is heated, the equation for the temperature distribution in the wake (equation (3), with $\overline{vT''}$ as in (8), or equation (17) with molecular conduction neglected) becomes

$$U_0 \frac{\partial T}{\partial x} = \frac{\partial}{\partial y} \left( \overline{Lv} \frac{\partial T}{\partial y} \right). \tag{32}$$

If (31) is accepted from experiment as a valid result for the velocity distribution, the temperature distribution can be calculated without making any assumption as to $\overline{Lv}$. On the vorticity transfer theory

† Taylor, *Proc. Roy. Soc.* A, **135** (1932), 685–696.

(either the two-dimensional theory or the modified theory) the equation for the velocity is

$$U_0 \frac{\partial U}{\partial x} = \overline{Lv} \frac{\partial^2 U}{\partial y^2}, \tag{33}$$

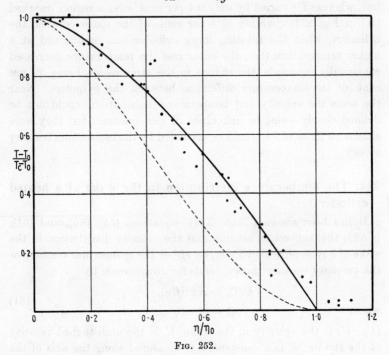

FIG. 252.

so that the equation for $f(\eta)$ is

$$-\tfrac{1}{2}U_0(f + \eta f') = \overline{Lv}f''. \tag{34}$$

If $T_0$ is the temperature outside the wake and if

$$T - T_0 = \text{const.}\, x^{-\frac{1}{2}}\phi(\eta),$$

then from (32)

$$-\tfrac{1}{2}U_0(\phi + \eta\phi') = \frac{\partial}{\partial \eta}(\overline{Lv}\phi'). \tag{35}$$

We substitute for $\overline{Lv}$ from (34) and integrate with respect to $\eta$. Then

$$\eta\phi = \phi'(f + \eta f')/f'', \tag{36}$$

no additive constant being necessary since $\phi' = 0$ at $\eta = 0$. Hence

$$\phi = \exp\left\{\int \frac{\eta f''}{f + \eta f'}\, d\eta\right\} = \exp\left\{-\frac{3}{2\eta_0} \int \frac{(\eta/\eta_0)^{\frac{1}{2}}}{1 - (\eta/\eta_0)^{\frac{3}{2}}}\, d\eta\right\}$$

$$= 1 - (\eta/\eta_0)^{\frac{3}{2}}. \tag{37}$$

If we put $\overline{Lv} = l^2|dU/dy|$, and make the same assumptions for $l$ as in calculating the velocity distributions (Chap. XIII, equation (56)), (35) integrates at once to (37); but (37) may be obtained without any assumptions as to $\overline{Lv}$, if (31) is taken as an empirical result.[†]

On the momentum transfer theory the temperature distribution is similar to the velocity distribution, so that $\phi = f$.

The results on the momentum and vorticity transfer theories are compared with the experiments of Fage and Falkner[‡] in Fig. 252, in which $(T-T_0)/(T_c-T_0)$ is plotted against $\eta/\eta_0$, $T_c$ being the value of $T$ in the middle of the wake. The full line is $\phi(\eta)$ and the dotted line is $f(\eta)$ (which is the value of $(T-T_0)/(T_c-T_0)$ on the momentum transfer theory). The points are the experimental results. It will be seen that the agreement with the vorticity transfer theory is good except near the edge of the wake.

## 285. The temperature distribution in the wake behind a row of heated parallel rods.

We have seen in Chap. XIII, § 253, that on either the momentum or vorticity transfer theories the velocity distribution behind a row of parallel equally-spaced similar rods tends to

$$U/U_0 = Kx^{-1}f(y), \tag{38}$$

where $(U_0-U)$ is the velocity in the wake, $U_0$ is the mean velocity of the stream, $K$ is a constant, $x$ is measured downstream from an origin near the rods, and the axis $y = 0$ passes through the centre of the section of one of the rods, which are situated at a distance $\lambda$ apart; and that, with $\overline{Lv} \propto x^{-1}$ and constant across a section of the wake,

$$f(y) = \cos(2\pi y/\lambda), \tag{39}$$

while with $\overline{Lv} = l^2|\partial U/\partial y|$ and $l$ constant,

$$f(y) = F(y),$$

where $F(y)$ is the function defined in equation (70) of Chapter XIII.

When the rods are heated, the temperature distribution far downstream is given by (32). If the function $f(y)$ in (38) is assumed to be known, the temperature distribution can be calculated without any

---

† If we were to regard $f(\eta)$ as not given by an analytical expression, but simply by a curve drawn through the experimental points, we should write

$$\int \eta f'' \, d\eta/(f+\eta f') = \log(f+\eta f') - 2 \int f' \, d\eta/(f+\eta f')$$

to avoid the double differentiation.

‡ *Proc. Roy. Soc.* A, **135** (1932), 702–705.

assumption as to $\overline{Lv}$. On the momentum transfer theory the temperature distribution is the same as the velocity distribution. On the vorticity transfer theory the equation for $f(y)$ is

$$U_0 f/x = -\overline{Lv}f'' \tag{40}$$

(since the equation for $U$ is equation (33)). If

$$T-T_0 = \text{const.}\, x^{-1}\phi(y),$$

where $T_0$ is the temperature far downstream, then from (32)

$$-U_0\phi/x = \frac{\partial}{\partial y}(\overline{Lv}\phi').$$

Substituting for $\overline{Lv}$ from (40) we obtain

$$\phi = \frac{d}{dy}\left(\frac{f\phi'}{f''}\right).$$

The solution which satisfies the condition $\phi' = 0$ for $y = 0$ is

$$\phi = C\left[f'\int_0^y \frac{dy}{f^2} + \frac{1}{f}\right], \tag{41}$$

where $C$ is a constant.

This equation holds on the vorticity transfer theory for any value of $\overline{Lv}$. If the experimental velocity distribution were known sufficiently accurately this could be used to determine the temperature distribution. With $f$ as in (39), (41) gives

$$\phi = f = \cos(2\pi y/\lambda),$$

thus verifying that when $\overline{Lv}$ is independent of $y$ the temperature distribution is the same as the velocity distribution, which it must be since the vorticity transfer theory and momentum transfer theory are then identical.

With $f = F$, (41) gives

$$\phi = C\left[F'\int_1^F \frac{dF}{F'F^2} + \frac{1}{F}\right] \tag{42}$$

since $F = 1$ at $y = 0$. From the equation for $F$ (see Chap. XIII, equation (69)), (42) may be written

$$\phi = C\left[-(1-F^2)^{\frac{1}{2}}\int_F^1 \frac{dF}{F^2(1-F^2)^{\frac{1}{2}}} + \frac{1}{F}\right],$$

which is easily transformed into

$$\phi/C = \frac{1}{F} + (1-F^2)^{\frac{1}{3}} \int\limits_F^1 \frac{d}{dF}\left[\frac{(1-F^2)^{\frac{1}{3}}}{F}\right](1-F^2)^{\frac{1}{3}}\, dF$$

$$= \frac{1}{F} - (1-F^2)^{\frac{1}{3}}\left[\frac{(1-F^2)^{\frac{2}{3}}}{F} - \frac{1}{3}\int\limits_F^1 (1-F^2)^{-\frac{1}{3}}\, dF\right]$$

$$= F + \tfrac{1}{3}(1-F^2)^{\frac{1}{3}} \int\limits_F^1 (1-F^2)^{-\frac{1}{3}}\, dF.$$

But if
$$\beta = \int\limits_0^1 \frac{dF}{(1-F^2)^{\frac{1}{3}}} = 1\cdot2935$$

then
$$y = \frac{\lambda}{4\beta} \int\limits_F^1 \frac{dF}{(1-F^2)^{\frac{1}{3}}}$$

(Chap. XIII, equations (70) and (71)). Also $F = 1$ at $y = 0$, and hence to make $\phi = 1$ at $y = 0$ we take $C = 1$, so that

$$\phi = F + \frac{4\beta y}{3\lambda}(1-F^2)^{\frac{1}{3}}. \tag{43}$$

These results are shown in Fig. 253, and compared with some experimental results of Gran Olsson.† It may be noted that the temperature distribution given by (43) is asymmetrical, and that $\phi'$ is infinite at $y = \frac{1}{2}\lambda$. This defect is not necessarily due to the vorticity transfer theory, but is probably due to the assumption that $\overline{Lv}$ (which in this case is put equal to $l^2|\partial U/\partial y|$ in obtaining the formula for $F$) vanishes for $\partial U/\partial y = 0$; it might be expected that the solution would fail at $y = \frac{1}{2}\lambda$.

Owing to the rapid equalization of temperature in the wake of the grid it is possible to compare theory and experiment only at a distance of 25 cm. downstream of the rods (which were placed 6 cm. apart): this distance is not really sufficient for the assumptions made in developing the theory to be valid. Farther downstream, however, the temperature differences were hardly measurable.

Gran Olsson also investigated the temperature distribution in the wake of a single heated rod which forms part of a grid of equidistant rods. With Prandtl's assumption that $\overline{Lv} = l^2|dU/dy|$ the temperature distribution is the same as if all the rods are heated, since

† *Zeitschr. f. angew. Math. u. Mech.* **16** (1936), 257–274.

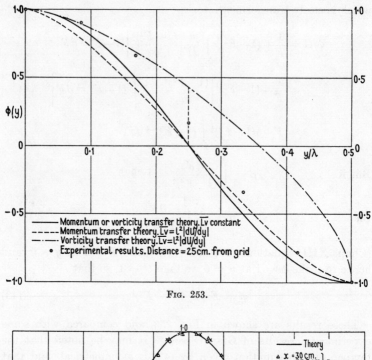

Fig. 253.

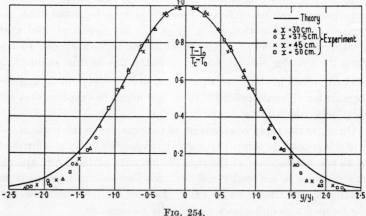

Fig. 254.

$dU/dy$ vanishes in the planes midway between the rods. With $\overline{Lv}$ constant across a section of the wake the temperature distribution at any section, given by (32), is shown to be

$$T - T_0 = (T_c - T_0)\exp[-(\log_e 2)(y/y_1)^2],$$

where $T_c$ is the temperature at the centre, $T_0$ is the temperature outside the wake, and $y_1$ is the value of $y$ for which $(T-T_0)/(T_c-T_0)$ is equal to $0.5$ at the section concerned. Gran Olsson's experimental results are compared with the temperature distribution given by this formula in Fig. 254, in which the scale of the experimental results has been chosen to make theory and experiment agree at $y = 0$ and $y = y_1$.

## 286. The wake behind a heated body of revolution.

The equation for the temperature far downstream in the wake of a heated body of revolution is

$$U_0\frac{\partial T}{\partial x} = \frac{1}{r}\frac{\partial}{\partial r}\left(r\overline{L_r v_r}\frac{\partial T}{\partial r}\right). \tag{44}$$

(Cf. equation (4); $x$ is written for $z$, and $U_0$ for $V_z$, and $\overline{L_r v_r}\,\partial T/\partial r$ has been substituted for $-\overline{v_r\,T'}$.)

On the momentum transfer theory the temperature distribution is the same as the velocity distribution, which (Chap. XIII, equation (84)) we take to be

$$U/U_0 = f(\eta)/x^{\frac{2}{3}}, \tag{45}$$

where $\eta = r/x^{\frac{1}{3}}$, so that if

$$T-T_0 = \text{constant}.x^{-\frac{2}{3}}\phi(\eta), \tag{46}$$

then $\phi = f$.

Expressions for $\phi(\eta)$ according to the vorticity transfer theory have been found by Goldstein[†] on the various assumptions on which the velocity was found in Chap. XIII, § 254. The results are shown in Fig. 255, where, with $T_c$ as the temperature at the centre of the wake, $(T-T_0)/(T_c-T_0)$ is plotted on the same scale as $U/U_{\max}$ in Fig. 234 of Chapter XIII, i.e. against $r/R$, where $R$ is the value of $r$ for which $U = \frac{1}{2}U_{\max}$. The figure also contains the mean experimental results of Hall and Hislop,[‡] which, apart from a small systematic discrepancy of doubtful significance, agree well with the results calculated on the modified vorticity transfer theory except near the edge of the wake.

Thus it appears from Figs. 234 and 255 that the experimental

[†] *Proc. Camb. Phil. Soc.* **34** (1938), 48–67 and 351–353. The calculation on the modified vorticity transfer theory with $l$ constant over a section of the wake has also been carried out by Tomotika, *Proc. Roy. Soc.* A, **165** (1938), 53–64.

[‡] *Proc. Camb. Phil. Soc.* **34** (1938), 345–350.

evidence is in favour of the validity of the modified vorticity transfer theory except in the outer parts of the wake.

The mean experimental velocity and temperature distributions are compared in Fig. 256 (p. 671).

When the velocity distribution—i.e. the function $f(\eta)$ in (45)—

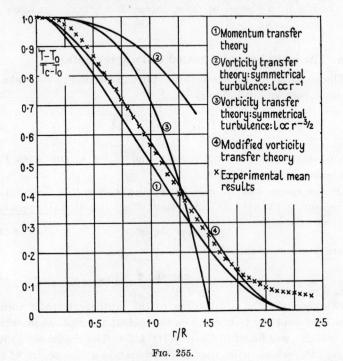

① Momentum transfer theory

② Vorticity transfer theory: symmetrical turbulence: $L \propto r^{-1}$

③ Vorticity transfer theory: symmetrical turbulence: $L \propto r^{-3/2}$

④ Modified vorticity transfer theory

× Experimental mean results

FIG. 255.

has been found experimentally, it is possible to find $\phi$ without any assumption as to $\overline{L_r v_r}$ by eliminating this quantity between the equations for the temperature and the velocity, as in the preceding section. It may be shown† that if the scales for $\eta$ and $f(\eta)$ are altered so that

$$\eta = r/R, \qquad U/U_{\max} = f(\eta), \tag{47}$$

then on the vorticity transfer theory with symmetrical turbulence

$$\frac{T-T_0}{T_c-T_0} = (f + \tfrac{1}{2}\eta f')\exp\left\{-4 \int_0^\eta \frac{f'}{\eta f'+2f}\,d\eta\right\}, \tag{48}$$

† Goldstein, *loc. cit.*

and on the modified vorticity transfer theory with isotropic turbulence

$$\frac{T-T_0}{T_c-T_0} = (f+\tfrac{1}{2}\eta f')\exp\left(-2\int_0^\eta \frac{f'}{\eta f'+2f}\,d\eta\right). \qquad (49)$$

On the momentum transfer theory $(T-T_0)/(T_c-T_0)$ is equal to
$U/U_{\max}$. All these results are independent of the value of the coeffi-

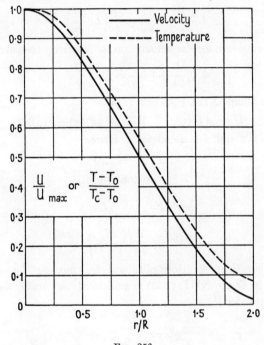

Fig. 256.

cient of turbulent diffusion; and the experimental results suffice to
show that for no value of that coefficient can either the momentum
transfer theory or the vorticity transfer theory with symmetrical
turbulence lead to satisfactory results, whereas the modified vorticity transfer theory does lead to satisfactory results over the inner
portion of the wake with Prandtl's assumption for the coefficient
of turbulent diffusion and with $l$ constant across a section of the
wake.

### 287. The temperature distribution in heated jets.

In Chap. XIII, § 255, pp. 597–599, we considered the two-dimensional problem of the turbulent mixing of a stream of fluid with fluid at rest, when the mixing, beginning at some value of $x$ (where the stream is moving parallel to the $x$-axis) takes place along a single boundary between the two portions of fluid. If one of the portions of fluid is hotter than the other, and $T$ is the temperature in the mixing region, then (equation (17) with molecular conduction neglected)

$$U\frac{\partial T}{\partial x}+V\frac{\partial T}{\partial y} = \frac{\partial}{\partial y}\Big(\overline{Lv}\frac{\partial T}{\partial y}\Big),$$

whilst the equation for the velocity on the vorticity transfer theory is

$$U\frac{\partial U}{\partial x}+V\frac{\partial U}{\partial y} = \overline{Lv}\frac{\partial^2 U}{\partial y^2}.$$

Now, as in Chap. XIII, equation (114),

$$U = AF'(\eta), \qquad V = A\{\eta F'(\eta)-F(\eta)\},$$

where $\eta = y/x$ and $A$ is a constant. Hence

$$-AxFF'' = \overline{Lv}F''',$$

and if

$$T = \text{constant}+g(\eta),$$

then

$$-AxFg' = \frac{\partial}{\partial \eta}(\overline{Lv}g').$$

Hence

$$Fg' = \frac{d}{d\eta}\Big(\frac{FF''}{F'''}g'\Big).$$

If the theoretical solution obtained by putting $\overline{Lv} = l^2\,\partial U/\partial y$, $l = cx$ (as in Chap. XIII, § 255) is sufficiently accurate, then on the vorticity transfer theory

$$F+c^2F''' = 0,$$

so that

$$Fg' = -\frac{d}{d\eta}(c^2F''g') = -c^2F'''g'-c^2F''g''$$

$$= Fg'-c^2F''g'',$$

and

$$g'' = 0,$$

$$g = A\eta+B.$$

Thus the temperature distribution, which on the momentum transfer theory is the same as the velocity distribution, is a straight line on the vorticity transfer theory. According to Ruden[†] this latter result is in much better agreement than the former with the temperature

† *Die Naturwissenschaften*, **21** (1933), 375–378.

distribution at the edge of a large heated symmetrical jet near the mouth of the nozzle.

Calculations of temperature distributions have been made by Howarth[†] for heated two-dimensional jets and heated jets symmetrical about an axis in three dimensions. For the plane jet, with $l \propto x$ as in Chap. XIII, § 255, equation (100), the temperature distribution on the vorticity transfer theory is the square root of the velocity distribution. No experiments have been reported. For the symmetrical jet all the calculations have been carried out which are analogous to those mentioned in the preceding section for symmetrical wakes. Small scale graphs of both velocity and temperature distributions have been published by Ruden (*loc. cit.*). The observed values of $(T-T_0)/(T_c-T_0)$ are rather greater than the observed values of $U/U_c$, so the momentum transfer theory does not give a satisfactory result. The vorticity transfer theory with symmetrical turbulence leads to a result completely at variance with experiment. The calculated velocity distribution according to the modified vorticity transfer theory with isotropic turbulence and with $l$ constant across a section of the jet does not agree well with the observed velocity distribution, but the corresponding calculated temperature distribution is in satisfactory agreement with the experimental results. Consequently an attempt to calculate the temperature distribution from the observed velocity distribution without any assumption concerning the coefficient of turbulent diffusion did not lead to a satisfactory result.

[†] *Proc. Camb. Phil. Soc.* **34** (1938), 185–203.

## ADDITIONAL REFERENCES FOR CHAPTERS XIV AND XV

*Treatises and Collected Works.*

FISHENDEN and SAUNDERS, *The Calculation of Heat Transmission* (London, 1932).

MCADAMS, *Heat Transmission* (New York, 1933).

GRÖBER and ERK, *Die Grundgesetze der Wärmeübertragung* (Berlin, 1933).

DRYDEN, Aerodynamics of Cooling (*Aerodynamic Theory*, edited by Durand, **6** (Berlin, 1936), 223–282).

ten BOSCH, *Die Wärmeübertragung* (Berlin, 1936).

*Heat Transfer in Potential Flow.*

PIERCY and WINNY, Convection of Heat from Isolated Plates and Cylinders in an Inviscid Stream, *Phil. Mag.* (7), **16** (1933), 390–408.

BICKLEY, Some Solutions of the Problem of Forced Convection, *Phil. Mag.* (7), **20** (1935), 322–343.

*Flow in Pipes.*

(a) *Laminar Flow.*

LÉVÊQUE, Résolution théorique du problème de l'échange de chaleur par circulation d'un fluide visqueux en mouvement tranquille à l'intérieur d'un tube cylindrique, *Comptes Rendus*, 185 (1927), 1190–1192.

KRAUSSOLD, Die Wärmeübertragung bei zähen Flüssigkeiten in Rohren, *Ver. deutsch. Ing.*, *Forschungsheft* 351 (1931).

SHERWOOD, KILEY, and MANGSEN, Heat Transmission to Oil Flowing in Pipes, *Ind. Eng. Chem.* 24 (1932), 273–277.

JACOB and ECK, Über den Wärmeaustausch bei der Strömung zäher Flüssigkeiten in Rohren, *Forsch. Ingwes.* 3 (1932), 121–126.

RIBAUD, Théorie thermique de la couche limite en régime laminaire, *Comptes Rendus*, 202 (1936), 32–35.

(b) *Turbulent Flow.*

NUSSELT, Der Wärmeübergang in Rohrleitungen, *Forschungsarbeiten des Ver. deutsch. Ing.*, No. 89 (1910).

MCADAMS and FROST, Heat Transfer by Conduction and Convection, *Ind. Eng. Chem.* 14 (1922), 13–18; 1101–1105.

BAILEY and COPE, Heat Transmission through Circular, Square and Rectangular Pipes, *A.R.C. Reports and Memoranda*, No. 1560 (1934).

GUCHMANN, ILJUCHIN, TARASSOWA, and WARSCHAWSKI, Untersuchung des Wärmeübergangs bei Bewegung eines Gases mit sehr grosser Geschwindigkeit in einem geraden Rohr, *Techn. Physics U.S.S.R.* 2 (1935), 375–413.

WINKLER, Wärmeübergang in Rohren bei hohen Reynoldschen Zahlen, *Forsch. Ingwes.* 6 (1935), 261–268.

JUNG, Wärmeübergang und Reibungswiderstand bei Gasströmung in Rohren bei hohen Geschwindigkeiten, *Ver. deutsch. Ing.*, *Forschungsheft* 380 (1936).

*Analogy between Heat Transfer and Skin-Friction.*

TAYLOR, Conditions at the Surface of a Hot Body exposed to the Wind, *Advisory Committee for Aeronautics, Reports and Memoranda*, No. 272 (1919).

STANTON and PANNELL, Heat Transmission over Surfaces, *Advisory Committee for Aeronautics, Reports and Memoranda*, No. 243 (1919).

RAYLEIGH, On the suggested Analogy between the Conduction of Heat and Momentum during the Turbulent Motion of a Fluid, *Advisory Committee for Aeronautics, Reports and Memoranda*, No. 497 (1919); *Scientific Papers*, 6, 486, 487.

MARSHALL, Further Experiments on the Relation between Skin Friction and Heat Transmission, *A.R.C. Reports and Memoranda*, No. 1004 (1926).

WHITE, Fluid Friction and its relation to Heat Transfer, *Trans. Inst. Chem. Eng.* 10 (1932), 66–80.

*Analogy between Heat Transfer and Diffusion.*

THOMA, *Hochleistungskessel* (Berlin, 1921).

LOHRISCH, Bestimmung von Wärmeübergangszahlen durch Diffusionsversuche, *Forschungsarbeiten des Ver. deutsch. Ing.*, No. 322 (1929).

NUSSELT, Wärmeübergang, Diffusion und Verdunstung, *Zeitschr. f. angew. Math. u. Mech.* **10** (1930), 105–121.

HOFFMANN, Wärmeübergang und Diffusion, *Forsch. Ingwes.* **6** (1935), 293–304.

*Forced Convection.*

DAVIS, On Convective Cooling in Liquids, *Phil. Mag.* (6), **47** (1924), 1057–1092.

BRYANT, OWER, HALLIDAY, and FALKNER, On the Convection of Heat from the Surface of an Aerofoil in a Wind Current, *A.R.C. Reports and Memoranda*, No. 1163 (1928).

PIERCY and SCHMIDT, On the Convection of Heat from Flat Plates in a Viscous Stream, *Phil. Mag.* (7), **17** (1934), 423–432.

KROUJILINE, La Théorie de la transmission de chaleur par un cylindre circulaire dans un courant fluide transversal, *Techn. Physics U.S.S.R.* **3** (1936), 311–320.

• PIERCY and PRESTON, A Simple Solution of the Flat Plate Problem of Skin Friction and Heat Transfer, *Phil. Mag.* (7), **21** (1936), 995–1005.

*Free Convection*

NUSSELT, Die Wärmeabgabe eines wagrecht liegenden Rohres oder Drahtes in Flüssigkeiten und Gasen, *Zeitschr. des Vereines deutscher Ingenieure*, **73** (1929), 1475–1478.

WEISE, Wärmeübergang durch freie Konvection an quadratischen Platten, *Forsch. Ingwes.* **6** (1935), 281–292.

JOUKOVSKY and KIREJEW, Optische Methode zur Untersuchung des Temperaturfeldes in der Umgebung umströmter Zylinder, *Techn. Physics U.S.S.R.* **3** (1936), 754–766.

*Temperature of Bodies immersed in Moving Fluids.*

KRUSHILIN, Die Bestimmung der Oberflächentemperatur eines wärmenichtleitenden Körpers, der sich in einem schnell bewegenden incompressibeln Flüssigkeitsstrom befindet, *Zeitschr. techn. Physik, Leningrad*, **6** (1936), 1574–1577.

A rather extensive list of references to works on heat transfer is given in Bulletin No. 84 of the National Research Council (Washington, 1932): *Hydrodynamics*, by Dryden, Murnaghan, and Bateman, 433–437.

# NOTE ON THE CONDITIONS AT THE SURFACE OF CONTACT OF A FLUID WITH A SOLID BODY

THE question of the conditions to be satisfied by a moving fluid in contact with a solid body was one of considerable difficulty for a long time, and its importance will justify a short historical note. At the present time it appears to be definitely settled that for practical purposes the fluid immediately in contact with a solid body may be taken as having no velocity relative to the solid, at any rate for nearly all fluids; but the exact conditions on a molecular scale remain still in doubt.

That a real fluid could not slip freely over the surface of a solid body was recognized by Daniel Bernoulli, who ascribed to this fact certain large discrepancies between the results he had calculated for a perfect fluid and those he had measured with a real one: 'Enormes has differentias maxima ex parte adhaesioni aquae ad latera tubi tribuo, quae certe adhaesio in hujusmodi casibus incredibilem exercere potest effectum.'† Du Buat‡ came to the conclusion that when the mean velocity of water flowing along a channel is sufficiently small, the fluid adjacent to the surface is at rest. Coulomb found that the resistance of an oscillating metallic disk in water was scarcely altered when the disk was smeared with grease, or when the grease was covered with powdered sandstone, so that the nature of the surface had practically no influence on the resistance;‖ and he also suggested that the molecules of the fluid in contact with an oscillating cylinder have the same velocity as the cylinder, that the molecules a short distance away have a smaller velocity, and that at a lateral distance of two or three millimetres the velocity becomes zero (*loc. cit.*, p. 296).

During the nineteenth century three different hypotheses were put forward by various authors at various times. According to the first, the velocity is the same at a solid wall as that of the solid itself, and changes continuously in the fluid, which has everywhere the same properties. This seems to have been Coulomb's belief. The second was put forward very clearly by Girard in the discussion of his experiments on the flow of liquids through tubes. He supposed that a very thin layer of fluid remains completely attached to the walls. The question then arises as to the conditions at the outer surface of this layer; Girard assumed that the rest of the fluid slips over it. He also supposed that if the walls are of the same material everywhere, the layer has a constant thickness, so that its surface presents to the current the same irregularities as those of the wall itself. Further, he assumed that the thickness of the layer depends on the curvature of the wall, and on the temperature.†† He took it to

† *Hydrodynamica* (Argentorati, 1738), p. 59.
‡ *Principes d'Hydraulique*, **1** (Paris, 1786), 92, 93.
‖ *Mémoires de l'Institut National des Sciences et des Arts: Sciences Mathématiques et Physiques*, **3**, prairial, an 9 (1800), 286.
†† *Mémoires de la Classe des Sciences Mathématiques et Physiques de l'Institut de France*, **14** (1813, 1814, 1815), 254, 324, 329. Similar ideas, but with a much thicker stagnant layer, had been expressed before, notably by Prony in his *Recherches physico-mathématiques sur la théorie des eaux courantes* (Paris, 1804).

be different for different liquids or different materials of the wall, and to become zero for liquids which do not wet the wall, as for mercury in glass tubes; in such cases he supposed that the liquid slips over the surface.[†] Thirdly, from the same molecular hypotheses which led him to the equations of motion of a viscous fluid, Navier deduced that there is slipping at a solid boundary, and this slipping is resisted by a force proportional to the relative velocity. Since the tangential stress on the solid wall at any point is the same as the stress at a neighbouring internal point of the fluid, this is equivalent to the boundary condition $\beta u = \mu \partial u/\partial n$ for flow in one direction along a plane wall, where $u$ is the velocity, the differentiation is along the normal away from the wall, and $\beta$ is a constant, such that $\mu/\beta$ is a length. This length is zero if there is no slip. Navier explained Girard's results by an application of this theory.[‡]

For some time confusion prevailed. Poisson obtained conditions essentially the same as Navier's, but suggested that these conditions are to be applied at the outer surface of a stagnant or quasi-solid layer similar to Girard's.[||] Stokes was initially inclined to the first hypothesis, but when calculations on flow through pipes gave results not in agreement with experiments then known to him (though they would have agreed with Hagen's or Poiseuille's), he hesitated between this hypothesis and Navier's.[††] In his report to the British Association in 1846 he mentioned all three hypotheses without deciding between them;[‡‡] finally he decided on the first, on the grounds that the existence of slip would imply that the friction between solid and fluid was of a different nature from, and infinitely less than, the friction between two layers of fluid, and also that the agreement with observation of results obtained on the assumption of no slip was highly satisfactory.[||||] Poiseuille, in his memoir on the movement of the blood, found a layer of stagnant blood at the walls of the containing vessel. He was led to examine the flow, through glass tubes, of liquids holding opaque bodies in suspension, and observed stagnant layers at the walls of thicknesses much less than any obtained by Girard.[†††] In his memoir on the flow of liquids through capillary tubes he remarked merely that the velocity cannot be uniform over a section of a tube, since it is known that the velocity of blood in a tube falls away from a maximum at the axis to a small value near the walls; but he added that hydraulic engineers should study the movement of particles in moving liquids with the aid of a microscope.[‡‡‡] Hagen, who obtained the laws of flow in capillary tubes experimentally a short time before Poiseuille, simply stated in his first paper that the velocity increases at a uniform rate from zero at the walls to a maximum in the middle; later he adopted the idea of a

† *Mémoires de l'Académie Royale des Sciences de l'Institut de France*, **1** (1816), 235, 247, 258.

‡ *Ibid.* **6** (1823), 414–416, 432–436.

|| *Journal de l'École Polytechnique*, **13** (1831), 161–169; see also *Mémoires de l'Académie Royale*, **11** (1832), 539.

†† *Trans. Camb. Phil. Soc.* **8** (1845), 299, 300; *Math. and Phys. Papers*, **1**, 96–98.

‡‡ *Papers*, **1**, 185, 186.

|||| *Trans. Camb. Phil. Soc.* **9** (1851), [17], [18]; *Math. and Phys. Papers*, **3**, 14, 15.

††† *Mémoires des Savants Étrangers*, **7** (1841), 150.

‡‡‡ *Ibid.* **9** (1846), 521. See the reference to Fage and Townend on p. 679.

stagnant layer near the walls, but without slip, and found that in his tubes the layer had to be thinner than the thinnest writing-paper.† Darcy, in his great memoir on the flow of water through pipes, substantially agreed with Girard's hypothesis.‡ Helmholtz, in his discussion of experiments by Piotrowski on the oscillations of spheres filled with liquid, adopted Navier's hypothesis and concluded that though there might be no slip in the case of water in contact with glass, or of ether or alcohol in contact with glass or a metal surface, there was a considerable slip for water in contact with a gilt surface.||

Gradually, however, the hypothesis finally adopted by Stokes, that there is no slip and that all parts of the fluid have the same properties, gained ground. The results of calculations on this hypothesis for the flow through tubes, begun by Stokes†† and carried out by various authors,‡‡ gave consistent results not only when applied to Poiseuille's experiments on the flow of water, but also—contrary to the opinion of most of these authors, and in disagreement with experiments carried out by Poiseuille himself||||—for the flow of mercury through tubes of glass (which is not wetted by the mercury).††† Moreover Whetham, who carried out experiments on the flow of water through silvered and copper tubes and repeated some of Piotrowski's experiments, came to the conclusion that there was no evidence of slip;‡‡‡ and Couette,|||||| after discussing at length various experiments on the determination of viscosity, including some of his own, came to the same conclusion, which Maxwell, in his experiments on the oscillations of glass disks in air, had also arrived at some time before.†††† All these experiments, being concerned in the main with methods of determining viscosity, relate almost entirely to non-turbulent flow; only Couette specially discussed the conditions at the boundaries for turbulent flow, and his conclusion was that the relative velocity is zero actually at the boundary, but changes very rapidly in its neighbourhood.

Any discussion of views held regarding the molecular phenomena at a surface of contact of a solid and a liquid would lead us too far astray; but a

† *Poggendorff's Annalen der Physik u. Chemie,* **46** (1839), 433; *Abhandlungen der Königlichen Akademie der Wissenschaften zu Berlin* (1854), *mathematische Abhandlungen,* pp. 57 and 62.

‡ *Mémoires des Savants Étrangers,* **16** (1858), 347.

|| *Sitzungsberichte der mathematisch-naturwissenschaftlichen Classe der K. Akademie der Wissenschaften zu Wien,* **40** (1860), 607–658; *Wissenschaftliche Abhandlungen,* **1,** 196–222.

†† *Trans. Camb. Phil. Soc.* **8** (1845), 304, 305; *Papers,* **1,** 104, 105.

‡‡ Wiedemann, *Poggendorff's Annalen der Physik u. Chemie,* **99** (1856), 217–221; Neumann (see Jacobson, *Archiv für Anatomie, Physiologie u. wissenschaftliche Medicin* (1860), pp. 88–91); Hagenbach, *Poggendorff's Annalen,* **109** (1860), 385–426; Mathieu, *Comptes Rendus des Séances de l'Académie des Sciences,* **57** (1863), 320–324; Boussinesq, *Comptes Rendus,* **65** (1867), 46–48.

|||| See p. 1186 of the report on Poiseuille's memoir by Messrs. Arago, Babinet, Piobert, and Regnault, *Comptes Rendus,* **15** (1842), 1167–1186.

††† Warburg, *Poggendorff's Annalen,* **140** (1870), 367–379; Villari, *Memorie della Accademia delle Scienze dell' Istituto di Bologna* (3), **6** (1875), 487–520; Koch, *Wiedemann's Annalen der Physik u. Chemie,* **14** (1881), 1–12.

‡‡‡ *Phil. Trans.* A, **181** (1890), 559–582. Warburg (*op. cit.* p. 370) had found no change in the discharge of water from a glass tube when the latter was silvered.

|||||| *Annales de Chimie et de Physique* (6), **21** (1890), 491, 508, 509.

†††† *Phil. Trans.* A, **156** (1866), 255, 256; *Scientific Papers,* **2,** 9, 10.

few words may be added on the results so far obtained for gases on the kinetic theory. Maxwell, though chary of attempting the investigation (because, as he said, 'it is almost certain that the stratum of gas nearest to a solid body is in a very different state from the rest of the gas') carried out some calculations at the request of a Royal Society referee, and arrived at the conclusion that if there are no inequalities of temperature, then slipping takes place according to Navier's equation, and the length $\mu/\beta$ is a moderate multiple of the mean free path, $L$, of a gas molecule,—probably about $2L$. Thus at atmospheric pressure the slip would be negligible: for rarefied gases, however, it would be considerable. This latter deduction is in agreement with experiment.† An investigation of the question has lately been undertaken by Rocard,‡ whose conclusion is that if the aggregate of molecules near a solid wall continues to have the properties of a gas, then not only the relative velocity, but also its normal derivative, must vanish at the wall, so that the velocity of the fluid is practically the same as that of the wall at some short distance away; he decided that, in fact, the stratum next to the wall does not have the properties of a gas.

After these brief historical and theoretical summaries, it remains to set out the reasons for the assumption—now generally accepted, and adopted throughout this book—that slip, if it takes place, is too small, or a quasi-solid layer of fluid, if there is one, is too thin, to be observed or to make any observable difference in the results of the theoretical deductions. In the first place we have the valuable evidence of direct observation, even though this can hardly be regarded as conclusive by itself. Thus in their examination with an ultramicroscope of the turbulent flow of water through a square pipe, the slowest particle observed by Fage and Townend had a mean velocity of 0·006 feet per second when the average velocity across a section of the pipe was 0·83 feet per second: on the assumption of no slip it was calculated that this particle would be at a distance of about $2·5 \times 10^{-5}$ inch from the surface,— much too small to measure.‖ Then we have the agreement with observation of calculations relating to the determination of viscosity, and to Stokes's and Oseen's theories of motion at small Reynolds numbers.†† More important are the agreement between Taylor's calculations and observations on the stability of flow between rotating cylinders‡‡ and the agreement between experiment and the results of several calculations mentioned in this book. These calculations relate, however, for the most part only to non-turbulent flow; and probably the most satisfactory evidence, covering both turbulent and non-turbulent flow, is provided by the numerous experiments which have verified the dependence, for geometrically similar systems, of non-dimensional quantities such as force coefficients on Reynolds number only. For if there is a quasi-solid layer at a wall, or if slip takes place according to Navier's

† *Phil. Trans.* **170** (1879), 249–256; *Scientific Papers,* **2,** 703–709.

‡ *L'Hydrodynamique et la Théorie Cinétique des Gaz* (Paris, 1932), chaps. ix and x.

‖ *Proc. Roy. Soc.* A, **135** (1932), 668, 669.

†† Lamb, *Hydrodynamics* (Cambridge, 1932), pp. 594–616.'

‡‡ *Phil. Trans.* A, **223** (1922), 289–343. See also Lewis, *Proc. Roy. Soc.* A, **117** (1928), 388–407. Taylor's experiments were with water and glass; Lewis used xylene, nitrobenzene, and a mixture of the two, with a range of $\nu$ from 0·008 to 0·018 c.g.s. units, and one silver and one lead surface.

equation, then another length, $l$, enters in addition to the length $d$ specifying the dimensions of the system,—namely, the thickness of the layer, or $\mu/\beta$; and when the dimensions of the system are changed, force-coefficients and other non-dimensional quantities would depend on $l/d$ as well as on the Reynolds number. Hence, unless in some curious way $l$ varies in proportion to $d$, the experiments mentioned above indicate that $l$ is zero (or so small that its effects are negligible).

# INDEX OF AUTHORS

# SUBJECT INDEX

Accelerated motion, in ideal fluid, 26.
  in viscous fluid, 39, 59–63, 186, 187.

Acceleration of fluid element, 96.

Acoustical phenomena, 189.

Aerofoil:
  boundary layer of: separation of, 445, 464, 465, 469–71; thickness of, 410, 411, 467; transition to turbulence in, 406, 410, 411, 461, 464–9, 483; turbulent, 490; velocity distribution in, 411, 412, 466–9.
  circulation around: 34, 36, 455–8; growth of, 41, 76, 458–61.
  drag of: 401; at negative incidences, 452; coefficient of, 35; effect of roughness on, 416, 417, 473–8; effect of shape on, 412–14; effect of thickness on, 401–3; effect of turbulence on, 440, 490; scale effect on, 412–16, 447–50; see also Induced drag.
  flaps, effect of, on drag and lift, 452–4.
  flow past: at stall, 69, 77, 469–71; of inviscid fluid with and without circulation, 34; of real fluid, 74, 76–8; stream-lines of, 454–8.
  form drag of, 403, 406–8, 416, 440, 448–50.
  'hysteresis' loop in lift-incidence curve for stalled flow of, 471, 472.
  Joukowski, generalized type, 401–12, 489, 490.
  lift of: 442–7; at negative incidences, 451, 452; calculations for, with boundary layer, 489; coefficient of, comparison of calculated with measured, 34, 35, 442; effect of roughness on, 473–8; effect of shape on, 464–6; fluctuations at stall of, 472; maximum, 443, (effect of roughness on) 474–8, (effect of turbulence on) 238, 240, 445–7, 465, 490.
  lift/drag ratio, 450, 451.
  pressure distribution on: 36, 75, 76, 403–5, 454, 455, 461; comparison with theory of inviscid flow, 36, 405; effect of shape on, 463.
  roughness, effect of: on drag, 416, 417, 473–8; on lift, 474–8.
  scale effect: on drag, 406, 412–16, 447–50, 452; on lift, 78, 442–5, 451, (effect of shape on) 464–6.
  separation of boundary layer of, 445, 464, 465, 469–71.

skin-friction on: 409, 410, 461; measurement of, 279, 280, 409.
skin-friction drag of, 403, 406–8, 440, 448–50, 463.
slotted: 79, 81, 530, 531; comparison with boundary layer control by pressure, 531, 543, 544; lift and drag of, 542, 543; quantity of air flowing through a slot, 543, 544; shadowgraphs of flow past, 289.
stagnation point on, flow near, 458.
stalling of, see Stalling.
stream-lines of flow past, 454–8.
tests on: 243, (references to) 441; full-scale, 466–9, 473, 474.
theory: potential, 34, 442, 458, 489, 490; three-dimensional, 42–6, 441, 444, 445, 450.
thickness, effect of, on drag, 401–3.
transition to turbulence in boundary layer of, 406, 410, 411, 461, 464–9, 483.
turbulence 'bubble' on, 469.
wake behind, 77, 580, 581.
with boundary layer control: 535–42; air quantities, 538, 539, 542, 544; effect of Reynolds number, 535, 539; lift and drag of, 538, 539, 541, 542; slots, optimum size and position of, 538, 540; theoretical lift exceeded, 542.

Aeroplane model, measurement of forces and moments on, 242–5, (maximum lift compared with full-scale) 444.

Air flow, visualization of, see Visualization of flow (air).

Airscrew, shadowgraphs of flow past, 289, 290.

Airship shapes, 21, 505 et seq.
  critical Reynolds numbers for boundary layer of, 511, 514, 515, 518, 519.
  drag of, 75, 507, 509–19, (effect of fineness ratio on) 507–9, 517, 518, (effect of roughness on) 524–6, (measurement of) 241, 243, 245, 246, 511, 514, 517, 518, 519, 522, 523.
  form drag of, 507.
  non-axial flow past, 527, 528.
  pressure distribution over, 24, 25, 523, 524.
  roughness of, effect of, 524–6.

Boundary layer control, 78–81, 529–49.
by pressure, 530, 531, (aerofoils) 536–42.
by suction: 80, 81, 84, 529, 530; aerofoils, 535–42; air quantities, 533, 534, 538, 539, 542, 544; diffusers, 534, 535; pressure gradients, 533, 534; type of flow, 531–3; velocity distribution with, final, 534.
deflexion of air-stream round cylinder, 530, 535.
effective drag coefficient with, 536.
moving surfaces, 531, 548, 549; see also Rotating cylinder; Rotating sphere.
pressure and suction systems compared, 541, 542.
'sink' effect, 532.
slotted wing, 79, 81, 530, 531, 542–4.
suction apertures: position of, 529, 532, 533, 538, 540; size of, 540.
Townend ring, 544, 545.
velocity distribution, final, with suction, 534.
Bound vorticity, bound vortices, 42, 43.
Buoyancy, force of, 12.

Calibration: of anemometers, 247; of pitot-static tube, 250, 251, 253; of hot wire anemometers, 266.
Capillarity, 19.
Cascades, see Guide vanes.
Cast-off vortices, 40–2, 76, 460.
Cavitation, 37, 62.
Channels, diverging and converging: solution for steady two-dimensional viscous flow in, 105–10, 143–5; turbulent flow in, 371–6; see also Diffuser.
open, flow in, 84, 87, 330, 400.
straight:
dissipation of energy in, distribution of, 395, 396.
heat transfer in, with constant temperature gradient along walls, 647–9.
inlet length of, (laminar) 309, 310, (turbulent) 325.
laminar pressure flow through, 308, 309, (effect of roughness on) 311.
resistance coefficient of, (laminar) 309, (turbulent, smooth surface) 336–9, (turbulent, rough surface) 376–80, 400.
stability or instability of steady flow through, 195, 198.

turbulent pressure flow through, 192, 193, 194, 340, 341, 344, 345, 352, 353.
transition to turbulence in, 319, 325.
velocity distribution in, (laminar) 308, 309, (effect of roughness on) 311, (turbulent, mean) 340 341, 344, 345, 350, 352, 353.
velocity fluctuations in, 400.
water (experimental), see Water channels.
see also Curved channels; Constriction in channel.
Chattock manometer, 274–6.
Chord, mean, of an aerofoil, 43.
Circular cylinder, see Cylinder.
Circular pipe, see Pipe.
Circulation, 26–31.
calculation of, in inviscid fluid for cylinder with section possessing salient point, 34.
growth of, round aerofoil, 458–61.
in moving circuit: constancy of, in inviscid fluid, 30; rate of change of, in viscous fluid, 30, 97.
motion of inviscid fluids with, 31–6.
production of, 41, 46, 66–9, 76, (round rotating cylinder) 81, (by suction) 84.
see also Aerofoil; Boundary layer control; Cylinder, elliptic, lift calculations; Flat plate; Rotating cylinder; Rotating sphere.
Closed circuit wind tunnel, see Wind tunnel.
Closed jet wind tunnel, see Wind tunnel.
Coating, chemical, of surfaces for examining flow, 281.
Coefficients, non-dimensional, 13, (for heat transfer) 19, 607–9.
Compensation of hot wire anemometers, see Hot wire anemometers.
Compressed air wind tunnel, see Wind tunnel.
tests in (N.P.L.), (aerofoils) 414, 442–5, 447, 448, 450–4, 475, 477, (airship models) 513, 514, (elliptic cylinder) 490, (Townend ring) 545.
Compressibility, 11, 17, 19, 602, 603.
Conduction of heat, 10, 19, 55, 604–6; see also Diffusion; Heat transfer; Thermal conductivity.
Constriction in channel, 88.
Continuity, equation of, 90, 102, 103, 104, 114, 192, 602, 611, (in Lagrangian system) 212.

Ellipsoids, 527.

Elliptic cylinder, see Cylinder, elliptic.

Energy criterion for turbulence, 194–6, (with density gradient) 229–32.

Energy, dissipation of, see Dissipation.

Energy equation 603–7.

for boundary la_er: 613–15; at a flat plate along the stream, 626, 627.

Energy spectrum, see Spectrum.

Entry length, 320, 321.

Equation of continuity, see Continuity.

Equation of state, 601.

Equations of motion, of viscous fluid, 95–7, 100, 104, 105, (with compressibility) 602; see also Boundary layer equations; Exact solutions.

mean, for turbulent flow, 192, 193, 210, 211.

Equiangular spiral flow of viscous fluid, 110.

Error law distribution of velocities in turbulent motion, 218, 219, 220.

Errors involved in boundary layer approximations, 121, 123.

Exact solutions of equations of motion of viscous fluid, 105–13, 117, 140, 143.

Expansion in series, solution of boundary layer equations by, 148–54, 178.

Filament lines: 280, 281; revealed by hot wire shadows, 289, 290, 292.

Filter circuits, see Hot wire anemometers.

Fineness ratio, of stream-line bodies, effect on drag of, 507–9, 517, 518.

of symmetrical cylinder, effect on drag of, 414–16.

Flaps, aerofoil, effect on drag and lift of, 452.

Flat plate at zero incidence, drag coefficient of: 136, 318; for turbulent flow, (smooth) 361–7, (rough) 380–2.

heat transfer from, (laminar flow) 623–7, 634–6, 675, (turbulent flow) 371, 660, 661, 674, 675.

laminar flow along: (descriptive) 50–3, (experimental) 316–18, (effect of roughness on) 318, 319; on Oseen theory (small Reynolds numbers), 550; solution of boundary layer equations for, 135–9, 157, 158, 167.

stability or instability of steady boundary layer flow along, 199, 200.

temperature distribution in turbulent boundary layer at, 371, 660, 661.

transition to turbulence in boundary layer at, 71, 137, 199, 200, 325–9, 330, 483.

turbulent flow along, (smooth) 361–71, (rough) 380–2.

velocity distribution in turbulent boundary layer at, 367–71.

wake behind, 571–4.

see also Convection of heat, free; Lamina; Plate thermometer; Rotating disk.

Flat plate normal to or inclined to stream: drag coefficient of, 37, 401, 403, 423, 568, 569; flow behind, 553 et seq., 564, 568–71; lift on and circulation round, 490; suction at rear of, 37; see also Disks.

Force coefficients, non-dimensional, 13, 14, 101.

Force measurements: in water tank, 247; in wind tunnels, 242–6; on whirling arm, 247.

Forced convection, see Convection, forced.

Form drag, 48, 241, 256, 257.

and position of separation, 65.

of aerofoil: 406–8, 440; dependence on thickness, 408, and on bluffness, 416; fraction of total drag, 408.

of circular cylinder, 424–7.

of stream-line bodies of revolution, 507.

Free convection, see Convection, free.

Free stream-lines, 36, 37, 49.

Frequency distribution, of particles in theory of discontinuous diffusion, 216.

of turbulent velocity, (in wind tunnel) 218, 219, (in atmosphere) 220.

Frequency of shedding of vortices, 421, 568–71.

Frictional resistance, see Skin-friction drag.

Friction velocity defined, 220.

Froude number, 18.

Full-scale, experiments, 443, 444, 466–9, 473, 474.

prediction, (from wind tunnel tests) 15, 17, 74, 78, 239, 443, 444, 466, 467, 506, (from water tank tests) 19, 367.

Gas constant, 601, 602.

Gases, kinetic theory of, 8–11, 17, 203, 679.

Gas law, 601.

General coordinates (orthogonal), 101–3, 114, 115.

Grashoff number, 607.

Gravity, effect of, on fluid of variable density, 229–32, 602, 603, 611.

Grids, turbulence behind, *see* Turbulence behind grids.

Guide vanes, 237, 238.

Heat, conduction of, *see* Conduction; Heat transfer.

convection of, *see* Convection; Heat transfer.

diffusion of, *see* Diffusion.

mechanical equivalent of, 605.

Heat transfer, 601–75.

boundary conditions for, 615, 616.

boundary layer equations with, 610–15.

coefficients for, 19, 608, 650.

diffusion, analogy with, 674.

eddy, 646, 647.

from circular cylinder: 636, 637, 675; free convection, 643, 644, 675.

from cylinder near forward stagnation point, 631, 632.

from flat plate, (laminar flow) 623–7, 634–6, 675, (turbulent flow) 660, 661, 674, 675.

in channels, 647–9.

in circular pipes: experimental results, 658–60; laminar flow, 616–23, 674; turbulent flow, 648, 649, 652–4, 674.

inlet length for, 654.

mixture length theory of, 648, 649.

Reynolds's analogy between, and skin-friction: 649–54; extension of, 654–7, 674.

*see also* Conduction of heat; Convection of heat; Diffusion in turbulent motion; Temperature distribution.

Honeycomb, 235, 237.

Honeycomb wall, 235.

Hot wire anemometers: 265–74; accuracy of, 265, 266; ageing of wires of, 265; compensation of, for thermal lag, 269; constant current type, 265; constant resistance type, 266; correlation measurements with, in turbulent flow, 269–73; direction measurements with, 267, 286; extension of range of, for high speeds, 266, 267; filter circuits for, 273, 274; measurement with, of energy spectrum in turbulent flow, 273, 274; need for frequent recalibration of, 266; response of, to high frequency

speed variations, 269; velocity measurements with, 265–7, (in turbulent flow) 239, 268, 269.

Hot wire shadows, *see* Visualization of flow (air).

House, flow past model of, 88, 89.

Hydraulic mean depth for pipes and channels, defined, 297.

Hydrographic measurements, 231, 232.

Hydrostatic pressure, 12.

Ideal fluid, 2.

theory, results of, 21–6, (with circulation) 31–6, (with concentrated vorticity) 36–46.

Impact pressure, 253.

Impulsive motion, in viscous fluid, 39, 59–63, 138, 181–6, 419, 420.

visualization of flow associated with, 293, 294, 459–61.

Inclined open channels, *see* Open channels.

Inclined tube manometer, 277.

Induced drag, 43, 46, 83, 441, 450.

Infinitesimal disturbances, 196–200, 230, 231.

Inflexion in graph of velocity in boundary layer, 125, (influence on stability) 200.

Initial motion of solid body relative to fluid, 40, 46, 55, 56, 59, 60, 181.

Inlet length, for heat transfer (turbulent flow), 654.

for laminar flow: in channels, 309, 310; in circular pipes, 21, 54, 299–308; in rectangular pipes, 330.

for turbulent flow: in channels, 325; in circular pipes, 323–5, 360, 361.

Inner and outer approximate solutions for boundary layer equations, 164–73.

Instability of steady flow, *see* Stability or instability.

Interference, effects of attachments, 73, 242, 243, 245, 246, 494; *see also* Wind tunnel interference.

Intersecting walls, flow at surface formed by two, 330.

Intrinsic energy, 604–6.

Inviscid fluid theory, 21–48.

compared with observation in real fluids, 48, 49.

Irrotational motion, defined, 27.

persistence of, in inviscid fluid, 46.

*see also* Ideal fluid theory.

# A CATALOGUE OF SELECTED
# DOVER SCIENCE BOOKS

# A CATALOGUE OF SELECTED
# DOVER SCIENCE BOOKS

Physics: The Pioneer Science, Lloyd W. Taylor. Very thorough non-mathematical survey of physics in a historical framework which shows development of ideas. Easily followed by laymen; used in dozens of schools and colleges for survey courses. Richly illustrated. Volume 1: Heat, sound, mechanics. Volume 2: Light, electricity. Total of 763 illustrations. Total of cvi + 847pp.

60565-5, 60566-3 Two volumes, Paperbound 5.50

THE RISE OF THE NEW PHYSICS, A. d'Abro. Most thorough explanation in print of central core of mathematical physics, both classical and modern, from Newton to Dirac and Heisenberg. Both history and exposition: philosophy of science, causality, explanations of higher mathematics, analytical mechanics, electromagnetism, thermodynamics, phase rule, special and general relativity, matrices. No higher mathematics needed to follow exposition, though treatment is elementary to intermediate in level. Recommended to serious student who wishes verbal understanding. 97 illustrations. Total of ix + 982pp.

20003-5, 20004-3 Two volumes, Paperbound $6.00

INTRODUCTION TO CHEMICAL PHYSICS, John C. Slater. A work intended to bridge the gap between chemistry and physics. Text divided into three parts: Thermodynamics, Statistical Mechanics, and Kinetic Theory; Gases, Liquids and Solids; and Atoms, Molecules and the Structure of Matter, which form the basis of the approach. Level is advanced undergraduate to graduate, but theoretical physics held to minimum. 40 tables, 118 figures. xiv + 522pp.

62562-1 Paperbound $4.00

BASIC THEORIES OF PHYSICS, Peter C. Bergmann. Critical examination of important topics in classical and modern physics. Exceptionally useful in examining conceptual framework and methodology used in construction of theory. Excellent supplement to any course, textbook. Relatively advanced.
Volume 1. Heat and Quanta. Kinetic hypothesis, physics and statistics, stationary ensembles, thermodynamics, early quantum theories, atomic spectra, probability waves, quantization in wave mechanics, approximation methods, abstract quantum theory. 8 figures. x + 300pp. 60968-5 Paperbound $2.50
Volume 2. Mechanics and Electrodynamics. Classical mechanics, electro- and magnetostatics, electromagnetic induction, field waves, special relativity, waves, etc. 16 figures, viii + 260pp. 60969-3 Paperbound $2.75

FOUNDATIONS OF PHYSICS, Robert Bruce Lindsay and Henry Margenau. Methods and concepts at the heart of physics (space and time, mechanics, probability, statistics, relativity, quantum theory) explained in a text that bridges gap between semi-popular and rigorous introductions. Elementary calculus assumed. "Thorough and yet not over-detailed," Nature. 35 figures. xviii + 537 pp.

60377-6 Paperbound $3.50

EINSTEIN'S THEORY OF RELATIVITY, Max Born. Relativity theory analyzed, explained for intelligent layman or student with some physical, mathematical background. Includes Lorentz, Minkowski, and others. Excellent verbal account for teachers. Generally considered the finest non-technical account. vii + 376pp.
60769-0 Paperbound $2.75

PHYSICAL PRINCIPLES OF THE QUANTUM THEORY, Werner Heisenberg. Nobel Laureate discusses quantum theory, uncertainty principle, wave mechanics, work of Dirac, Schroedinger, Compton, Wilson, Einstein, etc. Middle, non-mathematical level for physicist, chemist not specializing in quantum; mathematical appendix for specialists. Translated by C. Eckart and F. Hoyt. 19 figures. viii + 184pp.
60113-7 Paperbound $2.00

PRINCIPLES OF QUANTUM MECHANICS, William V. Houston. For student with working knowledge of elementary mathematical physics; uses Schroedinger's wave mechanics. Evidence for quantum theory, postulates of quantum mechanics, applications in spectroscopy, collision problems, electrons, similar topics. 21 figures. 288pp.
60524-8 Paperbound $3.00

ATOMIC SPECTRA AND ATOMIC STRUCTURE, Gerhard Herzberg. One of the best introductions to atomic spectra and their relationship to structure; especially suited to specialists in other fields who require a comprehensive basic knowledge. Treatment is physical rather than mathematical. 2nd edition. Translated by J. W. T. Spinks. 80 illustrations. xiv + 257pp.
60115-3 Paperbound $2.00

ATOMIC PHYSICS: AN ATOMIC DESCRIPTION OF PHYSICAL PHENOMENA, Gaylord P. Harnwell and William E. Stephens. One of the best introductions to modern quantum ideas. Emphasis on the extension of classical physics into the realms of atomic phenomena and the evolution of quantum concepts. 156 problems. 173 figures and tables. xi + 401pp.
61584-7 Paperbound $3.00

ATOMS, MOLECULES AND QUANTA, Arthur E. Ruark and Harold C. Urey. 1964 edition of work that has been a favorite of students and teachers for 30 years. Origins and major experimental data of quantum theory, development of concepts of atomic and molecular structure prior to new mechanics, laws and basic ideas of quantum mechanics, wave mechanics, matrix mechanics, general theory of quantum dynamics. Very thorough, lucid presentation for advanced students. 230 figures. Total of xxiii + 810pp.
61106-X, 61107-8 Two volumes, Paperbound $6.00

INVESTIGATIONS ON THE THEORY OF THE BROWNIAN MOVEMENT, Albert Einstein. Five papers (1905-1908) investigating the dynamics of Brownian motion and evolving an elementary theory of interest to mathematicians, chemists and physical scientists. Notes by R. Fürth, the editor, discuss the history of study of Brownian movement, elucidate the text and analyze the significance of the papers. Translated by A. D. Cowper. 3 figures. iv + 122pp.
60304-0 Paperbound $1.50

FUNDAMENTAL FORMULAS OF PHYSICS, edited by Donald H. Menzel. Most useful reference and study work, ranges from simplest to most highly sophisticated operations. Individual chapters, with full texts explaining formulae, prepared by leading authorities cover basic mathematical formulas, statistics, nomograms, physical constants, classical mechanics, special theory of relativity, general theory of relativity, hydrodynamics and aerodynamics, boundary value problems in mathematical physics, heat and thermodynamics, statistical mechanics, kinetic theory of gases, viscosity, thermal conduction, electromagnetism, electronics, acoustics, geometrical optics, physical optics, electron optics, molecular spectra, atomic spectra, quantum mechanics, nuclear theory, cosmic rays and high energy phenomena, particle accelerators, solid state, magnetism, etc. Special chapters also cover physical chemistry, astrophysics, celestian mechanics, meteorology, and biophysics. Indispensable part of library of every scientist. Total of xli + 787pp.

60595-7, 60596-5 Two volumes, Paperbound $6.00

INTRODUCTION TO EXPERIMENTAL PHYSICS, William B. Fretter. Detailed coverage of techniques and equipment: measurements, vacuum tubes, pulse circuits, rectifiers, oscillators, magnet design, particle counters, nuclear emulsions, cloud chambers, accelerators, spectroscopy, magnetic resonance, x-ray diffraction, low temperature, etc. One of few books to cover laboratory hazards, design of exploratory experiments, measurements. 298 figures. xii + 349pp.

(EBE) 61890-0 Paperbound $3.00

CONCEPTS AND METHODS OF THEORETICAL PHYSICS, Robert Bruce Lindsay. Introduction to methods of theoretical physics, emphasizing development of physical concepts and analysis of methods. Part I proceeds from single particle to collections of particles to statistical method. Part II covers application of field concept to material and non-material media. Numerous exercises and examples. 76 illustrations. x + 515pp. 62354-8 Paperbound $4.00

AN ELEMENTARY TREATISE ON THEORETICAL MECHANICS, Sir James Jeans. Great scientific expositor in remarkably clear presentation of basic classical material: rest, motion, forces acting on particle, statics, motion of particle under variable force, motion of rigid bodies, coordinates, etc. Emphasizes explanation of fundamental physical principles rather than mathematics or applications. Hundreds of problems worked in text. 156 figures. x + 364pp. 61839-0 Paperbound $2.75

THEORETICAL MECHANICS: AN INTRODUCTION TO MATHEMATICAL PHYSICS, Joseph S. Ames and Francis D. Murnaghan. Mathematically rigorous introduction to vector and tensor methods, dynamics, harmonic vibrations, gyroscopic theory, principle of least constraint, Lorentz-Einstein transformation. 159 problems; many fully-worked examples. 39 figures. ix + 462pp. 60461-6 Paperbound $3.50

THE PRINCIPLE OF RELATIVITY, Albert Einstein, Hendrick A. Lorentz, Hermann Minkowski and Hermann Weyl. Eleven original papers on the special and general theory of relativity, all unabridged. Seven papers by Einstein, two by Lorentz, one each by Minkowski and Weyl. "A thrill to read again the original papers by these giants," *School Science and Mathematics*. Translated by W. Perret and G. B. Jeffery. Notes by A. Sommerfeld. 7 diagrams. viii + 216pp.

60081-5 Paperbound $2.25

MATHEMATICAL FOUNDATIONS OF STATISTICAL MECHANICS, A. I. Khinchin. Introduction to modern statistical mechanics: phase space, ergodic problems, theory of probability, central limit theorem, ideal monatomic gas, foundation of thermodynamics, dispersion and distribution of sum functions. Provides mathematically rigorous treatment and excellent analytical tools. Translated by George Gamow. viii + 179pp. 60147-1 Paperbound $2.50

INTRODUCTION TO PHYSICAL STATISTICS, Robert B. Lindsay. Elementary probability theory, laws of thermodynamics, classical Maxwell-Boltzmann statistics, classical statistical mechanics, quantum mechanics, other areas of physics that can be studied statistically. Full coverage of methods; basic background theory. ix + 306pp. 61882-X Paperbound $2.75

DIALOGUES CONCERNING TWO NEW SCIENCES, Galileo Galilei. Written near the end of Galileo's life and encompassing 30 years of experiment and thought, these dialogues deal with geometric demonstrations of fracture of solid bodies, cohesion, leverage, speed of light and sound, pendulums, falling bodies, accelerated motion, etc. Translated by Henry Crew and Alfonso de Salvio. Introduction by Antonio Favaro. xxiii + 300pp. 60099-8 Paperbound $2.25

FOUNDATIONS OF SCIENCE: THE PHILOSOPHY OF THEORY AND EXPERIMENT, Norman R. Campbell. Fundamental concepts of science examined on middle level: acceptance of propositions and axioms, presuppositions of scientific thought, scientific law, multiplication of probabilities, nature of experiment, application of mathematics, measurement, numerical laws and theories, error, etc. Stress on physics, but holds for other sciences. "Unreservedly recommended," *Nature* (England). Formerly *Physics: The Elements.* ix + 565pp. 60372-5 Paperbound $4.00

THE PHASE RULE AND ITS APPLICATIONS, Alexander Findlay, A. N. Campbell and N. O. Smith. Findlay's well-known classic, updated (1951). Full standard text and thorough reference, particularly useful for graduate students. Covers chemical phenomena of one, two, three, four and multiple component systems. "Should rank as the standard work in English on the subject," *Nature.* 236 figures. xii + 494pp. 60091-2 Paperbound $3.50

THERMODYNAMICS, Enrico Fermi. A classic of modern science. Clear, organized treatment of systems, first and second laws, entropy, thermodynamic potentials, gaseous reactions, dilute solutions, entropy constant. No math beyond calculus is needed, but readers are assumed to be familiar with fundamentals of thermometry, calorimetry. 22 illustrations. 25 problems. x + 160pp.
60361-X Paperbound $2.00

TREATISE ON THERMODYNAMICS, Max Planck. Classic, still recognized as one of the best introductions to thermodynamics. Based on Planck's original papers, it presents a concise and logical view of the entire field, building physical and chemical laws from basic empirical facts. Planck considers fundamental definitions, first and second principles of thermodynamics, and applications to special states of equilibrium. Numerous worked examples. Translated by Alexander Ogg. 5 figures. xiv + 297pp. 60219-2 Paperbound $2.50

MICROSCOPY FOR CHEMISTS, Harold F. Schaeffer. Thorough text; operation of microscope, optics, photomicrographs, hot stage, polarized light, chemical procedures for organic and inorganic reactions. 32 specific experiments cover specific analyses: industrial, metals, other important subjects. 136 figures. 264pp.
61682-7 Paperbound $2.50

OPTICKS, Sir Isaac Newton. A survey of 18th-century knowledge on all aspects of light as well as a description of Newton's experiments with spectroscopy, colors, lenses, reflection, refraction, theory of waves, etc. in language the layman can follow. Foreword by Albert Einstein. Introduction by Sir Edmund Whittaker. Preface by I. Bernard Cohen. cxxvi + 406pp.
60205-2 Paperbound $4.00

LIGHT: PRINCIPLES AND EXPERIMENTS, George S. Monk. Thorough coverage, for student with background in physics and math, of physical and geometric optics. Also includes 23 experiments on optical systems, instruments, etc. "Probably the best intermediate text on optics in the English language," *Physics Forum.* 275 figures. xi + 489pp.
60341-5 Paperbound $3.50

PHYSICAL OPTICS, Robert W. Wood. A classic in the field, this is a valuable source for students of physical optics and excellent background material for a study of electromagnetic theory. Partial contents: nature and rectilinear propagation of light, reflection from plane and curved surfaces, refraction, absorption and dispersion, origin of spectra, interference, diffraction, polarization, Raman effect, optical properties of metals, resonance radiation and fluorescence of atoms, magneto-optics, electro-optics, thermal radiation. 462 diagrams, 17 plates. xvi + 846pp.
61808-0 Paperbound $4.50

MIRRORS, PRISMS AND LENSES: A TEXTBOOK OF GEOMETRICAL OPTICS, James P. C. Southall. Introductory-level account of modern optical instrument theory, covering unusually wide range: lights and shadows, reflection of light and plane mirrors, refraction, astigmatic lenses, compound systems, aperture and field of optical system, the eye, dispersion and achromatism, rays of finite slope, the microscope, much more. Strong emphasis on earlier, elementary portions of field, utilizing simplest mathematics wherever possible. Problems. 329 figures. xxiv + 806pp.
61234-1 Paperbound $5.00

THE PSYCHOLOGY OF INVENTION IN THE MATHEMATICAL FIELD, Jacques Hadamard. Important French mathematician examines psychological origin of ideas, role of the unconscious, importance of visualization, etc. Based on own experiences and reports by Dalton, Pascal, Descartes, Einstein, Poincaré, Helmholtz, etc. xiii + 145pp.
20107-4 Paperbound $1.50

INTRODUCTION TO CHEMICAL PHYSICS, John C. Slater. A work intended to bridge the gap between chemistry and physics. Text divided into three parts: Thermodynamics, Statistical Mechanics, and Kinetic Theory; Gases, Liquids and Solids; and Atoms, Molecules and the Structure of Matter, which form the basis of the approach. Level is advanced undergraduate to graduate, but theoretical physics held to minimum. 40 tables, 118 figures. xiv + 522pp.
62562-1 Paperbound $4.00

CONTRIBUTIONS TO THE FOUNDING OF THE THEORY OF TRANSFINITE NUMBERS, Georg Cantor. The famous articles of 1895-1897 which founded a new branch of mathematics, translated with 82-page introduction by P. Jourdain. Not only a great classic but still one of the best introductions for the student. ix + 211pp.
60045-9 Paperbound $2.50

ESSAYS ON THE THEORY OF NUMBERS, Richard Dedekind. Two classic essays, on the theory of irrationals, giving an arithmetic and rigorous foundation; and on transfinite numbers and properties of natural numbers. Translated by W. W. Beman. iii + 115pp. 21010-3 Paperbound $1.75

GEOMETRY OF FOUR DIMENSIONS, H. P. Manning. Part verbal, part mathematical development of fourth dimensional geometry. Historical introduction. Detailed treatment is by synthetic method, approaching subject through Euclidean geometry. No knowledge of higher mathematics necessary. 76 figures. ix + 348pp.
60182-X Paperbound $3.00

AN INTRODUCTION TO THE GEOMETRY OF N DIMENSIONS, Duncan M. Y. Sommerville. The only work in English devoted to higher-dimensional geometry. Both metric and projectiv. Oproperties of n-dimensional geometry are covered. Covers fundamental ideas of incidence, parallelism, perpendicularity, angles between linear space, enumerative geometry, analytical geometry, polytopes, analysis situs, hyperspacial figures. 60 diagrams. xvii + 196pp. 60494-2 Paperbound $2.00

THE THEORY OF SOUND, J. W. S. Rayleigh. Still valuable classic by the great Nobel Laureate. Standard compendium summing up previous research and Rayleigh's original contributions. Covers harmonic vibrations, vibrating systems, vibrations of strings, membranes, plates, curved shells, tubes, solid bodies, refraction of plane waves, general equations. New historical introduction and bibliography by R. B. Lindsay, Brown University. 97 figures. lviii + 984pp.
60292-3, 60293-1 Two volumes, Paperbound $6.00

ELECTROMAGNETIC THEORY: A CRITICAL EXAMINATION OF FUNDAMENTALS, Alfred O'Rahilly. Critical analysis and restructuring of the basic theories and ideas of classical electromagnetics. Analysis is carried out through study of the primary treatises of Maxwell, Lorentz, Einstein, Weyl, etc., which established the theory. Expansive reference to and direct quotation from these treatises. Formerly *Electromagnetics*. Total of xvii + 884pp.
60126-9, 60127-7 Two volumes, Paperbound $6.00

ELEMENTARY CONCEPTS OF TOPOLOGY, Paul Alexandroff. Elegent, intuitive approach to topology, from the basic concepts of set-theoretic topology to the concept of Betti groups. Stresses concepts of complex, cycle and homology. Shows how concepts of topology are useful in math and physics. Introduction by David Hilbert. Translated by Alan E. Farley. 25 figures. iv + 57pp.
60747-X Paperbound $1.25

THE ELEMENTS OF NON-EUCLIDEAN GEOMETRY, Duncan M. Y. Sommerville. Presentation of the development of non-Euclidean geometry in logical order, from a fundamental analysis of the concept of parallelism to such advanced topics as inversion, transformations, pseudosphere, geodesic representation, relation between parataxy and parallelism, etc. Knowledge of only high-school algebra and geometry is presupposed. 126 problems, 129 figures. xvi + 274pp.
60460-8 Paperbound $2.50

NON-EUCLIDEAN GEOMETRY: A CRITICAL AND HISTORICAL STUDY OF ITS DEVELOPMENT, Roberto Bonola. Standard survey, clear, penetrating, discussing many systems not usually represented in general studies. Easily followed by non-specialist. Translated by H. Carslaw. Bound in are two most important texts: Bolyai's "The Science of Absolute Space" and Lobachevski's "The Theory of Parallels," translated by G. B. Halsted. Introduction by F. Enriques. 181 diagrams. Total of 431pp.
60027-0 Paperbound $3.00

ELEMENTS OF NUMBER THEORY, Ivan M. Vinogradov. By stressing demonstrations and problems, this modern text can be understood by students without advanced math backgrounds. "A very welcome addition," *Bulletin, American Mathematical Society.* Translated by Saul Kravetz. Over 200 fully-worked problems. 100 numerical exercises. viii + 227pp.
60259-1 Paperbound $2.50

THEORY OF SETS, E. Kamke. Lucid introduction to theory of sets, surveying discoveries of Cantor, Russell, Weierstrass, Zermelo, Bernstein, Dedekind, etc. Knowledge of college algebra is sufficient background. "Exceptionally well written," *School Science and Mathematics.* Translated by Frederick Bagemihl. vii + 144pp.
60141-2 Paperbound $1.75

A TREATISE ON THE DIFFERENTIAL GEOMETRY OF CURVES AND SURFACES, Luther P. Eisenhart. Detailed, concrete introductory treatise on differential geometry, developed from author's graduate courses at Princeton University. Thorough explanation of the geometry of curves and surfaces, concentrating on problems most helpful to students. 683 problems, 30 diagrams. xiv + 474pp.
60667-8 Paperbound $3.50

AN ESSAY ON THE FOUNDATIONS OF GEOMETRY, Bertrand Russell. A mathematical and physical analysis of the place of the a priori in geometric knowledge. Includes critical review of 19th-century work in non-Euclidean geometry as well as illuminating insights of one of the great minds of our time. New foreword by Morris Kline. xx + 201pp.
60233-8 Paperbound $2.50

INTRODUCTION TO THE THEORY OF NUMBERS, Leonard E. Dickson. Thorough, comprehensive approach with adequate coverage of classical literature, yet simple enough for beginners. Divisibility, congruences, quadratic residues, binary quadratic forms, primes, least residues, Fermat's theorem, Gauss's lemma, and other important topics. 249 problems, 1 figure. viii + 183pp.
60342-3 Paperbound $2.00

AN ELEMENTARY INTRODUCTION TO THE THEORY OF PROBABILITY, B. V. Gnedenko and A. Ya. Khinchin. Introduction to facts and principles of probability theory. Extremely thorough within its range. Mathematics employed held to elementary level. Excellent, highly accurate layman's introduction. Translated from the fifth Russian edition by Leo Y. Boron. xii + 130pp.

60155-2 Paperbound $2.00

SELECTED PAPERS ON NOISE AND STOCHASTIC PROCESSES, edited by Nelson Wax. Six papers which serve as an introduction to advanced noise theory and fluctuation phenomena, or as a reference tool for electrical engineers whose work involves noise characteristics, Brownian motion, statistical mechanics. Papers are by Chandrasekhar, Doob, Kac, Ming, Ornstein, Rice, and Uhlenbeck. Exact facsimile of the papers as they appeared in scientific journals. 19 figures. v + 337pp. $6\frac{1}{8}$ x $9\frac{1}{4}$.

60262-1 Paperbound $3.50

STATISTICS MANUAL, Edwin L. Crow, Frances A. Davis and Margaret W. Maxfield. Comprehensive, practical collection of classical and modern methods of making statistical inferences, prepared by U. S. Naval Ordnance Test Station. Formulae, explanations, methods of application are given, with stress on use. Basic knowledge of statistics is assumed. 21 tables, 11 charts, 95 illustrations. xvii + 288pp.

60599-X Paperbound $2.50

MATHEMATICAL FOUNDATIONS OF INFORMATION THEORY, A. I. Khinchin. Comprehensive introduction to work of Shannon, McMillan, Feinstein and Khinchin, placing these investigations on a rigorous mathematical basis. Covers entropy concept in probability theory, uniqueness theorem, Shannon's inequality, ergodic sources, the E property, martingale concept, noise, Feinstein's fundamental lemma, Shanon's first and second theorems. Translated by R. A. Silverman and M. D. Friedman. iii + 120pp.

60434-9 Paperbound $1.75

INTRODUCTION TO SYMBOLIC LOGIC AND ITS APPLICATION, Rudolf Carnap. Clear, comprehensive, rigorous introduction. Analysis of several logical languages. Investigation of applications to physics, mathematics, similar areas. Translated by Wiliam H. Meyer and John Wilkinson. xiv + 214pp.

60453-5 Paperbound $2.50

SYMBOLIC LOGIC, Clarence I. Lewis and Cooper H. Langford. Probably the most cited book in the literature, with much material not otherwise obtainable. Paradoxes, logic of extensions and intensions, converse substitution, matrix system, strict limitations, existence of terms, truth value systems, similar material. vii + 518pp.

60170-6 Paperbound $4.50

VECTOR AND TENSOR ANALYSIS, George E. Hay. Clear introduction; starts with simple definitions, finishes with mastery of oriented Cartesian vectors, Christoffel symbols, solenoidal tensors, and applications. Many worked problems show applications. 66 figures. viii + 193pp.

60109-9 Paperbound $2.50

GUIDE TO THE LITERATURE OF MATHEMATICS AND PHYSICS, INCLUDING RELATED WORKS ON ENGINEERING SCIENCE, Nathan Grier Parke III. This up-to-date guide puts a library catalog at your fingertips. Over 5000 entries in many languages under 120 subject headings, including many recently available Russian works. Citations are as full as possible, and cross-references and suggestions for further investigation are provided. Extensive listing of bibliographical aids. 2nd revised edition. Complete indices. xviii + 436pp.
60447-0 Paperbound $3.00

INTRODUCTION TO ELLIPTIC FUNCTIONS WITH APPLICATIONS, Frank Bowman. Concise, practical introduction, from familiar trigonometric function to Jacobian elliptic functions to applications in electricity and hydrodynamics. Legendre's standard forms for elliptic integrals, conformal representation, etc., fully covered. Requires knowledge of basic principles of differentiation and integration only. 157 problems and examples, 56 figures. 115pp. 60922-7 Paperbound $1.50

THEORY OF FUNCTIONS OF A COMPLEX VARIABLE, A. R. Forsyth. Standard, classic presentation of theory of functions, stressing multiple-valued functions and related topics: theory of multiform and uniform periodic functions, Weierstrass's results with additiontheorem functions. Riemann functions and surfaces, algebraic functions, Schwarz's proof of the existence-theorem, theory of conformal mapping, etc. 125 figures, 1 plate. Total of xxviii + 855pp. 6⅛ x 9¼
61378-X, 61379-8 Two volumes, Paperbound $6.00

THEORY OF THE INTEGRAL, Stanislaw Saks. Excellent introduction, covering all standard topics: set theory, theory of measure, functions with general properties, and theory of integration emphasizing the Lebesgue integral. Only a minimal background in elementary analysis needed. Translated by L. C. Young. 2nd revised edition. xv + 343pp. 61151-5 Paperbound $3.00

THE THEORY OF FUNCTIONS, *Konrad Knopp. Characterized as "an excellent introduction . . . remarkably readable, concise, clear, rigorous" by the* Journal of the American Statistical Association *college text.*

A COURSE IN MATHEMATICAL ANALYSIS, Edouard Goursat. *The entire "Cours d'analyse" for students with one year of calculus, offering an exceptionally wide range of subject matter on analysis and applied mathematics. Available for the first time in English. Definitive treatment.*

VOLUME I: Applications to geometry, expansion in series, definite integrals, derivatives and differentials. Translated by Earle R. Hedrick. 52 figures. viii + 548pp. 60554-X Paperbound $5.00

VOLUME II, PART I: Functions of a complex variable, conformal representations, doubly periodic functions, natural boundaries, etc. Translated by Earle R. Hedrick and Otto Dunkel. 38 figures. x + 259pp. 60555-8 Paperbound $3.00

VOLUME II, PART II: Differential equations, Cauchy-Lipschitz method, non-linear differential equations, simultaneous equations, etc. Translated by Earle R. Hedrick and Otto Dunkel. 1 figure. viii + 300pp. 60556-6 Paperbound $3.00

VOLUME III, PART I: Variation of solutions, partial differential equations of the second order. Poincaré's theorem, periodic solutions, asymptotic series, wave propagation, Dirichlet's problem in space, Newtonian potential, etc. Translated by Howard G. Bergmann. 15 figures. x + 329pp. 61176-0 Paperbound $3.50

VOLUME III, PART II: Integral equations and calculus of variations: Fredholm's equation, Hilbert-Schmidt theorem, symmetric kernels, Euler's equation, transversals, extreme fields, Weierstrass's theory, etc. Translated by Howard G. Bergmann. Note on Conformal Representation by Paul Montel. 13 figures. xi + 389pp.
61177-9 Paperbound $3.00

ELEMENTARY STATISTICS: WITH APPLICATIONS IN MEDICINE AND THE BIOLOGICAL SCIENCES, Frederick E. Croxton. Presentation of all fundamental techniques and methods of elementary statistics assuming average knowledge of mathematics only. Useful to readers in all fields, but many examples drawn from characteristic data in medicine and biological sciences. vii + 376pp.
60506-X Paperbound $2.50

ELEMENTS OF THE THEORY OF FUNCTIONS. A general background text that explores complex numbers, linear functions, sets and sequences, conformal mapping. Detailed proofs. Translated by Frederick Bagemihl. 140pp.
60154-4 Paperbound $1.50

THEORY OF FUNCTIONS, PART I. Provides full demonstrations, rigorously set forth, of the general foundations of the theory: integral theorems, series, the expansion of analytic functions. Translated by Federick Bagemihl. vii + 146pp.
60156-0 Paperbound $1.50

INTRODUCTION TO THE THEORY OF FOURIER'S SERIES AND INTEGRALS, Horatio S. Carslaw. A basic introduction to the theory of infinite series and integrals, with special reference to Fourier's series and integrals. Based on the classic Riemann integral and dealing with only ordinary functions, this is an important class text. 84 examples. xiii + 368pp. 60048-3 Paperbound $3.00

AN INTRODUCTION TO FOURIER METHODS AND THE LAPLACE TRANSFORMATION, Philip Franklin. Introductory study of theory and applications of Fourier series and Laplace transforms, for engineers, physicists, applied mathematicians, physical science teachers and students. Only a previous knowledge of elementary calculus is assumed. Methods are related to physical problems in heat flow, vibrations, eletcrical transmission, electromagnetic radiation, etc. 828 problems with answers. Formerly *Fourier Methods*. x + 289pp. 60452-7 Paperbound $2.75

INFINITE SEQUENCES AND SERIES, Konrad Knopp. Careful presentation of fundamentals of the theory by one of the finest modern expositors of higher mathematics. Covers functions of real and complex variables, arbitrary and null sequences, convergence and divergence. Cauchy's limit theorem, tests for infinite series, power series, numerical and closed evaluation of series. Translated by Frederick Bagemihl. v + 186pp. 60153-6 Paperbound $2.00

INTRODUCTION TO THE DIFFERENTIAL EQUATIONS OF PHYSICS, Ludwig Hopf. No math background beyond elementary calculus is needed to follow this classroom or self-study introduction to ordinary and partial differential equations. Approach is through classical physics. Translated by Walter Nef. 48 figures. v + 154pp.
60120-X Paperbound $1.75

DIFFERENTIAL EQUATIONS FOR ENGINEERS, Philip Franklin. For engineers, physicists, applied mathematicians. Theory and application: solution of ordinary differential equations and partial derivatives, analytic functions. Fourier series, Abel's theorem, Cauchy Riemann differential equations, etc. Over 400 problems deal with electricity, vibratory systems, heat, radio; solutions. Formerly *Differential Equations for Electrical Engineers*. 41 illustrations. vii + 299pp.
60601-5 Paperbound $2.50

THEORY OF FUNCTIONS, PART II. Single- and multiple-valued functions; full presentation of the most characteristic and important types. Proofs fully worked out. Translated by Frederick Bagemihl. x + 150pp.
60157-9 Paperbound $1.50

PROBLEM BOOK IN THE THEORY OF FUNCTIONS, I. More than 300 elementary problems for independent use or for use with "Theory of Functions, I." 85pp. of detailed solutions. Translated by Lipman Bers. viii + 126pp.
60158-7 Paperbound $1.50

PROBLEM BOOK IN THE THEORY OF FUNCTIONS, II. More than 230 problems in the advanced theory. Designed to be used with "Theory of Functions, II" or with any comparable text. Full solutions. Translated by Frederick Bagemihl. 138pp.
60159-5 Paperbound $1.75

INTRODUCTION TO THE THEORY OF EQUATIONS, Florian Cajori. Classic introduction by leading historian of science covers the fundamental theories as reached by Gauss, Abel, Galois and Kronecker. Basics of equation study are followed by symmetric functions of roots, elimination, homographic and Tschirnhausen transformations, resolvents of Lagrange, cyclic equations, Abelian equations, the work of Galois, the algebraic solution of general equations, and much more. Numerous exercises include answers. ix + 239pp. 62184-7 Paperbound $2.75

LAPLACE TRANSFORMS AND THEIR APPLICATIONS TO DIFFERENTIAL EQUATIONS, N. W. McLachlan. Introduction to modern operational calculus, applying it to ordinary and partial differential equations. Laplace transform, theorems of operational calculus, solution of equations with constant coefficients, evaluation of integrals, derivation of transforms, of various functions, etc. For physics, engineering students. Formerly *Modern Operational Calculus*. xiv + 218pp.
60192-7 Paperbound $2.50

PARTIAL DIFFERENTIAL EQUATIONS OF MATHEMATICAL PHYSICS, Arthur G. Webster. Introduction to basic method and theory of partial differential equations, with full treatment of their applications to virtually every field. Full, clear chapters on Fourier series, integral and elliptic equations, spherical, cylindrical and ellipsoidal harmonics, Cauchy's method, boundary problems, method of Riemann-Volterra, many other basic topics. Edited by Samuel J. Plimpton. 97 figures. vii + 446pp.
60263-X Paperbound $3.00

ASTRONOMY AND COSMOGONY, Sir James Jeans. Modern classic of exposition, Jean's latest work. Descriptive astronomy, atrophysics, stellar dynamics, cosmology, presented on intermediate level. 16 illustrations. Preface by Lloyd Motz. xv + 428pp. 60923-5 Paperbound $3.50

EXPERIMENTAL SPECTROSCOPY, Ralph A. Sawyer. Discussion of techniques and principles of prism and grating spectrographs used in research. Full treatment of apparatus, construction, mounting, photographic process, spectrochemical analysis, theory. Mathematics kept to a minimum. Revised (1961) edition. 110 illustrations. x + 358pp. 61045-4 Paperbound $3.50

THEORY OF FLIGHT, Richard von Mises. Introduction to fluid dynamics, explaining fully the physical phenomena and mathematical concepts of aeronautical engineering, general theory of stability, dynamics of incompressible fluids and wing theory. Still widely recommended for clarity, though limited to situations in which air compressibility effects are unimportant. New introduction by K. H. Hohenemser. 408 figures. xvi + 629pp. 60541-8 Paperbound $5.00

AIRPLANE STRUCTURAL ANALYSIS AND DESIGN, Ernest E. Sechler and Louis G. Dunn. Valuable source work to the aircraft and missile designer: applied and design loads, stress-strain, frame analysis, plates under normal pressure, engine mounts; landing gears, etc. 47 problems. 256 figures. xi + 420pp. 61043-8 Paperbound $3.50

PHOTOELASTICITY: PRINCIPLES AND METHODS, H. T. Jessop and F. C. Harris. An introduction to general and modern developments in 2- and 3-dimensional stress analysis techniques. More advanced mathematical treatment given in appendices. 164 figures. viii + 184pp. 6⅛ x 9¼. (USO) 60720-8 Paperbound $2.50

THE MEASUREMENT OF POWER SPECTRA FROM THE POINT OF VIEW OF COMMUNICATIONS ENGINEERING, Ralph B. Blackman and John W. Tukey. Techniques for measuring the power spectrum using elementary transmission theory and theory of statistical estimation. Methods of acquiring sound data, procedures for reducing data to meaningful estimates, ways of interpreting estimates. 36 figures and tables. Index. x + 190pp. 60507-8 Paperbound $2.50

GASEOUS CONDUCTORS: THEORY AND ENGINEERING APPLICATIONS, James D. Cobine. An indispensable reference for radio engineers, physicists and lighting engineers. Physical backgrounds, theory of space charges, applications in circuit interrupters, rectifiers, oscillographs, etc. 83 problems. Over 600 figures. xx + 606pp. 60442-X Paperbound $3.75

*Prices subject to change without notice.*

Available at your book dealer or write for free catalogue to Dept. Sci, Dover Publications, Inc., 180 Varick St., N.Y., N.Y. 10014. Dover publishes more than 150 books each year on science, elementary and advanced mathematics, biology, music, art, literary history, social sciences and other areas.